Rotating Electrical Machines

Rotating Electrical Machines

Edited by Elsa Hughes

| STATES |
ACADEMIC PRESS
www.statesacademicpress.com

States Academic Press,
109 South 5th Street,
Brooklyn, NY 11249, USA

Visit us on the World Wide Web at:
www.statesacademicpress.com

ISBN: 978-1-63989-744-5

Cataloging-in-Publication Data

Rotating electrical machines / edited by Elsa Hughes.
   p. cm.
Includes bibliographical references and index.
ISBN 978-1-63989-744-5
1. Electric machinery. 2. Electric machines. 3. Electric motors. 4. Turbomachines.
5. Rotors. 6. Electric controllers. I. Hughes, Elsa.
TK2000 .R68 2023
621.310 42--dc23

Table of Contents

Permissions

List of Contributors

Index

Preface

An electric machine is a device that is used to convert electrical energy into mechanical energy or vice versa. They are categorized into two classes, namely, static electric machines and rotating electric machines. Static electric machines include transformers. Electric motor and electric generator are important examples of rotating electric machines. Electric motor converts electric energy into mechanical energy whereas electric generator does the opposite. The three major types of rotating electric machines are DC electric machines, synchronous electric machines, and induction motors. Rotating electric machines consist of two components: the rotor and the stator, which are made up of highly permeable magnetic materials such as silicon steel. This book traces various advancements in the study of rotating electric machines. A number of latest researches have been included to keep the readers up-to-date with the global concepts in this area of study. It will serve as a valuable source of reference for graduate and post graduate students.

This book is a result of research of several months to collate the most relevant data in the field.

When I was approached with the idea of this book and the proposal to edit it, I was overwhelmed. It gave me an opportunity to reach out to all those who share a common interest with me in this field. I had 3 main parameters for editing this text:

1. Accuracy – The data and information provided in this book should be up-to-date and valuable to the readers.

2. Structure – The data must be presented in a structured format for easy understanding and better grasping of the readers.

3. Universal Approach – This book not only targets students but also experts and innovators in the field, thus my aim was to present topics which are of use to all.

Thus, it took me a couple of months to finish the editing of this book.

I would like to make a special mention of my publisher who considered me worthy of this opportunity and also supported me throughout the editing process. I would also like to thank the editing team at the back-end who extended their help whenever required.

Editor

Novel Diagnostic Techniques for Rotating Electrical Machines

Lucia Frosini

Department of Electrical, Computer and Biomedical Engineering, University of Pavia, Via Ferrata 5, 27100 Pavia, Italy; lucia.frosini@unipv.it

† This paper is an extended version of my paper published in the 2019 IEEE Workshop on Electrical Machines Design, Control and Diagnosis (WEMDCD), Athens, Greece, 22–23 April 2019.

Abstract: This paper aims to update the review of diagnostic techniques for rotating electrical machines of different type and size. Each of the main sections of the paper is focused on a specific component of the machine (stator and rotor windings, magnets, bearings, airgap, load and auxiliaries, stator and rotor laminated core) and divided into subsections when the characteristics of the component are different according to the type or size of the machine. The review considers both the techniques currently applied on field for the diagnostics of the electrical machines and the novel methodologies recently proposed by the researchers in the literature.

Keywords: diagnostics; fault detection; electrical machine; electromagnetic signal; vibration

1. Introduction

Recently, diagnostic techniques for the condition monitoring of rotating electrical machines have experienced an extraordinary growth and development, as can be observed in the increased number and quality of papers on this topic published in scientific journals, in their special issues and in related conferences of recent years, e.g., Symposium on Diagnostics for Electric Machines, Power Electronics and Drives (SDEMPED), Workshop on Electrical Machines Design, Control and Diagnosis (WEMDCD), International Conference on Electrical Machines (ICEM), etc. This happened for several reasons. First of all, predictive maintenance is increasingly widespread in the industrial sector and in many other fields, such as electric traction, and this strategy requires the use of advanced diagnostic techniques, possibly non-intrusive and on-line. Secondly, the world is becoming more and more electric [1] and therefore electrical machines are progressively more present in any application. Lastly, the sensors needed to measure the quantities to be monitored to predict the condition of the machines are now more reliable, miniaturized and cheaper than in the past, thanks to the progress in electronic and computer engineering, and are consequently easily applicable not only to large power machines, but also to those of small and medium power.

A further evolution of diagnostics (or diagnosis) is prognostics (or prognosis); diagnostics consist in identifying the fault that will lead to a failure and estimating its severity, while prognostics rely on continuous monitoring of the variables and parameters of the system and use this information to predict the time until a failure occurs, known as remaining useful life (RUL) [2]. In other words, prognostics aim to forecast the evolution of a fault and to predict when the machine will no longer operate as designed or desired [3]. Prognostics may be used to evaluate the rate of degradation and may permit the machine to continue to operate regularly until the moment of failure; it may diminish unnecessary expensive maintenance and unexpected failures [4].

Due to the huge amount of papers on both diagnostics and prognostics and their applications to engineering systems, e.g., [3–7], this review focuses on the first theme, with application to rotating electrical machines, therefore excluding mechanical systems, such as gas turbines and engines. Electrical machines obviously include those supplied by electronic converters, but the faults analyzed in this paper do not comprise those that occur in electronic devices.

In recent literature, there are some excellent reviews on diagnostics of electrical machines, published in journals five and six years ago [8–10]. Furthermore, a very interesting review was published in the proceedings of a conference last year, but it is limited to the use of the stray flux analysis as a diagnostic technique [11]. As research in this field has recently grown exponentially, this paper aims to update the state of the art, adding the most significant outcomes reported in the literature of the last five years. Unlike previous reviews, this paper is structured with two first sections (the second and the third) that report broad-spectrum considerations on diagnostics of electrical machines, coming from an overall evaluation of the literature on this topic and from the author's experience. Each of the following seven sections (from Section 4 to Section 10) is dedicated to one of the main components of the machines prone to fault, as graphically shown in Figure 1; in each of these sections, the techniques appropriate to all major types of electrical machines are reviewed, with particular attention to the problems arising from the supply of the machines through electronic converters.

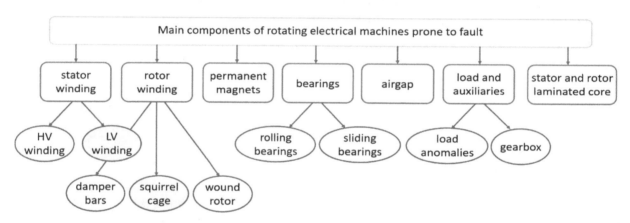

Figure 1. Main components of rotating electrical machines prone to fault, as divided into sections and subsections of this paper (from Section 4 to Section 10); in yellow, the subsections with a sub-subsection dedicated to power supply by electronic converter.

2. The Targets of Diagnostics

Diagnostics are the procedure for translating information deriving from the measurement of parameters relating to a machine into information concerning actual or incipient faults of the machine itself. In other words, diagnostics are the complex of analysis and synthesis activities which—using the acquisition of some physical quantities, characteristic of the monitored machine—allows for collecting significant information on the condition of the machine and on its trend over time, for evaluation of its short- and long-term reliability.

The targets of the diagnostics are: (i) detection (whether the fault is present or not); (ii) isolation (in which part of the machine); (iii) identification (which kind of fault). The general problem of diagnostics is to detect whether or not a specific fault is present based on the available information, preferably without an intrusive inspection of the machine [12]. This problem can be described with a statistical approach as a hypothesis-testing problem: the null hypothesis H_0 affirms that the fault is present and the alternative hypothesis H_1 asserts that the fault is not present. Hypothesis testing is subject to two types of error. A type I error occurs when the null hypothesis H_0 is true and it is rejected, i.e., when, on the basis of the available information, it is decided that the fault is not present, but really it is. Therefore, the machine is not stopped and repaired before the actual manifestation of the failure, with possible catastrophic consequences. A type II error occurs when the null hypothesis is false and one fails to reject it, i.e., when it is decided that the fault is present, but in reality it is not. Then, the machine is stopped and repaired in vain, with useless economic costs.

A diagnostic program is developed according to the following steps: (i) data acquisition; (ii) data processing; (iii) decision-making. The methods employed for data acquisition and processing and for choosing the threshold which separates the faulty condition from the healthy condition of a component can heavily influence the probability of committing an error during the decision-making phase. The term "fault detection" generally refers to a decision made by an individual, while the term "diagnosis" or "diagnostics" usually refers to a decision given by an automatic algorithm.

If the object of diagnostics is a rotating electrical machine, its operation cannot be considered separately from: the operation of the mechanical machine connected along its axis line (pump, fan or other load for an electrical motor; turbine or other prime mover for an electrical generator); the type of the mechanical coupling (joint, gears, belts, etc.); the possible control system (inverter, etc.). All these mechanical and electronic systems can: induce faults in the electrical machine; arouse variations in the parameters of the electrical machine, even in the absence of a fault; suffer faults induced by the electrical machine.

For these reasons, it is important to separate the following situations: (i) a rotating electrical machine directly connected to the grid; (ii) a rotating electrical machine connected to the drive in open loop; (iii) a rotating electrical machine with closed-loop control.

3. Diagnostics of Rotating Electrical Machines

The main types of rotating electrical machines include: induction machine (IM) with a squirrel cage rotor (SCIM) or wound rotor (WRIM); synchronous machine (SM) with salient poles or a wound cylindrical rotor; permanent magnet synchronous machine (PMSM); synchronous reluctance machine (SynRM); switched reluctance machine (SRM); brush DC machine (BDCM). There are many variants of these machines, which can be realized as: single-phase, three-phase or multi-phase; with radial or axial flux; connected directly online (DOL) to the grid or driven by an electronic converter. Further variants comprise: doubly fed induction machine (DFIM), brushless doubly fed induction machine (BDFIM). Each variant determines peculiar features in the behavior of the machine and in its condition monitoring.

Faults can occur in the following components: (i) stator winding; (ii) rotor winding; (iii) magnets; (iv) bearings; (v) airgap; (vi) load and auxiliaries; (vii) stator and rotor laminated core; (viii) other components (commutator, collector rings, slip rings, etc.) [13].

Diagnostic methods can be founded in the analysis of electromagnetic, mechanical and other parameters: vibration, current, external stray flux, internal airgap flux, voltage, electric power (instantaneous, active, reactive), temperature (also with infrared thermography), partial discharges and acoustic noise. Some quantities, such as current, can be measured in steady state or during the starting transient. The methods to analyze the current in steady state are mainly divided into motor current signature analysis (MCSA) and negative sequence (Park current).

Vibration measurement by means of transducers positioned on the bearings is generally employed for the detection of mechanical faults, such as bearing defects, mechanical imbalance (even due to the load) and malfunctions in the transmission system. However, an anomalous vibration can also be a symptom of electrical faults, since any electrical fault produces an asymmetry in the distribution of the magnetic flux at the airgap. In turn, this asymmetry produces an asymmetrical distribution of the electromagnetic forces inside the electrical machine, which affects its vibrations. For this reason, the vibration measured in the casing and in the end-windings can help in the detection of electrical faults, such as non-uniform airgap (static or dynamic eccentricity), faults in stator or rotor windings and imbalance in the power supply.

On the other hand, electromagnetic signals can help in the detection of mechanical faults, since most mechanical faults produce a radial displacement between rotor and stator and this displacement produces an asymmetry in the magnetic flux at the airgap. In turn, this asymmetry affects the inductances of the machine which determine harmonics in the stator current and in the stray magnetic flux. Therefore, not only does the primary effect of each fault need to be evaluated as a possible diagnostic indicator, but even its secondary effect.

In recent years, traditional diagnostic techniques, based on the vibration measurement, have been progressively abandoned in favor of the analysis of electromagnetic signals, in particular the stator current and the external stray flux around the motor [11]. The aim is to use fewer sensors, possibly already existing in the electrical drive for machine control.

Diagnostics of Electrical Motors Fed by Electronic Converter

Generally, diagnostics through the measurement of electromagnetic signals are heavily influenced by the power supply of the electronically commutated machines [13]. The main problems examined in the literature are: (i) the presence of harmonics in the electromagnetic signals coming from the inverter, which diminishes the likelihood of fault detection; (ii) the compensation effect due to the control system, if the drive works in closed loop, which masks the effects induced by the fault; (iii) the operations at variable frequency (and hence at variable speed), since the characteristic harmonics of the fault are distributed over a wide frequency range and may be undetectable [14,15]. Moreover, when a motor is working in a closed field-oriented control system, the fault could change the values of stator and rotor winding parameters and this modification could cause disturbances in the proper functioning of the frequency control system, due to the inaccurate value of the estimated flux. An uncontrolled growth in the damage degree can lead to unstable operation of the drive [16,17].

These concerns are common to all faults that require electromagnetic signal analysis for their detection; several researchers have tried to solve them by applying filters and algorithms to these signals, in steady state and during the transient. Since, in the case of stator and rotor winding faults of low voltage machines, the influence of electronic converters is particularly relevant, both for their diagnostics and for the frequency of failure occurrence, a sub-subsection of the following sections is dedicated to this topic.

4. Stator Winding

Insulation is one of the most fault-prone electrical machine components: for example, the proportion of stator winding insulation faults is between 21% and 40% of total faults in IMs, according to the type and dimension of the machine [8]. Stator short circuits may be classified as: (i) turn-to-turn or inter-turns (between turns of the same phase); (ii) phase-to-phase (between turns of different phases); (iii) phase-to-ground (between turns and stator core). The construction of the winding insulation for low (<700 V) and high (≥700 V) voltage machines is very different; therefore, even the phenomenon which leads to the fault and the diagnostic procedures which can be used to detect it are different, as explained in the following subsections. In Figure 2, two examples of a short circuit in stator winding at high voltage (a) and low voltage (b) are reported.

(a)   (b)

Figure 2. Stator winding faults: (**a**) short circuit in a high voltage winding; (**b**) short circuit in a low voltage winding.

4.1. Stator Winding at High Voltage (HV)

For machines with nominal voltages ≥700 V, the insulation system is manufactured as "form-wound" to prevent partial discharges (PDs) through high dielectric strength materials, such as mica, and by means of a vacuum pressure impregnation (VPI) process [18]. For this type of insulation, the most stressed area is located between the turns and the stator core (phase-to-ground) and its diagnostics are implemented through periodic measurements: off-line (insulation resistance, polarization index, tan delta and, in some cases, AC or DC hipot test) and on-line (PD analysis). Furthermore, to diagnose turn-to-turn short circuits, an off-line insulation measurement may be applied between turns of the same phase. From the interesting reviews of these techniques described in [19,20], it arises that the tan delta and PD measurements may not provide a correct diagnosis of the insulation conditions for machines with voltages lower than 6 kV.

Recently, the issues coming from the supply of motors in the range of 3–13.8 kV by means of pulse-width modulation (PWM) drives have been highlighted and solutions to correctly measure the PDs in these systems have been proposed. These drives generate high-voltage impulses in the kilovolt range with rise times in the sub-microsecond range; therefore, these impulses represent a noticeable electrical interference that can make the on-line detection of PDs difficult, which have magnitudes 1000 times smaller, owing to the overlapping frequency content between PDs and these impulses. For this reason, PD detection in these systems has become a challenge. Additionally, PWM drives may cause more serious stator winding insulation aging. In [21], a method was proposed to measure the stator winding PDs during the operation of medium-voltage motors fed from PWM drives. This method is suitable to diagnose both the normal aging processes in stator windings and the aging processes that can be accelerated in variable-speed drives.

In addition to these tests based on electrical quantities, the condition monitoring of HV stator insulation can be completed with the analysis of winding temperature, the chemical composition of gases and vibrations [19]. Moreover, a new technique based on electromagnetic interference (EMI) seems promising; it may detect insulation deterioration and conductor-related defects, along with some mechanical problems, such as bearing rub or shaft misalignment [22].

4.2. Stator Winding at Low Voltage (LV)

The insulation system for nominal voltages <700 V is manufactured as "random-wound", because up to this voltage level, PDs are not expected with a sinusoidal power supply [23]. The failure in this kind of winding generally begins as shorts between turns and rapidly progresses to phase-to-phase or phase-to-ground shorts, in just a few minutes or hours. This failure is quickly progressive, unlike the HV stator insulation failure, which is slowly progressive. An early warning of an LV stator insulation fault can only be accomplished if shorts within a few turns can be detected through a continuous on-line diagnostic procedure. Hence, the techniques of the diagnosis of inter-turn short circuits must be able to detect them very fast, in their initial phase, when the fault current is still low or, at least, before the conventional protection system acts by interrupting the power supply [10].

Important reviews of the methods for detecting on-line faults between stator turns in LV electrical motors are described in [24,25], especially regarding IMs and PMSMs. The insulation fault diagnostics are essentially founded on the measurement of stator current and external stray flux. In particular, the stator current analysis can be based on the assessment of the harmonics arising in its spectrum or on the evaluation of its negative sequence component.

The methods founded on the evaluation of the harmonics in the current spectrum of the actual machine with respect to the same machine in a healthy condition are known as MCSA. These methods can also be used with the external stray flux spectrum, because it provides similar diagnostic information, though its signal is normally weaker and may be affected by the sensor location on the motor. Several typical frequencies of inter-turn short circuits, that can be detected in current and flux spectra of IMs, have been proposed in the literature [26–31]. The first important work on this topic identified the following frequencies in the axial leakage flux [27]:

$$\left(n_1 \pm k_1 \frac{1-s}{p}\right)f_s = n_1 f_s \pm k_1 f_r \tag{1}$$

where f_s is the supply frequency, f_r the rotational frequency of the rotor, p the number of pole pairs, s the slip, n_1 an odd positive integer (1, 3) and k_1 a positive integer (1, 2, 3, ... $2p-1$). The same group of researchers, some years later, discovered two further sets of frequencies in the phase current [28]:

$$\left(j_r N_r \frac{1-s}{2p} \pm 2j_s + i_s\right)f_s \tag{2}$$

$$\left((j_r N_r \pm k)\frac{1-s}{2p} \pm 2j_s + i_s s\right)f_s \tag{3}$$

where N_r is the number of rotor bars, k and i_s are positive integers (1, 2, 3,...) and j_r and j_s are positive integers or null numbers (0, 1, 2, 3, ...). In the same year, another author pointed out a group of frequencies similar to (1) in the current spectrum [29]:

$$\left(k\frac{1-s}{p} \pm n\right)f_s = kf_r \pm nf_s \tag{4}$$

where n is an odd positive integer (1, 3, 5, ...). Later, in [26], the following set of frequencies was highlighted in the external stray flux:

$$\left(\gamma\frac{N_r}{p}(1-s) \pm v\right)f_s \tag{5}$$

where γ is a positive integer or null number (0, 1, 2, 3, ...) and v is a harmonic index of the stator current (1, 5, 7, 11, ...). Furthermore, some multiples of the supply frequency have been identified as characteristics of this fault; precisely, the third one in the current [30]:

$$3f_s \tag{6}$$

and the following multiples in the external stray flux [31]:

$$15f_s \; ; \; 17f_s \tag{7}$$

The implementation of this methodology must take into account that the current spectrum of a healthy motor always contains harmonic components, which are exhaustively described in [32] for a SCIM. It is worth noting that, if the simple fast Fourier transform (FFT) is applied to analyze the current or flux signals, a high-frequency resolution is necessary and this requires long sampling periods, which may surpass the time sufficient to produce a catastrophic failure. For this purpose, novel signal analysis techniques have been suggested in [33]. However, even these methods may implicate long computation times. Hence, further studies on more efficient signal processing methods have been carried out to detect this fault on time.

An interesting development of these methods comprises the external stray flux analysis through discrete wavelet transform (DWT) at standstill and during the starting transient of an IM. This technique is based on the computation of the energy of the detail decomposition in a healthy condition and during a short circuit [34]. An upgrading of this technique is described in [35,36], where a statistical procedure is shown to detect stator winding short circuits in IMs and SMs: it requires the computation of a correlation coefficient achieved by examining the evolution of magnetic flux harmonics collected in the motor case at no load and in load operation. An adapted wavelet transform (WT) is employed in [37] to identify an intermittent inter-turn fault in a PMSM that induces particular distortions in the stator currents and reference voltages. It is important to note that stator shorts are more dangerous for PMSMs, as they may cause magnetic field strengths greater than the coercivity of magnets, permanently demagnetizing the magnets [25].

From the review of the literature up to this point, it is evident that the processing of the stator current and external stray flux signals with simple techniques, like FFT, or more complex methods, such as WT, could be effective in detecting stator short circuits in low-voltage machines. Nevertheless, other approaches have been suggested, as reported in the following.

Always considering the stator current, the analysis of its negative sequence component has been proposed, starting from the fact that the defective phase has fewer turns and a different impedance than the healthy phases: therefore, it produces lower electromotive force (EMF), causing an imbalance in the three-phase currents. A disadvantage of the methods based on this analysis is that other causes may induce a negative sequence component in the current, for example, imbalanced supply voltages or constructive asymmetries [38].

Other signals and the related sensors needed to collect them also appear to be useful for the same purpose. For example, in [39], the voltage transformers usually installed for a multifunction protection relays in an SM are employed to detect shorts between turns. The measurement of the phase-to-phase voltage to monitor an SM is instead proposed in [40], together with the collection of the airgap magnetic field through a special coil inserted in the stator slots; the authors demonstrate that the 3rd and 9th harmonics in the voltage and the 3rd component in the magnetic field are clearly sensitive to stator inter-turn short circuits. The tests were carried out in open loop, but the authors are confident that the voltage imbalance created by a stator short would remain, even if an increase in excitation was performed by the voltage regulator to maintain the three-phase voltage at the reference level.

For the monitoring of a WRIM, even the rotor current can be measured and, therefore, in [41], a method focused on FFT and WT of the rotor current is evaluated to identify stator shorts between turns. This technique does not practically depend on the load and may quickly diagnose the fault and its seriousness.

A different approach is proposed in [42], starting from the consideration that the origin of the external signals (current and stray flux) is the magnetic flux moving in time and space inside the airgap, which contains indications of any occurring imbalance or fault. However, when transformed into a signal available for external measurements, this information is significantly reduced and distorted. Moreover, it loses one of its dimensions, i.e., the space. Therefore, an array of miniature Hall effect flux sensors (HEFSs), in the order of 1 mm high, was installed within the airgap of a 11 kW IM supplied by a two-level IGBT drive. This technique has been proven to be suitable for detecting both stator short circuits and static eccentricity and for determining the fault location. It is worth noticing that normally the airgap of a small- or medium-power IM is equal to or smaller than 1 mm. Only very high-power IMs have an airgap in the order of 2 or 3 mm, which is sufficient to insert this instrumentation; in a few cases, these large motors could be supplied at low voltage, but over a certain power (about 1300 kW) they need to be supplied at voltage higher than 700 V. In the latter case, the insulation of the winding is different and techniques like those presented in Section 4.1 have to be considered. Therefore, this technique seems practically infeasible in most cases.

The proposal to use a thermographic analysis is evaluated in [43] by means of a finite element method (FEM) simulation, considering that a substantial increment of the temperature in the stator windings is experienced when an inter-turn short circuit occurs. This analysis, based on a thermal camera, can be employed in machines that do not have any temperature sensors inside, while enabling an online noninvasive monitoring.

Stator Winding at Low Voltage Supplied by Electronic Converter

While some works cited in the above subsection consider motors fed by an electronic converter, it is worth dedicating a sub-subsection on this topic.

First of all, it is significant to note that, though with a sinusoidal power supply the winding insulation of LV machines does not experience PDs, they can occur if these machines are driven by inverters; hence, the monitoring and testing of an inverter-fed LV machine should be performed according to IEC 60034-18-41, as described in [44].

Some papers demonstrated the possibility of continuing to use stator current and stray flux to detect stator short circuits, by means of different techniques to process these signals.

In [14], an approach based on WT combined with the power spectral density of the stator current was shown to be able to detect stator inter-turns short circuits and broken rotor bars in IMs during variable load torque operations. In [45], a WT-based pre-processing method proved effective in filtering the stator current and stray flux signals in order to diagnose stator shorts in an IM supplied by an electronic converter, through the typical harmonics of this fault. A subsequent paper [46] has evaluated a procedure to detect stator shorts and bearing defects (even in the case of simultaneous presence), based on the high sampling frequency and filtering process of stator current and stray flux. This procedure allows for the discrimination not only of the existence of single and multiple faults, but also their progression from an early stage to more serious conditions.

Other diagnostic methodologies for inverter-fed IMs are founded on the measurement of their signals during the transients, e.g., the start-up [47]; in particular, the Hilbert–Huang transform results is suitable for non-stationary and non-linear signals and the adaptive slope transform allows for a better adaptation to the time-varying harmonic content of the signal.

In [48], an analysis of the stator current and torque in the frequency domain for a direct torque control (DTC)-driven IM shows that inter-turn shorts cause numerous harmonic components in these signals and even inter-harmonic components that mimic or mask the typical signatures of a DOL faulty IM; this happens due to the DTC reaction, the switching of the inverter, the saturation effect and the noise signal coming from the industrial environment. An update to this study investigates the stray radiated magnetic field measured around the same motor through a magnetic loop antenna at different distances and angles, to examine its influence on the diagnostics of stator shorts [49].

In [50], a real-time procedure is presented for the early detection of stator shorts between turns in an IM supplied by a voltage source inverter (VSI). The methodology is based on the fact that both non-sinusoidal input voltage and short circuits cause harmonics in the stator current; this combination of harmonic components complicates the diagnostics based on spectral analysis. Therefore, the paper aimed to study the effect of the fundamental and switching frequencies of the inverter on the detection and classification of incipient inter-turn shorts. DWT-based analysis is implemented on stator currents; the results are promising, but the disadvantage of the method is that it needs the measurement of all three-phase currents.

A particular machine is considered in [51], i.e., the BDFIM. The diagnostic proposal starts from the reflection that the existing techniques to detect inter-turn short circuits consider rotor slot harmonics in stator current spectra as fault indicators for conventional DFIMs. Nevertheless, these techniques cannot be used for a BDFIM, owing to its different stator and rotor winding structure. Therefore, the paper presents a novel analytical formulation for the nested-loop rotor slot harmonics as inter-turn short circuit indicators in BDFIMs. Another particular machine has been evaluated in [52], i.e., a SynRM with a dual three-phase stator configuration; the novel method proposed for its stator fault detection is based on the reactive power measurement of both three-phase systems.

5. Rotor Winding

Differently from the stator winding, which is manufactured in a similar manner for most electrical machines, the rotor can be wound in various modes, it can consist of a squirrel cage, it can comprise permanent magnets or it can be made only by iron sheets.

A wide and accurate review of the rotor failure diagnostic techniques in IMs is described in [53]. These methods start from the fact that any electrical rotor failure leads to an asymmetry in the equivalent winding impedances, which in turn produces an asymmetry in current distribution. Consequently, as a first effect, a rotor failure modifies the airgap flux and the stator current; thus, the current can be successfully employed to detect this fault, as well as the external stray flux, which provides similar diagnostic information.

For high-power SMs, rotor winding diagnostic techniques have been well established for many years and include: the recurrent surge oscillation (RSO) test, dynamic impedance measurement, shaft voltage measurement and internal rotor stray flux monitoring. Nevertheless, the rotors of these machines also often contain damper bars; the detection of faults in these bars is still not simple and some recent papers have proposed new techniques to solve this problem, as reported in the following subsection.

An additional component that may be present on SM rotors is the rotating rectifier of the brushless excitation system; this component is quite vulnerable and its failure rate is affected by centrifugal force and thermal stress. Due to the ageing phenomenon, rectifier diodes can fail; the two major failure conditions are open circuit and short circuit. This issue is rarely mentioned in reviews on rotating electrical machines diagnostics, although it is relevant in several applications. The main approach to the early detection of diode faults is based on the spectral analysis of the induced electromotive force measured by a search coil, which is often already installed in the exciter stator slots. When this device is not incorporated in the SM, alternative methods can be applied, based on the output voltage analysis [54] or on the total harmonic distortion (THD) and polarity of the asynchronous exciter armature current [55].

5.1. Damper Bars

Damper bars are often used in rotors of high-power SMs, both with salient poles and cylindrical wound rotors, as reported in Figure 3. The damper winding acts like the cage of an SCIM and it consists of short-circuited copper bars embedded in the face of the salient poles or in the slots containing the rotor winding.

(a)   (b)

Figure 3. Damper bars in SMs: (**a**) damper bars in a salient pole; (**b**) damper bars in a cylindrical rotor.

When an SM is used as a generator, these bars intervene during the transient following an abrupt change of electromagnetic torque and the consequent variation of the load angle; their purpose is to dampen the transient oscillations by means of an asynchronous torque. When an SM is used as a motor, they also allow the starting of the machine, without any other device.

Several cases of broken damper bars in SMs have recently been reported in the literature [56]. The detection of damper bar failures is difficult, because these bars are active only during the starting or the load transients. The traditional diagnostic tests on damper bars consist in off-line visual inspections.

For this reason, in [56], an on-line method has been proposed, based on flux measurements through an airgap search coil during the starting transient. This kind of airgap search coil is increasingly being installed in high-power SMs to detect field-winding short circuits, therefore, the method can be implemented without any additional hardware.

In [57,58], two further methodologies for identifying damper bar faults in salient pole SMs without disassembly have been proposed. The first one is an off-line test able to detect the change in asymmetry by injecting a pulsating field from a low-power three-phase inverter. The second method is on-line and it is based on the analysis of the starting current, by means of time–frequency transforms to extract the fault-related component.

5.2. Squirrel Cage Rotor

The SCIM rotor winding is made up of several bars and two frontal short-circuit rings on opposite sides; it may be produced with two techniques, known as die-cast and fabricated. With the first technique, the complete cage is shaped in a unique piece by pouring molten metal (aluminum or copper) into a mold. With the second technique, copper bars are inserted into the rotor slots and welded to the frontal rings. A fabricated rotor can be used at a high power, generally above 250 kW or less, when the number of poles is large and the rotor diameter is large. For fabricated rotors, the most common fault is caused by the breakage of a bar close to the welding between the bar and ring. Conversely, die-cast rotors are more long-lasting and robust, even if some defects can be introduced during production, e.g., porosity or blowholes, which may worsen the performance and reliability of the motor [59]. The growth in rotor resistance and rotor cage asymmetry caused by porosity give rise to motor efficiency degradation, torque pulsation and imbalanced magnetic pull. Porosity can also affect the starting performance and the torque–speed characteristics, which can differ considerably from those indicated by the manufacturer; this deterioration cannot be tolerated for high-power motors. For this reason, besides the traditional quality assurance tests, a novel off-line test has been proposed in [59], based on a flux injection probe, which can excite each rotor bar to achieve porosity information during the post-production balancing of rotors.

Although die-cast rotors can rarely exhibit breakage of the bars, they are often used in laboratories to validate diagnostic procedures for fabricated rotors [53].

Broken bars and cracked end rings represent only the 5–10% of SCIM failures, nevertheless, their detection is fundamental as they may cause serious secondary effects: broken fragments of the bar may impact the stator winding, severely damaging its insulation, causing an expensive repair and a lost production. Furthermore, the mechanical and thermal stresses increase in the bars adjacent to the broken one, due to the redistribution of the currents in the healthy bars. This produces a slowly progressive propagation of the fault; hence, diagnostics may be useful in preventing it.

Once the rotor is injured, its equivalent winding impedances become asymmetrical and a reverse rotating magnetic field is present. As a consequence, a group of harmonic components arises in the stator current and stray flux spectra at the following frequencies [53]:

$$(1 \pm 2\gamma s)f_s \tag{8}$$

where γ is a positive integer or null number, as defined in Section 4.2.

The monitoring of these harmonic components at steady state is often sufficient to detect this fault, but the literature has highlighted some conditions which may lead to false positive or false negative diagnostic alarms. False positives are present in the case of load torque oscillations, the existence of rotor cooling axial ducts and magnetic anisotropy issues. This happens because an oscillating load torque induces sidebands that can sometimes be localized near the typical frequencies of the breaking of the rotor bars; moreover, the amplitude of the sidebands caused by a broken bar may be of the order of magnitude of that produced by inherent manufacturing asymmetries or by special rotor structures. False negatives may happen in the case of non-adjacent broken bars, outer bar breakage in double cage rotors and diagnosis under light load or no load [60]. Furthermore, the existence of rotor inter-bar currents may decrease the sideband amplitude.

Several methods have been suggested in the literature to overcome these difficulties, on the basis of various signals (e.g., start-up current and external stray flux, active and reactive power, vibrations, torque, etc.), on sophisticated signal processing methods and artificial intelligent systems or on a combination of the above [53].

The analysis of the stator current during motor starting has been proved to offer satisfying results because the harmonic components typical of the fault are amplified due to the high rotor (and stator) current in high-slip conditions. A time–frequency approach based on WT can be applied to this signal, obtaining an effective diagnosis in double cage rotors, as in [61,62], and in the case of the presence of rotor axial air ducts [63]. Precisely, the left sideband during the starting is examined in [61], while in [62], both sidebands are analyzed during a general transient. Despite the diagnostic reliability offered by this analysis during the transient, in many applications, large induction motors do not experience frequent start-ups. They have a low rotor resistance to keep the steady-state operation very efficient, which reduces the starting capabilities. Therefore, in these applications, diagnostic methods depending on the starting current are not feasible.

A detailed review of the methods to detect load anomalies and to distinguish their effects from those produced by rotor failures is reported in [64,65]; besides, these papers suggest new methodologies to avoid possible false alarms in diagnosing rotor faults.

A recent study is presented in [66,67], which deals with broken bar detection in large IMs, where the sideband components are usually located near the fundamental frequency, owing to the low value of slip at steady state. A time–frequency analysis of the external stray flux is proposed, through short time Fourier transform (STFT), and the sideband signatures of higher harmonics are considered, focusing on the sidebands around the 5th and 7th harmonics, since they stand at the distances $(-4s\,f_s)$ and $(-6s\,f_s)$ for the 5th and at $(-6s\,f_s)$ and $(-8s\,f_s)$ for the 7th harmonic. Further investigations on the diagnostic capability of the stray flux to detect rotor faults are reported in [68], showing that the mechanical frequency-associated harmonics can be purely related to broken bars and totally independent from rotor eccentricity and rotor imbalance. Therefore, these specific harmonics can be good indicators of broken bars at low slip operation.

The feasibility of the stray flux analysis during the start-up of an IM has been investigated in [69], to detect two types of faults (broken bars and misalignment), even when they coexist. Two signal processing techniques were applied for the analysis of the stray flux: a continuous tool (STFT) is used to track the evolutions of the fault harmonics during the start-up and a DWT is employed to calculate a new rotor fault severity indicator.

In [70], an approach focused on the analysis of the axial vibrations, arising in the presence of inter-bar currents, is proposed in combination with MCSA to overcome the problems due to the inter-bar currents in detecting broken rotor bars.

Squirrel Cage Rotor of Motors Fed by Electronic Converter

As mentioned, the supply by electronic converter adds further uncertainties and complications to the diagnostics of the motors. For this reason, different processing techniques have been evaluated in the literature to make more robust the broken rotor bar detection in these systems.

A method based on the standard current sensors already existing in modern industrial inverters is proposed in [71] to identify rotor bar defects at no load and almost at standstill. A new fault indicator is achieved by energizing with voltage pulses generated by the switching of the inverter and measuring the resulting current slope.

An approach based on the combination of complete ensemble empirical mode decomposition (CEEMD) and the multiple signal classification (MUSIC) is presented in [72]; it was applied to the stator current of an inverter-fed IM during a starting transient followed by a steady-state period and its ability to recognize a single broken rotor bar, a mixed eccentricity due to a motor-load misalignment and the coexistence of both defects was proven.

A further improvement of the time–frequency analysis is presented in [73], where the Dragon Transform, applied to the stator current of an IM with a broken bar, represents the harmonic evolutions as very thin lines. Its high time–frequency resolution allows for the detection and quantification of the bar breakage harmonics during the start-up of an inverter-fed IM.

Another technique to diagnose broken bars in IMs supplied by inverters during non-stationary regimes is proposed in [74]; it is based on the combination of two methods, time-corrected instantaneous frequency and a spectrogram, and it is called "reassigned spectrogram". The effectiveness of this approach to detect one broken bar has been shown during a start-up, followed by a steady-state period. A novel technique to identify broken bars in an IM powered by three different types of inverter is focused on the amplitude of the stator current in the time domain [75]. This paper examines the accuracy in detecting broken bars over a wide frequency range and in various load conditions by evaluating the effectiveness of analog and digital filters applied to the current signal of the inverter-fed IM. Moreover, the performance of four pattern classification methods is evaluated and experimental tests confirm the effectiveness of this methodology. In [76], three types of analysis (FFT, Hilbert transform and modulus of Park's vector) are implemented on the current signals coming from an IM driven by an inverter at nominal load and with broken bars, and an evaluation of the effectiveness of these methods is shown. Another appropriate methodology for broken bar diagnosis in both DOL and inverter-fed IMs is founded on the Goertzel algorithm [77].

5.3. Wound Rotor

Less research work is available in the literature for WRIMs, compared to SCIMs. However, it is reasonable to presume that the rotor winding failure rate for WRIMs is higher than that for SCIMs, because WRIM rotor windings are generally less protected when auxiliary components (slip ring connections and resistors) are also present. Not many papers have been published on this subject, owing to the few industrial applications of WRIMs, particularly in the past. Recently, renewed interest in WRIMs has arisen as double fed induction generators (DFIGs) for wind power generation, in which the rotor windings are fed by an electronic converter to control the active and reactive power flows from the generator to the electrical network [53,78–80]. For this machine, a rotor failure is similar to a stator failure and can manifest itself in an increase in rotor resistance, a short circuit or an open circuit. In the first case, the WRIM may continue to operate, whereas in the event of short or open circuits, the operation is limited to a short period after the fault [53].

Similar to SCIMs, an electrical failure of the rotor in a WRIM induces a phase dissymmetry and MCSA can be useful to detect it. As explained in [80], a rotor dissymmetry produces an inverse rotating magnetic field and an inverse sequence component in the rotor currents at frequency sf_s. This inverse sequence is mirrored in the stator currents and causes the frequency component $(1 - 2s)f_s$, which determines both electromagnetic and mechanical interactions between rotor and stator. Consequently, a set of components at frequencies defined by (8) appears in the stator current spectrum and a set of components at the following frequencies in the spectrum of the rotor currents:

$$\pm (1 + 2\gamma)sf_s \tag{9}$$

where γ is defined as above and the main component is at frequency sf_s.

However, in closed-loop and time-varying operations, MCSA presents some problems, as the control system may compensate for current harmonics caused by rotor failure and hide the fault signatures. Hence, as the rotor winding is accessible, rotor currents can also be measured and a group of harmonic components of these currents at frequencies (9) can be examined.

In [79], a diagnostic method based on the combination of WT and a pre-processing of rotor voltages in time-varying operations is investigated to identify rotor imbalances. The paper focuses on tracing the most significant fault frequencies in rotor voltages; since the machine-load inertia produces a damping effect on higher-order fault harmonics, the authors examined only the signature of the sideband sf_s on a 5.5 kW WRIM with two pole pairs, connected with a PWM back-to-back converter on the rotor side.

A method to identify a rotor short circuit between turns in a WRIM operating as a generator, through spectral and bispectral analysis, is described in [78]. Stator currents have been shown to contain interesting signatures, because an inter-turn short of the rotor induces novel harmonics in them:

$$[1 \mp (\lambda + \gamma)(1 - s)]f_s \tag{10}$$

where $\lambda = 6\gamma \pm 1$ and γ is defined as above. Spectral analysis demonstrated that these harmonics are sensitive to this failure and their magnitude depends on the seriousness of the fault. In [81], a DWT-based slip-independent technique for detecting stator and rotor asymmetries in WRIMs is presented. Both the space vector magnitude and the instantaneous magnitude of the stator current were proven to be efficient signatures for diagnosing rotor asymmetry. The proposed technique was validated with experimental results on a 7.5 kW WRIM. In [82], the effects of rotor inter-turn shorts and imbalanced rotor resistance on fault indicators were evaluated. The diagnosis of these anomalies was carried out by means of suitable signals present in the DFIG control system (stator current, reactive power and rotor modulating voltage). The capability of the fault indicators was investigated for different degrees of failure and for various DFIG operating conditions.

In [83], a novel approach is introduced to detect rotor asymmetry faults in a WRIM, based on the measurement of the stray flux in three different positions. The level of electrical asymmetry was induced by means of an external rheostat inserted in series with one rotor winding. Similar to the diagnostics of electrical rotor asymmetry in SCIMs, the most significant components amplified by this fault in WRIMs are expected at the frequencies of the main sidebands around the fundamental:

$$(1 \pm 2s)f_s \tag{11}$$

Moreover, additional components at frequencies sf_s and $3sf_s$ are expected to be excited in the FFT spectrum of the stray flux signals; the nature of these components is axial, and, consequently, they are more likely to be detected in those sensor locations where a greater part of axial flux is measured. The amplitude of the component at sf_s can also be increased by the existence of eccentricity/misalignment. The proposed approach is based on the detection of patterns which arise in the time–frequency maps during the motor start-up. These patterns can be detected by means of different signal processing methods applied to the signal during the transient: the STFT, to track the evolution of fault harmonics, and the DWT, to calculate a fault severity indicator.

6. Permanent Magnets

The demagnetization of the permanent magnets can considerably decrease the torque produced by a PMSM. Consequently, a larger stator current is required to obtain an equal value of torque. Besides, demagnetization gives rise to a growth in Joule losses and in temperature. In turn, this stimulates greater demagnetization, which further increases current and diminishes efficiency [84]. Demagnetization has been recognized as one of the main causes for PMSM failure, because it induces a flux disturbance, which modifies several parameters of the machine [53,85]. A typical signature occurs in the stator current and, therefore, MCSA may be able to identify this problem, although this signature is strongly influenced by the winding configuration, both in case of local and uniform demagnetization. Some research revealed that partial demagnetization produces harmonic components in the stator currents at the following frequencies [84,86,87]:

$$f_{demag} = (1 \pm k/p)f_s \tag{12}$$

However, according to the configuration of the stator windings and the type of demagnetization, sometimes no new harmonic or subharmonic components arise in the case of demagnetization other than those normally existing in a healthy motor [87]. Furthermore, other rotor defects, like dynamic eccentricity, may be identified through the same frequencies in the stator currents [84]. Therefore, in some instances, MCSA does not discriminate between demagnetization and other rotor faults and this is why other techniques have been investigated. In [84], the early rotor demagnetization in a surface-mounted PMSM was detected through an on-line monitoring of the zero-sequence voltage component (ZSVC). By means of simulations and tests, it was proven that local demagnetization reduces the amplitude of the ZSVC and this can allow for fault identification; nevertheless, this method needs windings to have an accessible neutral and an artificial neutral point; the latter is realized with a three-phase balanced resistor network connected to the motor terminals. In [86], time–frequency WT-based techniques were effectively used to diagnose demagnetization in PMSMs during non-stationary operations.

In [88], the Hall sensors already existing in an interior PMSM for its control are employed to measure the flux variation inside the machine caused by a magnetic asymmetry of the rotor; FEM analysis and experimental tests demonstrate that this technique may permit a correct diagnosis of local demagnetization and dynamic/mixed eccentricity. Furthermore, the same research group proposed the zero-sequence component of the magnetic flux density to detect demagnetization [89]. The authors proved that this method considerably decreases the sensitivity of the diagnostics to various problems, such as the position of the sensors, the permanent magnet temperature, etc. This technique can be used for condition monitoring, but also for torque ripple compensation techniques.

7. Bearings

The main classification of bearings used to support the rotor of electrical machines distinguishes them between rolling and plain (or sliding) bearings. Rolling bearings are more common, while sliding bearings are generally employed for high-power SMs or special applications. According to the type and size of the machine, the bearing failure distribution is variable between 40% and 90% of the total faults, from large to small machines [90].

7.1. Rolling Bearings

Condition monitoring of rolling bearings is traditionally performed by analyzing the vibration of the machine frame and/or the bearings themselves. A comprehensive review of this argument is described in [91], even if in recent years further variants of the vibration analysis have been proposed, as, for example, in [92]. Most bearing failures are slowly progressive and then they may normally be prevented through periodic vibration monitoring, along with periodical replacement and/or lubrication after a defined period. This kind of maintenance is usually effective, but: (i) it is costly and it can require the use of resources outside the company; (ii) it may involve premature unnecessary substitution of bearings; (iii) it may be unsuccessful when quickly progressive faults happen in the case of shaft currents due to electronic converters. Therefore, in the last twenty years, a research work has been concentrated on implementing a predictive condition monitoring procedure capable of identifying bearing failures in their early phase through a continuous on-line analysis of current and stray flux [92–94].

This detection approach categorizes bearing faults as: (i) single-point defects; (ii) generalized roughness. The first causes an impact between ball and raceway and produces vibrations at specific frequencies, which depend on the surface of the bearing affected by the fault:

$$\text{Outer race}: \; f_o = N/2 \cdot (1 - d/D \cdot \cos\alpha) f_r \tag{13}$$

$$\text{Inner race}: \; f_i = N/2 \cdot (1 + d/D \cdot \cos\alpha) f_r \tag{14}$$

$$\text{Ball}: \; f_b = D/2d \cdot \left(1 - (d/D)^2 \cos^2\alpha\right) f_r \tag{15}$$

$$\text{Cage}: \quad f_c = 1/2 \cdot (1 + d/D \cdot \cos\alpha) f_r \qquad (16)$$

where N is the number of balls, d the ball diameter, D the bearing pitch diameter and α the ball contact angle.

The radial movement between rotor and stator due to a single-point bearing defect causes stator currents at the following frequencies:

$$f_p = |f_s \pm k f_v| \qquad (17)$$

where f_v is one of the specific vibration frequencies, as defined in (13)–(16). Other typical frequencies have been defined in [90]. Although ideally these harmonics are not present in the event of generalized roughness, the analysis of stator current and external stray flux has been assessed to identify this type of fault in rolling bearings, with promising results [95]. However, it is obvious that the primary effect of bearing failures is in vibration, while in electromagnetic signals, the effect is secondary and weaker, so it could hardly be noticed, especially in inverter-fed motors. Hence, MCSA has been considered a suitable method for integrating other techniques, e.g., vibration or thermography [96].

A different approach, although based on the measurement of the stator current, starts from the consideration that the current spectrum is not effective for bearing fault detection. Therefore, a method based on vibration envelope analysis is applied to the stator current, obtaining promising results by means of the squared envelope analysis. Moreover, the fast kurtogram algorithm has been proved effective in identifying the frequency bands in which the fault impulses are concentrated [97].

In recent years, further alternative methodologies to detect rolling bearing faults, based on different signals, have been assessed. In [98], a method based on rotor speed signal was proposed, which is advantageous in terms of cost and simplicity. This technique was tested under both constant and variable speed, at constant load.

In [99], rolling bearing faults were detected by means of the acoustic signal collected with a mobile phone. This signal was analyzed through a combination of spectral kurtosis and Hilbert transform post-processing methods, therefore using the high-frequency content of the signal. In [100], a different technique for bearing diagnosis in DOL IMs is investigated, based on the voltage and current signals in the time domain; a neural network scheme is proven to be capable of identifying bearing failures by examining a half-cycle sampling of the IM supply voltages and stator currents. In [101], a new diagnostic technique for PMSMs is presented; it is based on a speed sensorless observer, which aims to acquire the rotor angle and speed. The angle signal is utilized to resample the non-stationary speed signal into a stationary signal in the angular domain for order spectrum analysis. The presence of excitation in the bearing fault characteristic order in the resampled signal spectrum is utilized for the diagnosis of the fault.

It is important to note that all papers reviewed until this point in this subsection are aimed at investigating the ability of sensors other than traditional accelerometers to detect various types of bearing defects in different working conditions and in particular when variable speed induction motors are supplied by electronic converters. These papers often have the purpose of identifying the potential of sensors that are cheaper than the traditional ones and/or already present in the overall system in which the electrical machine is inserted (e.g., current sensors for closed-loop control). Another purpose of this group of papers is to discriminate a bearing fault in the event of the coexistence of multiple faults [46].

However, there is another group of papers which is focused on machine learning (ML) and deep learning (DL) techniques [102]; to implement effective ML and DL algorithms for bearing fault detection, good data collection is needed and, therefore, these papers sometimes refer to datasets available on-line [103], but also to wide campaigns of laboratory measurements [104,105]. In [106], a very recent survey of these papers is reported, together with a comparative study of the classification accuracy of various algorithms that use the open-source Case Western Reserve University (CWRU) bearing dataset.

7.2. Plain Bearings

Plain bearing diagnostics require sensors able to measure the relative motion of the rotor with respect to the bearings. Unlike rolling bearings, which present almost no damping, the oil film of plain bearings considerably dampens rotor vibrations. Consequently, rotor excursions must be measured through relative sensors positioned close to both bearings. High-power rotating machines normally operate with permanent monitoring of the relative rotor excursions. The best solution for sliding bearing vibration monitoring is given by two relative sensors positioned 90° from each other at each bearing, since with this method, it is possible to observe not only the vibrations in the time domain, but even the trajectory of the rotor movement [107]. As the fault detection of these bearings is mainly based on mechanical signals, in this paper, the review of the diagnostic techniques in this field is limited to the above considerations.

8. Airgap

Airgap eccentricity causes a force on the rotor, known as unbalanced magnetic pull (UMP), which pulls the rotor towards the stator bore, along the minimum airgap. When the level of eccentricity exceeds definite limits (usually <10%), eccentricity may induce excessive stress on the machine and may increase bearing wear. A strong airgap eccentricity may eventually cause rubbing between rotor and stator, with a consequent damage to the stator and rotor core and windings. This may determine insulation failure of the windings, broken rotor bars (for SCIMs) and shorts between the laminations, as reported in Figure 4.

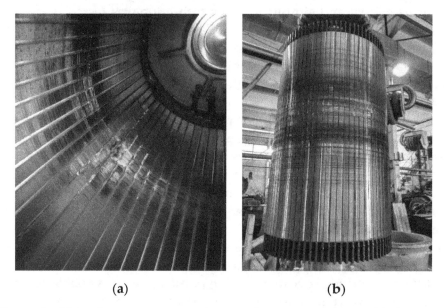

(a)   (b)

Figure 4. Effects of rubbing between rotor and stator due to eccentricity: (**a**) damage of the stator; (**b**) damage of the rotor.

Eccentricity is generally classified into static and dynamic, which may coexist. It mainly affects the airgap flux and then the stator current; for this reason, most of the literature has used stator current as a fault indicator, along with external stray flux, which provides equivalent diagnostic information. Nevertheless, this fault also significantly affects the vibration of the stator frame and then even this signal can be successfully employed to detect eccentricity, particularly if dynamic. The typical frequencies due to this fault to be identified in the stator current and in the frame vibration of an IM are [108]:

$$\left[(N_r \pm n_d) \frac{(1-s)}{p} \pm n_\omega \right] f_s \tag{18}$$

where $n_d = 0$ for static eccentricity, $n_d = 1$ for dynamic eccentricity, $n_\omega = 1, 3, 5, \ldots$ for current analysis, and $n_\omega = 0, 2, 4, \ldots$ for vibration analysis. Beside these frequencies, the literature points out the following sidebands due to eccentricity in the current spectrum of an IM [109]:

$$f_s \pm f_r \tag{19}$$

and further harmonics excited by eccentricity in the vibration spectrum at frequencies [108,109]:

$$2f_s \ ; \ f_r \ ; \ 2f_s \pm f_r \tag{20}$$

In addition, in [110], the coexistence of static and dynamic eccentricities in IMs was evaluated and a novel fault severity index for mixed eccentricity was developed, based on the stator current harmonic components. This index can be computed by no-load or low-load tests and it does not need any preliminary knowledge of the machine in a healthy condition.

A very recent paper, [111], starts with the considerations that: (i) often it is not possible to distinguish rotor eccentricity and load defects; (ii) false rotor cage fault indications are frequent; (iii) a new trend aims to integrate smart self-diagnostics into electrical machines by means of embedded sensors, especially in applications where motor inspection is difficult (e.g., submersible pumps, nuclear plants, etc.). Therefore, the paper evaluates the airgap flux measurement, by means of a search coil, as an alternative for the diagnostics of motor and load defects; a novel technique based on the analysis of the airgap flux during the start-up transient has been proven to be effective in detecting and classifying broken rotor bars, eccentricity and load mechanical anomalies. The method has been shown to be insensitive to load imbalance, misalignment and axial duct influence, which can cause MCSA-based detection of eccentricity and broken bars to fail. This technique should reliably detect rotor failures with an inexpensive airgap search coil, when used together with MCSA.

A different technique described in [112] analyzes the transient current response of an IM to an excitation with voltage pulses provided by the inverter switching and proves that the indicator achieved from the transient current response does not depend on the number of slots per pole, while the performance of the MCSA is affected by this parameter.

Concerning PMSMs, these frequencies have been recognized as typical of eccentricity (static, dynamic and mixed) [113,114]:

$$[1 \pm (2k - 1)/p]f_s \tag{21}$$

In [113], the analysis of the current spectrum of faulty PMSMs caused by eccentricity, open circuits, short circuits and demagnetization shows that only eccentricity produces these components. In [114], a new index is defined for static and dynamic eccentricity diagnostics in PMSMs; it is calculated through a linear combination of energy, shape factor, peak, head angle of the peak, area below the peak, gradient of the peak of the detail signals in wavelet decomposition and coefficients of the autoregressive model, which are derived from the stator current. Always with reference to PMSMs, a recent paper considers in-wheel motors for automotive machines and proposes a new method to directly measure the airgap width, in either static or dynamic eccentricity, through an optical sensor integrated in the airgap, which acquires the reflection of the infrared radiation between rotor and stator [115]. The technique is validated by means of a parallel measuring system with an analog Hall sensor which measures the change in the magnetic flux density.

Regarding salient-pole SMs, in [116], a new off-line technique was proposed to diagnose static eccentricity through the measurement of the three-phase currents in a locked rotor condition, in three rotor positions; the spectrum of a pseudo zero-sequence current provides reliable indicators to identify this fault.

9. Load and Auxiliaries

9.1. Load Anomalies

Even in the absence of rotor eccentricity, some mechanical load anomalies may cause sidebands in the stator current spectrum. These anomalies are generally divided into two main types. The former is usually due to a speed reduction coupling with speed reduction ratio r. For an IM, the mechanical load oscillations (f_r/r) are transmitted to the motor rotor across the coupling and revealed in the stator current at frequencies [64,65]:

$$f_{coup} = f_s \pm k\frac{f_r}{r} = f_s\left(1 \pm k\frac{1-s}{p \cdot r}\right) \tag{22}$$

The second type of mechanical anomaly is caused by periodic low-frequency oscillations of the load around a constant torque and induces specific sidebands in the stator current at frequencies:

$$f_{oscil} = f_s \pm kf_o \tag{23}$$

where f_o is the fundamental frequency of the torque ripple [117]. As regards the possibility of detecting torque oscillations in IMs powered by inverters, an interesting, although complex, approach is presented in [118]; the stator current and the torque are measured and three fault indicators are calculated through the Wigner distribution and the instantaneous frequency estimation; all these indicators are capable of recognizing small torque oscillations. In [119], a new approach is described to detect load oscillations by means of the harmonic multiple of the fundamental. The paper examines the effects of these oscillations in current and external stray flux signals and proves that the characteristic signatures of this anomaly are also evident in an IM supplied by an inverter. Furthermore, for an automatic diagnosis of this anomaly, a technique focused on linear discriminant analysis and on the calculation of the first odd harmonics of the measured signals is suggested.

9.2. Gearbox

Similarly to bearings, even the diagnostics of the gearbox are mainly handled through the vibration analysis, as, for example, in [120]. Nevertheless, in [121], a comparative study of vibration, acoustic pressure and stator current analysis abilities is proposed for the diagnosis of a gear tooth wear fault and is tested on a 250 W SCIM shaft connected to a single-stage gearbox. Vibrational and acoustic analyses appear as the most suitable techniques to detect these faults, while current analysis is more challenging because the amplitudes of the fault-related harmonics are close to the noise level and long-term data acquisition and high-resolution systems are required to correctly identify some of these amplitudes. To overcome this problem, in [122], a noninvasive approach for the detection of gear tooth surface damage faults, focused on the stator current space vector analysis, is investigated, providing good results. In [123], both electrical and mechanical signatures of a WRIM are used for gear fault diagnosis. Numerical and experimental evidence shows that localized gear tooth defects can be identified by both the mechanical torque and the WRIM stator current signature.

10. Stator and Rotor Laminated Core

Whenever a stator (or rotor) core is crossed by a variable magnetic flux, it is manufactured with thin silicon–steel (Si–Fe) insulated laminations to minimize eddy-current losses. Each lamination sheet is deburred and coated with insulation material to avoid conduction between the sheets and to reduce the risk of inter-laminar eddy currents. The inter-laminar insulation can be damaged due to several causes, e.g., poor or damaged lamination coating, rubbing in a loose core, excessive burrs during processing, mechanical damage from foreign objects or lamination burning in the area of a winding failure. If this damage is unattended, it can spread and cause, in extreme cases, catastrophic failure of the machine [124].

The traditional off-line tests to detect local damages in a stator core are: (i) the core-ring test or loop test; (ii) the low-energy core test, often commercially known as electromagnetic core imperfection detection (EL-CID). The first one requires an external winding around the yoke of the stator core, after the rotor removal, to energize the stator core yoke at 80–100% of the rated flux. After the excitation, an infrared camera is used to discover possible hot spots in the stator bore due to inter-laminar fault currents. The second one needs a similar excitation configuration, but it requires only 3–4% of the rated flux. Both tests have been used for many years and are generally effective, but they involve rotor removal and specialized test equipment. For these reasons, these tests are generally applied to rotating electrical machines above tens of megawatts. For this kind of machine, there is also an on-line diagnostic technique to detect core damage, based on chemical monitoring to detect hot spots in the stator core.

Some papers, about ten years ago, proposed different diagnostic techniques for the stator core faults. In [125], a methodology applicable to inverter-fed IMs without disassembly is suggested, focused on the measurement and calculation of the input power as a function of flux vector angle, which reveals a different trend in the case of a healthy and faulty stator core. In [126,127], the proposal starts from the consideration that surface currents caused by a short circuit between laminations have a significant influence on the external magnetic field; therefore, an on-line diagnostic technique based on the analysis of the external magnetic field was theoretically and experimentally investigated.

It is worth noting that the diagnostics in this field have not proposed particular novelties in recent years, due to the fact that stator core damage alone is unlikely as a failure, but generally it arises as a secondary effect of other faults.

11. Discussion

This paper describes the state of the art in the diagnostics of rotating electrical machines and drives, dividing the different methodologies according to the main parts of the machine where a fault can occur. It is nearly impossible to be comprehensive on this topic, but efforts have focused on highlighting new techniques proposed at conferences and journal papers in recent years, along with established methods still used in the industry. In addition, some older papers, which represent the milestones of electrical machine fault detection, are reported as key references.

This review has elaborated some interesting suggestions on the modern diagnostic trend of electrical machines.

In the first place, the effort made by the research groups in the study of methods applicable on-line, during the normal operation of the machine in steady state or under a starting transient, is evident; the advantage is twofold, since an on-line method permits the continuous operation of the entire process in which the machine is involved and, moreover, it can evaluate the conditions of the machine under its normal stresses (thermal, electrical, mechanical, environmental, etc.).

In addition, as electronic converters are increasingly present in electrical systems, a significant amount of research has been devoted to solving the problems derived from the supply by the electronic converter, which can dampen or mask the typical signatures induced by faults in the signals measured for diagnostic purposes. For the same reason, the literature of recent years has been extended to new types of electrical drives, such as PMSMs, SynRMs and DFIGs, and it is not limited to IMs and SMs, as in the past.

The massive use of electronic converters has even induced the use of electromagnetic signals as much as possible for diagnostic purposes in research, to detect faults not only of electrical origin, but also of mechanical origin, such as bearing failure and load anomalies, which in the past required vibration measurement. The analysis of electromagnetic signals in electrical drives allows the use of transducers already present in the system, avoiding the insertion of additional sensors.

The outcomes of recent papers are promising, although, in some cases, the new proposed techniques must be optimized and become more robust. Therefore, research in the diagnostic field is still open to further theoretical and experimental studies, with the aim of achieving methodologies to be applied on-line and able to detect all types of faults in their early stage, by means of fewer sensors, possibly already present in the electrical drive for the control of the machine.

References

1. Nøland, J.K.; Leandro, M.; Suul, J.A.; Molinas, M. High-power machines and starter-generator topologies for more electric aircraft: A technology outlook. *IEEE Access* **2020**, *8*, 130104–130123. [CrossRef]
2. Strangas, E.G.; Aviyente, S.; Neely, J.D.; Zaidi, S.S.H. The effect of failure prognosis and mitigation on the reliability of permanent-magnet AC motor drives. *IEEE Trans. Ind. Electron.* **2013**, *60*, 3519–3528. [CrossRef]
3. Muetze, A.; Strangas, E.G. The useful life of inverter-based drive bearings: Methods and research directions from localized maintenance to prognosis. *IEEE Ind. Appl. Mag.* **2016**, *22*, 63–73. [CrossRef]
4. Jensen, W.R.; Strangas, E.G.; Foster, S.N. A method for online stator insulation prognosis for inverter-driven machines. *IEEE Trans. Ind. Appl.* **2018**, *54*, 5897–5906. [CrossRef]
5. Zaidan, M.A.; Harrison, R.F.; Mills, A.R.; Fleming, P.J. Bayesian hierarchical models for aerospace gas turbine engine prognostics. *Expert Syst. Appl.* **2015**, *42*, 539–553. [CrossRef]
6. Zaidan, M.A.; Mills, A.R.; Harrison, R.F.; Fleming, P.J. Gas turbine engine prognostics using Bayesian hierarchical models: A variational approach. *Mech. Syst. Signal Process.* **2016**, *70*, 120–140. [CrossRef]
7. Jin, X.; Que, Z.; Sun, Y.; Guo, Y.; Qiao, W. A data-driven approach for bearing fault prognostics. *IEEE Trans. Ind. Appl.* **2019**, *55*, 3394–3401. [CrossRef]
8. Henao, H.; Capolino, G.-A.; Fernandez-Cabanas, M.; Filippetti, F.; Bruzzese, C.; Strangas, E.; Pusca, R.; Estima, J.; Riera-Guasp, M.; Hedayati-Kia, S. Trends in fault diagnosis for electrical machines. *IEEE Ind. Electron. Mag.* **2014**, *8*, 31–42. [CrossRef]
9. Capolino, G.-A.; Antonino-Daviu, J.A.; Riera-Guasp, M. Modern diagnostics techniques for electrical machines, power electronics, and drives. *IEEE Trans. Ind. Electron.* **2015**, *62*, 1738–1745. [CrossRef]
10. Riera-Guasp, M.; Antonino-Daviu, J.A.; Capolino, G.-A. Advances in electrical machine, power electronic and drive condition monitoring and fault detection: State of the art. *IEEE Trans. Ind. Electron.* **2015**, *62*, 1746–1759. [CrossRef]
11. Capolino, G.-A.; Romary, R.; Hénao, H.; Pusca, R. State of the art on stray flux analysis in faulted electrical machines. In Proceedings of the 2019 IEEE WEMDCD, Athens, Greece, 22–23 April 2019.
12. Tavner, P.; Ran, L.; Penman, J.; Sedding, H. *Condition Monitoring of Rotating Electrical Machines*, 1st ed.; The Institution of Engineering and Technology: London, UK, 2008.
13. Frosini, L. Monitoring and diagnostics of electrical machines and drives: A state of the art. In Proceedings of the 2019 IEEE WEMDCD, Athens, Greece, 22–23 April 2019.
14. Cusido, J.; Romeral, L.; Ortega, J.; Rosero, J.; Garcia Espinosa, A. Fault detection in induction machines using power spectral density in wavelet decomposition. *IEEE Trans. Ind. Electron.* **2008**, *55*, 633–643. [CrossRef]
15. Zarri, L.; Gritli, Y.; Rossi, C.; Bellini, A.; Filippetti, F. Fault detection based on closed-loop signals for induction machines. In Proceedings of the 2015 IEEE WEMDCD, Torino, Italy, 26–27 March 2015.
16. Wolkiewicz, M.; Tarchala, G.; Orlowska-Kowalska, T. Diagnosis of stator and rotor faults of an induction motor in closed-loop control structure. In Proceedings of the 2018 SPEEDAM, Amalfi, Italy, 20–22 June 2018.
17. Wolkiewicz, M.; Tarchała, G.; Orłowska-Kowalska, T.; Kowalski, C.T. Online stator interturn short circuits monitoring in the DFOC induction-motor drive. *IEEE Trans. Ind. Electron.* **2016**, *63*, 2517–2528. [CrossRef]
18. IEC 60034-18-42. *Rotating Electrical Machines—Part 18-42: Partial Discharge Resistant Electrical Insulation Systems (Type II) Used in Rotating Electrical Machines Fed from Voltage Converters—Qualification Tests*; International Electrotechnical Commission: Geneva, Switzerland, 2017.
19. Cabanas, M.F.; Norniella, J.G.; Melero, M.G.; Rojas, C.H.; Cano, J.M.; Pedrayes, F.; Orcajo, G.A. Detection of stator winding insulation failures: On-line and off-line tests. In Proceedings of the 2013 IEEE WEMDCD, Paris, France, 11–12 March 2013.
20. Verginadis, D.; Antonino-Daviu, J.; Karlis, A.; Danikas, M.G. Diagnosis of stator faults in synchronous generators: Short review and practical case. In Proceedings of the 2020 ICEM, Gothenburg, Sweden, 23–26 August 2020. (Virtual Conference).

21. Stone, G.C.; Sedding, H.G.; Chan, C. Experience with online partial-discharge measurement in high-voltage inverter-fed motors. *IEEE Trans. Ind. Appl.* **2018**, *54*, 866–872. [CrossRef]

22. Timperley, J.E.; Vallejo, J.M. Condition assessment of electrical apparatus with EMI diagnostics. *IEEE Trans. Ind. Appl.* **2017**, *53*, 693–699. [CrossRef]

23. IEC 60034-18-41. *Rotating Electrical Machines—Part 18-41: Partial Discharge Free Electrical Insulation Systems (Type I) Used in Rotating Electrical Machines Fed from Voltage Converters—Qualification and Quality Control Tests*; International Electrotechnical Commission: Geneva, Switzerland, 2014.

24. Grubic, S.; Aller, J.; Lu, B.; Habetler, T. A survey on testing and monitoring methods for stator insulation systems of low-voltage induction machines focusing on turn insulation problems. *IEEE Trans. Ind. Electron.* **2008**, *55*, 4127–4136. [CrossRef]

25. Gandhi, A.; Corrigan, T.; Parsa, L. Recent advances in modeling and online detection of stator interturn faults in electrical motors. *IEEE Trans. Ind. Electron.* **2011**, *58*, 1567–1575. [CrossRef]

26. Henao, H.; Demian, C.; Capolino, G.-A. A frequency-domain detection of stator winding faults in induction machines using an external flux sensor. *IEEE Trans. Ind. Appl.* **2003**, *39*, 1272–1279. [CrossRef]

27. Penman, J.; Sedding, H.G.; Lloyd, B.A.; Fink, W.T. Detection and location of interturn short circuits in the stator windings of operating motors. *IEEE Trans. Energy Convers.* **1994**, *9*, 652–658. [CrossRef]

28. Stavrou, A.; Sedding, H.G.; Penman, J. Current monitoring for detecting inter-turn short circuits in induction motors. *IEEE Trans. Energy Convers.* **2001**, *16*, 32–37. [CrossRef]

29. Thomson, W.T. On-line MCSA to diagnose shorted turns in low voltage stator windings of 3-phase induction motors prior to failure. In Proceedings of the 2001 IEMDC, Cambridge, MA, USA, 17–20 June 2001.

30. Cruz, S.M.A.; Cardoso, A.J.M. Diagnosis of stator inter-turn short circuits in DTC induction motor drives. *IEEE Trans. Ind. Appl.* **2004**, *40*, 1349–1360. [CrossRef]

31. Romary, R.; Pusca, R.; Lecointe, J.P.; Brudny, J.F. Electrical machines fault diagnosis by stray flux analysis. In Proceedings of the 2013 WEMDCD, Paris, France, 11–12 March 2013.

32. Joksimovic, G.M.; Riger, J.; Wolbank, T.M.; Peric, N.; Vašak, M. Stator-current spectrum signature of healthy cage rotor induction machines. *IEEE Trans. Ind. Electron.* **2013**, *60*, 4025–4033. [CrossRef]

33. Kia, S.H.; Henao, H.; Capolino, G.-A. Efficient digital signal processing techniques for induction machine fault diagnosis. In Proceedings of the 2013 WEMDCD, Paris, France, 11–12 March 2013.

34. Cherif, H.; Menacer, A.; Romary, R.; Pusca, R. Dispersion field analysis using discrete wavelet transform for inter-turn stator fault detection in induction motors. In Proceedings of the 2017 SDEMPED, Tinos, Greece, 29 August–1 September 2017.

35. Irhoumah, M.; Pusca, R.; Lefevre, E.; Mercier, D.; Romary, R.; Demian, C. Information fusion with belief functions for detection of interturn short-circuit faults in electrical machines using external flux sensors. *IEEE Trans. Ind. Electron.* **2018**, *65*, 2642–2652. [CrossRef]

36. Irhoumah, M.; Pusca, R.; Lefevre, E.; Mercier, D.; Romary, R. Detection of the stator winding inter-turn faults in asynchronous and synchronous machines through the correlation between harmonics of the voltage of two magnetic flux sensors. *IEEE Trans. Ind. Appl.* **2019**, *55*, 2682–2689. [CrossRef]

37. Obeid, N.H.; Battiston, A.; Boileau, T.; Nahid-Mobarakeh, B. Identification and localization of incipient intermittent inter-turn fault in the stator of a three phase permanent magnet synchronous motor. In Proceedings of the 2017 SDEMPED, Tinos, Greece, 29 August–1 September 2017.

38. Bakhri, S.; Ertugrul, N.; Soong, W.L. Negative sequence current compensation for stator shorted turn detection in induction motors. In Proceeding of the 2012 IECON, Montreal, QC, Canada, 25–28 October 2012.

39. Redondo, M.; Platero, C.A.; Gyftakis, K.N. Turn-to-turn fault protection technique for synchronous machines without additional voltage transformers. In Proceedings of the 2017 SDEMPED, Tinos, Greece, 29 August–1 September 2017.

40. Filleau, C.; Picot, A.; Maussion, P.; Manfé, P.; Jannot, X. Stator short-circuit diagnosis in power alternators based on Flux2D/Matlab co-simulation. In Proceedings of the 2017 SDEMPED, Tinos, Greece, 29 August–1 September 2017.

41. Keravand, M.; Faiz, J.; Soleimani, M.; Ghasemi-Bijan, M.; Bandar-Abadi, M.; Cruz, S.M.Â. A fast, precise and low cost stator inter-turn fault diagnosis technique for wound rotor induction motors based on wavelet transform of rotor current. In Proceedings of the 2017 SDEMPED, Tinos, Greece, 29 August–1 September 2017.

42. Mirzaeva, G.; Imtiaz Saad, K. Advanced diagnosis of stator turn-to-turn faults and static eccentricity in induction motors based on internal flux measurement. *IEEE Trans. Ind. Appl.* **2018**, *54*, 3961–3970. [CrossRef]

43. Muxiri, A.C.P.; Bento, F.; Fonseca, D.S.B.; Marques Cardoso, A.J. Thermal analysis of an induction motor subjected to inter-turn short-circuit failures in the stator windings. In Proceedings of the 2019 ICIEAM, Sochi, Russia, 25–29 March 2019.

44. Tozzi, M.; Cavallini, A.; Montanari, G.C. Monitoring off-line and on-line PD under impulsive voltage on induction motors—part 1: Standard procedure. *IEEE Elect. Insul. Mag.* **2010**, *26*, 16–26. [CrossRef]

45. Frosini, L.; Zanazzo, S.; Albini, A. A wavelet-based technique to detect stator faults in inverter-fed induction motors. In Proceedings of the 2016 ICEM, Lausanne, Switzerland, 4–7 September 2016.

46. Frosini, L.; Minervini, M.; Ciceri, L.; Albini, A. Multiple faults detection in low voltage inverter-fed induction motors. In Proceedings of the 2019 SDEMPED, Toulouse, France, 27–30 August 2019.

47. Fernandez-Cavero, V.; Morinigo-Sotelo, D.; Duque-Perez, O.; Pons-Llinares, J. Fault detection in inverter-fed induction motors in transient regime: State of the art. In Proceedings of the 2015 SDEMPED, Guarda, Portugal, 1–4 September 2015.

48. Eldeeb, H.H.; Berzoy, A.; Mohammed, O. Comprehensive investigation of harmonic signatures resulting from inter-turn short-circuit faults in DTC driven IM operating in harsh environments. In Proceeding of the 2018 ICEM, Alexandroupoli, Greece, 3–6 September 2018.

49. Eldeeb, H.H.; Berzoy, A.; Mohammed, O. Stator fault detection on DTC-driven IM via magnetic signatures aided by 2-D FEA co-simulation. *IEEE Trans. Magn.* **2019**, *55*. [CrossRef]

50. Akhil Vinayak, B.; Anjali Anand, K.; Jagadanand, G. Wavelet-based real-time stator fault detection of inverter-fed induction motor. *IET Electr. Power Appl.* **2020**, *14*, 82–90. [CrossRef]

51. Afshar, M.; Tabesh, A.; Ebrahimi, M.; Khajehoddin, S.A. Stator short-circuit fault detection and location methods for brushless DFIMs using nested-loop rotor slot harmonics. *IEEE Trans. Power Electr.* **2020**, *35*, 8559–8568. [CrossRef]

52. Bianchini, C.; Torreggiani, A.; Davoli, M.; Bellini, A.; Babetto, C.; Bianchi, N. Stator fault diagnosis by reactive power in dual three-phase reluctance motors. In Proceedings of the 2019 SDEMPED, Toulouse, France, 27–30 August 2019.

53. Filippetti, F.; Bellini, A.; Capolino, G.-A. Condition monitoring and diagnosis of rotor faults in induction machines: State of art and future perspectives. In Proceedings of the 2013 WEMDCD, Paris, France, 11–12 March 2013.

54. Salah, M.; Bacha, K.; Chaari, A.; Benbouzid, M.E.H. Brushless three-phase synchronous generator under rotating diode failure conditions. *IEEE Trans. Energy Convers.* **2014**, *29*, 594–601. [CrossRef]

55. Wei, Z.; Liu, W.; Pang, J.; Sun, C.; Zhang, Z.; Ma, P. Fault diagnosis of rotating rectifier based on waveform distortion and polarity of current. *IEEE Trans. Ind. Appl.* **2019**, *55*, 2356–2367. [CrossRef]

56. Yun, J.; Park, S.W.; Yang, C.; Lee, S.B.; Antonino-Daviu, J.A.; Sasic, M.; Stone, G.C. Airgap search coil-based detection of damper bar failures in salient pole synchronous motors. *IEEE Trans. Ind. Appl.* **2019**, *55*, 3640–3648. [CrossRef]

57. Antonino-Daviu, J.; Fuster-Roig, V.; Park, S.; Park, Y.; Choi, H.; Park, J.; Lee, S.B. Electrical monitoring of damper bar condition in salient-pole synchronous motors without motor disassembly. *IEEE Trans. Ind. Appl.* **2020**, *56*, 1423–1431. [CrossRef]

58. Castro-Coronado, H.; Antonino-Daviu, J.; Quijano-Lopez, A.; Llovera-Segovia, P.; Fuster-Roig, V. Stray-flux and current analyses under starting for the detection of damper failures in cylindrical rotor synchronous machines. In Proceedings of the 2020 ICEM, Gothenburg, Sweden, 23–26 August 2020. (Virtual Conference).

59. Jeong, M.; Yun, J.; Park, Y.; Lee, S.B.; Gyftakis, K.N. Quality assurance testing for screening defective aluminum die-cast rotors of squirrel cage induction machines. *IEEE Trans. Ind. Appl.* **2018**, *54*, 2246–2254. [CrossRef]

60. Gyftakis, K.N.; Spyropoulos, D.V.; Arvanitakis, I.; Panagiotou, P.A.; Mitronikas, E.D. Induction motors torque analysis via frequency extraction for reliable broken rotor bar detection. In Proceedings of the 2020 ICEM, Gothenburg, Sweden, 23–26 August 2020. (Virtual Conference).

61. Antonino-Daviu, J.; Riera-Guasp, M.; Pons-Llinares, J.; Park, J.; Lee, S.B.; Yoo, J.; Kral, C. Detection of broken outer-cage bars for double-cage induction motors under the startup transient. *IEEE Trans. Ind. Appl.* **2012**, *48*, 1539–1548. [CrossRef]

62. Gritli, Y.; Lee, S.B.; Filippetti, F.; Zarri, L. Advanced diagnosis of outer cage damage in double squirrel cage induction motors under time-varying condition based on wavelet analysis. *IEEE Trans. Ind. Appl.* **2014**, *50*, 1791–1800. [CrossRef]

63. Yang, C.; Kang, T.J.; Hyun, D.; Lee, S.B.; Antonino-Daviu, J.A.; Pons-Llinares, J. Reliable detection of induction motor rotor faults under the rotor axial air duct influence. *IEEE Trans. Ind. Appl.* **2014**, *50*, 2493–2502. [CrossRef]

64. Kim, H.; Lee, S.B.; Park, S.; Kia, S.H.; Capolino, G.-A. Reliable detection of rotor faults under the influence of low-frequency load torque oscillations for applications with speed reduction couplings. *IEEE Trans. Ind. Appl.* **2016**, *52*, 1460–1468. [CrossRef]

65. Drif, M.; Kim, H.; Kim, J.; Lee, S.B.; Cardoso, A.J.M. Active and reactive power spectra-based detection and separation of rotor faults and low frequency load torque oscillations. *IEEE Trans. Ind. Appl.* **2017**, *53*, 2702–2710. [CrossRef]

66. Panagiotou, P.A.; Arvanitakis, I.; Lophitis, N.; Antonino-Daviu, J.A.; Gyftakis, K.N. On the broken rotor bar diagnosis using time-frequency analysis: Is one spectral representation enough for the characterization of monitored signals? *IET Electr. Power Appl.* **2019**, *11*, 932–942. [CrossRef]

67. Panagiotou, P.A.; Arvanitakis, I.; Lophitis, N.; Antonino-Daviu, J.A.; Gyftakis, K.N. A new approach for broken rotor bar detection in induction motors using frequency extraction in stray flux signals. *IEEE Trans. Ind. Appl.* **2019**, *55*, 3501–3511. [CrossRef]

68. Gyftakis, K.N.; Panagiotou, P.A.; Lee, S.B. The role of the mechanical speed frequency on the induction motor fault detection via the stray flux. In Proceedings of the 2019 SDEMPED, Toulouse, France, 27–30 August 2019.

69. Ramirez-Nunez, J.A.; Antonino-Daviu, J.A.; Climente-Alarcón, V.; Quijano-López, A.; Razik, H.; Osornio-Rios, R.A.; Romero-Troncoso, R.J. Evaluation of the detectability of electromechanical faults in induction motors via transient analysis of the stray flux. *IEEE Trans. Ind. Appl.* **2018**, *54*, 4324–4332. [CrossRef]

70. Concari, C.; Franceschini, G.; Tassoni, C. Differential diagnosis based on multivariable monitoring to assess induction machine rotor conditions. *IEEE Trans. Ind. Electron.* **2008**, *55*, 4156–4166. [CrossRef]

71. Wolbank, T.M.; Nussbaumer, P.; Chen, H.; Macheiner, P.E. Monitoring of rotor-bar defects in inverter-fed induction machines at zero load and speed. *IEEE Trans. Ind. Electron.* **2011**, *58*, 1468–1478. [CrossRef]

72. Romero-Troncoso, R.J.; Garcia-Perez, A.; Moringo-Sotelo, D.; Duque-Perez, O.; Osornio-Rios, R.A.; Ibarra-Manzano, M.A. Rotor unbalance and broken rotor bar detection in inverter-fed induction motors at start-up and steady-state regimes by high-resolution spectral analysis. *Elsevier Electr. Power Syst. Res.* **2016**, *133*, 142–148. [CrossRef]

73. Fernandez-Cavero, V.; Pons-Llinares, J.; Duque-Perez, O.; Moringo-Sotelo, D. Detection of broken rotor bars in non-linear startups of inverter-fed induction motors. In Proceedings of the 2019 SDEMPED, Toulouse, France, 27–30 August 2019.

74. Garcia-Perez, A.; Romero-Troncoso, R.J.; Camarena-Martinez, D.; Osornio-Rios, R.A.; Amezquita-Sanchez, J.P. Broken rotor bar detection in inverter-fed induction motors by time-corrected instantaneous frequency spectrogram. In Proceedings of the 2017 SDEMPED, Tinos, Greece, 29 August–1 September 2017.

75. Godoy, W.F.; da Silva, I.N.; Goedtel, A.; Palácios, R.H.C.; Scalassara, P.; Moriñnigo-Sotelo, D.; Duque-Perez, O. Detection of broken rotor bars faults in inverter-fed induction motors. In Proceeding of the 2018 ICEM, Alexandroupoli, Greece, 3–6 September 2018.

76. Asad, B.; Vaimann, T.; Belahcen, A.; Kallaste, A. Broken rotor bar fault diagnostic of inverter fed induction motor using FFT, Hilbert and Park's vector approach. In Proceeding of the 2018 ICEM, Alexandroupoli, Greece, 3–6 September 2018.

77. Spyropoulos, D.V.; Mitronikas, E.D.; Dermatas, E.S. Broken rotor bar fault diagnosis in induction motors using a Goertzel algorithm. In Proceeding of the 2018 ICEM, Alexandroupoli, Greece, 3–6 September 2018.

78. Yazidi, A.; Henao, H.; Capolino, G.-A.; Betin, F. Rotor inter-turn short circuit fault detection in wound rotor induction machines. In Proceedings of the 2010 ICEM, Rome, Italy, 6–8 September 2010.

79. Gritli, Y.; Zarri, L.; Mengoni, M.; Rossi, C.; Filippetti, F.; Casadei, D. Rotor fault diagnosis of wound rotor induction machine for wind energy conversion system under time-varying conditions based on optimized wavelet transform analysis. In Proceedings of the 2013 EPE, Lille, France, 2–6 September 2013.

80. Gritli, Y.; Zarri, L.; Rossi, C.; Filippetti, F.; Capolino, G.-A.; Casadei, D. Advanced diagnosis of electrical faults in wound-rotor induction machines. *IEEE Trans. Ind. Electr.* **2013**, *60*, 4012–4024. [CrossRef]

81. Kia, S.H. Monitoring of wound rotor induction machines by means of discrete wavelet transform. *Electr. Power Compon. Syst.* **2018**, *46*, 2021–2035. [CrossRef]

82. Moosavi, S.-M.M.; Faiz, J.; Abadi, M.B.; Cruz, S.M.A. Comparison of rotor electrical fault indices owing to inter-turn short circuit and unbalanced resistance in doubly-fed induction generator. *IET Electr. Power Appl.* **2019**, *13*, 235–242. [CrossRef]

83. Zamudio-Ramirez, I.; Antonino-Daviu, J.A.; Osornio-Rios, R.A.; Romero-Troncoso, R.J.; Razik, H. Detection of winding asymmetries in wound-rotor induction motors via transient analysis of the external magnetic field. *IEEE Trans. Ind. Electron.* **2020**, *67*, 5050–5059. [CrossRef]

84. Urresty, J.; Riba, J.R.; Delgado, M.; Romeral, L. Detection of demagnetization faults in surface-mounted permanent magnet synchronous motors by means of the zero-sequence voltage component. *IEEE Trans. Energy Convers.* **2012**, *27*, 42–51. [CrossRef]

85. Zarate, S.; Almandoz, G.; Ugalde, G.; Poza, J.; Escalada, A.J. Effects of demagnetization on torque ripples in permanent magnet synchronous machines with manufacturing tolerances. In Proceeding of the 2018 ICEM, Alexandroupoli, Greece, 3–6 September 2018.

86. Riba, J.R.; Rosero, J.A.; Garcia, A.; Romeral, L. Detection of demagnetization faults in permanent-magnet synchronous motors under nonstationary conditions. *IEEE Trans. Magn.* **2009**, *45*, 2961–2969.

87. Casadei, D.; Filippetti, F.; Rossi, C.; Stefani, A. Magnets faults characterization for permanent magnet synchronous motors. In Proceeding of the 2009 SDEMPED, Cargese, France, 31 August–3 September 2009.

88. Park, Y.; Fernandez, D.; Lee, S.B.; Hyun, D.; Jeong, M.; Kommuri, S.K.; Cho, C.; Diaz, D.; Briz, F. On-line detection of rotor eccentricity and demagnetization faults in PMSMs based on Hall-effect field sensor measurements. *IEEE Trans. Ind. Appl.* **2019**, *55*, 2499–2509. [CrossRef]

89. Reigosa, D.; Fernández, D.; Martínez, M.; Park, Y.; Lee, S.B.; Briz, F. Permanent magnet synchronous machine non-uniform demagnetization detection using zero-sequence magnetic field density. *IEEE Trans. Ind. Appl.* **2019**, *55*, 3823–3833. [CrossRef]

90. Immovilli, F.; Bianchini, C.; Cocconcelli, M.; Bellini, A.; Rubini, R. Bearing fault model for induction motor with externally induced vibration. *IEEE Trans. Ind. Electron.* **2013**, *60*, 3408–3418. [CrossRef]

91. Randall, R.B.; Antoni, J. Rolling element bearing diagnostics—A tutorial. *Mech. Syst. Signal Process.* **2011**, *25*, 485–520. [CrossRef]

92. Schmidt, S.; Heyns, P.S.; Gryllias, K.C. A discrepancy analysis methodology for rolling element bearing diagnostics under variable speed conditions. *Mech. Syst. Signal Process.* **2019**, *116*, 40–61. [CrossRef]

93. Frosini, L.; Bassi, E. Stator current and motor efficiency as indicators for different types of bearing faults in induction motors. *IEEE Trans. Ind. Electron.* **2010**, *57*, 244–251. [CrossRef]

94. Frosini, L.; Harlişca, C.; Szabó, L. Induction machine bearing faults detection by means of statistical processing of the stray flux measurements. *IEEE Trans. Ind. Electron.* **2015**, *62*, 1846–1854. [CrossRef]

95. Frosini, L.; Magnaghi, M.; Albini, A.; Magrotti, G. A new diagnostic instrument to detect generalized roughness in rolling bearings for induction motors. In Proceedings of the 2015 SDEMPED, Guarda, Portugal, 1–4 September 2015.

96. Martínez-Montes, E.; Jiménez-Chillarón, L.; Gilabert-Marzal, J.; Antonino-Daviu, J.; Quijano-López, A. Evaluation of the detectability of bearing faults at different load levels through the analysis of stator currents. In Proceeding of the 2018 ICEM, Alexandroupoli, Greece, 3–6 September 2018.

97. Leite, V.C.M.N.; Borges da Silva, J.G.; Cintra Veloso, G.F.; Borges da Silva, L.E.; Lambert-Torres, G.; Bonaldi, E.L.; de Lacerda de Oliveira, L.E. Detection of localized bearing faults in induction machines by spectral kurtosis and envelope analysis of stator current. *IEEE Trans. Ind. Electron.* **2015**, *62*, 1855–1865. [CrossRef]

98. Hamadache, M.; Lee, D.; Veluvolu, K.C. Rotor speed-based bearing fault diagnosis (RSB-BFD) under variable speed and constant load. *IEEE Trans. Ind. Electron.* **2015**, *62*, 6486–6495. [CrossRef]

99. Rzeszucinski, P.; Orman, M.; Pinto, C.T.; Tkaczyk, A.; Sulowicz, M. Bearing health diagnosed with a mobile phone: Acoustic signal measurements can be used to test for structural faults in motors. *IEEE Ind. Appl. Mag.* **2018**, *24*, 17–23. [CrossRef]

100. Gongora, W.S.; Goedtel, A.; Favoretto Castoldi, M.; Oliveira da Silva, S.A.; Nunes da Silva, I. Embedded system to detect bearing faults in line-connected induction motors. In Proceeding of the 2018 ICEM, Alexandroupoli, Greece, 3–6 September 2018.

101. Ye, M.; Huang, J. Bearing fault diagnosis under time-varying speed and load conditions via speed sensorless algorithm and angular resample. In Proceeding of the 2018 ICEM, Alexandroupoli, Greece, 3–6 September 2018.

102. He, M.; He, D. Deep learning based approach for bearing fault diagnosis. *IEEE Trans. Ind. Appl.* **2017**, *53*, 3057–3065. [CrossRef]

103. Guedidi, A.; Guettaf, A.; Cardoso, A.J.M.; Laala, W.; Arif, A. Bearing faults classification based on variational mode decomposition and artificial neural network. In Proceedings of the 2019 SDEMPED, Toulouse, France, 27–30 August 2019.

104. Khlaief, A.; Nguyen, K.; Medjaher, K.; Picot, A.; Maussion, P.; Tobon, D.; Chauchat, B.; Cheron, R. Feature engineering for ball bearing combined-fault detection and diagnostic. In Proceedings of the 2019 SDEMPED, Toulouse, France, 27–30 August 2019.

105. Immovilli, F.; Lippi, M.; Cocconcelli, M. Automated bearing fault detection via long short-term memory networks. In Proceedings of the 2019 SDEMPED, Toulouse, France, 27–30 August 2019.

106. Zhang, S.; Zhang, S.; Wang, B.; Habetler, T.G. Deep learning algorithms for bearing fault diagnostics— A comprehensive review. *IEEE Access* **2020**, *8*, 29857–29881. [CrossRef]

107. Klempner, G.; Kerszenbaum, I. *Handbook of Large Turbo-Generators Operation and Maintenance*, 3rd ed.; Wiley-IEEE Press: Piscataway, NJ, USA, 2018.

108. Salah, A.A.; Dorrell, D.G.; Guo, Y. A review of the monitoring and damping unbalanced magnetic pull in induction machines due to rotor eccentricity. *IEEE Trans. Ind. Appl.* **2019**, *55*, 2569–2580. [CrossRef]

109. Dorrell, D.G.; Thomson, W.T. Analysis of airgap flux, current, and vibration signals as a function of the combination of static and dynamic airgap eccentricity in 3-phase induction motors. *IEEE Trans. Ind. Appl.* **1997**, *33*, 24–34. [CrossRef]

110. Concari, C.; Franceschini, G.; Tassoni, C. Toward practical quantification of induction drive mixed eccentricity. *IEEE Trans. Ind. Appl.* **2011**, *47*, 1232–1239. [CrossRef]

111. Park, Y.; Choi, H.; Shin, J.; Park, J.; Lee, S.-B.; Jo, H. Airgap flux based detection and classification of induction motor rotor and load defects during the starting transient. *IEEE Trans. Ind. Electron.* **2020**, *67*, 10075–10084. [CrossRef]

112. Samonig, M.A.; Wolbank, T.M. Exploiting rotor slotting harmonics to determine and separate static and dynamic air-gap eccentricity in induction machines. In Proceedings of the 2017 SDEMPED, Tinos, Greece, 29 August–1 September 2017.

113. Ebrahimi, B.M.; Faiz, J.; Roshtkhari, M.J. Static-, dynamic-, and mixed-eccentricity fault diagnoses in permanent-magnet synchronous motors. *IEEE Trans. Ind. Electron.* **2009**, *56*, 4727–4739. [CrossRef]

114. Ebrahimi, B.M.; Roshtkhari, M.J.; Faiz, J.; Khatam, S.V. Advanced eccentricity fault recognition in permanent magnet synchronous motors using stator current signature analysis. *IEEE Trans. Ind. Electron.* **2014**, *61*, 2041–2052. [CrossRef]

115. Herman, J.; Beguš, S.; Mihalič, P.; Bojkovski, J. Novel method for direct measurement of air gap anomalies in direct-drive electrical motors. *IEEE Trans. Ind. Electron.* **2020**, *67*, 2422–2429. [CrossRef]

116. Gyftakis, K.N.; Platero, C.A.; Bernal, S. Off-line detection of static eccentricity in salient-pole synchronous machines. In Proceeding of the 2018 ICEM, Alexandroupoli, Greece, 3–6 September 2018.

117. Salles, G.; Filippetti, F.; Tassoni, C.; Grellet, G.; Franceschini, G. Monitoring of induction motor load by neural network techniques. *IEEE Trans. Power Electron.* **2000**, *15*, 762–768. [CrossRef]

118. Blodt, M.; Regnier, J.; Chabert, M.; Faucher, J. Fault indicators for stator current based detection of torque oscillations in induction motors at variable speed using time-frequency analysis. In Proceedings of the 2006 IET PEMD, Dublin, Ireland, 4–6 April 2006.

119. Frosini, L.; Beccarisi, F.; Albini, A. Detection of torque oscillations in induction motor drives by linear discriminant analysis. In Proceedings of the 2017 SDEMPED, Tinos, Greece, 29 August–1 September 2017.

120. D'Elia, G.; Mucchi, E.; Cocconcelli, M. On the identification of the angular position of gears for the diagnostics of planetary gearboxes. *Mech. Syst. Signal Process.* **2017**, *83*, 305–320. [CrossRef]

121. Kia, S.H.; Henao, H.; Capolino, G.-A. A comparative study of acoustic, vibration and stator current signatures for gear tooth fault diagnosis. In Proceedings of the 2012 ICEM, Marseille, France, 2–5 September 2012.

122. Kia, S.H.; Henao, H.; Capolino, G.-A. Gear tooth surface damage fault detection using induction machine stator current space vector analysis. *IEEE Trans. Ind. Electron.* **2015**, *62*, 1866–1878. [CrossRef]

123. Kia, S.H.; Marzebali, M.H.; Henao, H.; Capolino, G.-A.; Faiz, J. Simulation and experimental analyses of planetary gear tooth defect using electrical and mechanical signatures of wound rotor induction generators. In Proceedings of the 2017 SDEMPED, Tinos, Greece, 29 August–1 September 2017.

124. Bertenshaw, D.R.; Smith, A.C.; Ho, C.W.; Chan, T.; Sasic, M. Detection of stator core faults in large electrical machines. *IET Electr. Power Appl.* **2012**, *6*, 295–301. [CrossRef]

125. Lee, K.; Hong, J.; Lee, K.-W.; Lee, S.B.; Wiedenbrug, E.J. A stator-core quality-assessment technique for inverter-fed induction machines. *IEEE Trans. Ind. Appl.* **2010**, *46*, 213–221.

126. Romary, R.; Jelassi, S.; Brudny, J.F. Stator-interlaminar-fault detection using an external-flux-density sensor. *IEEE Trans. Ind. Electron.* **2010**, *57*, 237–243. [CrossRef]

127. Romary, R.; Demian, C.; Schlupp, P.; Roger, J.-Y. Offline and online methods for stator core fault detection in large generators. *IEEE Trans. Ind. Electron.* **2013**, *60*, 4084–4092. [CrossRef]

Rotating Electrical Machine Condition Monitoring Automation

Mallikarjun Kande [1,2,*,†], **Alf J. Isaksson** [3], **Rajeev Thottappillil** [1] and **Nathaniel Taylor** [1]

[1] KTH—Royal Institute of Technology, 100 44 Stockholm, Sweden; rajeev@kth.se (R.T.); taylor@kth.se (N.T.)
[2] ABB Power Generation, Wickliffe, OH 44092, USA
[3] ABB Corporate Research, 721 78 Västerås, Sweden; alf.isaksson@se.abb.com
*   Correspondence: mskande@kth.se
†   Current address: 29801 Euclid Ave, Wickliffe, OH 44092, USA.

Abstract: We review existing machine condition monitoring techniques and industrial automation for plant-wide condition monitoring of rotating electrical machines. Cost and complexity of a condition monitoring system increase with the number of measurements, so extensive condition monitoring is currently mainly restricted to the situations where the consequences of poor availability, yield or quality are so severe that they clearly justify the investment in monitoring. There are challenges to obtaining plant-wide monitoring that includes even small machines and non-critical applications. One of the major inhibiting factors is the ratio of condition monitoring cost to equipment cost, which is crucial to the acceptance of using monitoring to guide maintenance for a large fleet of electrical machinery. Ongoing developments in sensing, communication and computation for industrial automation may greatly extend the set of machines for which extensive monitoring is viable.

Keywords: condition monitoring; rotating electrical machine; motor fleet; plant-wide

1. Introduction

Condition Monitoring (CM) is a process of acquiring equipment health status and predicting the operational ability of a system in a given environment: the health of the system is evaluated during its operation, and possible failures associated with it are recognized at an early stage. Motivations for condition monitoring (CM) in industrial automation include reductions in downtime, maintenance activity and related faults, and increases in energy efficiency, yield, and quality. Predictive diagnostics based on CM permit a well-informed plant maintenance activity. Condition based maintenance (CBM) using the equipment condition assessment has several benefits as compared to scheduled cyclic or reactive plant maintenance, mainly in terms of reduced downtime and replacement cost [1]. The methods used in system health assessment depend on plant infrastructure, operational criticality, process work flow, and ease of repair and service. A cost effective CBM can in some cases be achieved even with low cost condition monitoring equipment, for example by statistical treatment to reduce false positives and negatives in spite of the uncertainty of measurements or fault-trigger thresholds [2]. The success of CBM depends on the overall cost of the condition monitoring system in a plant, its arrangement, and on the relative benefits compared to the operation and maintenance costs over the life of the plant.

The infrastructure chain in the power and process industries commonly has power generation, substation, power distribution, process control and plant operation equipment. Most of these subsystems are integrated with the distributed control system (DCS) and can publish their health status periodically for predictive diagnostics. The successful deployment of condition monitoring in process and power industries also relies upon the combined analysis of interrelated elements, such as structural, stochastic, resource, economic, etc, [3]. Generators, motors, turbines, and other rotating electrical machines are usually distributed in large numbers across the plant. The monitoring of the distributed rotating electrical machines such as smart motors sending fault information wirelessly may demand modification of the legacy CM practices. This paper reviews the traditional condition monitoring techniques, recent activities, and deployment of the intelligent machines in an industrial automation system. It covers the evolution and recent trends regarding rotating electrical machines, as well as deployment use case for plant-wide operations and future directions.

2. Machine Condition Monitoring: An Evolution

One of the first references to machine condition monitoring can be found in 1924 by Walker [4], who proposed a measurement system that used data and theoretically based analysis for diagnosis of motor faults. Vibration analysis for machine diagnosis started about 1950 [5], with accelerometers proposed in place of magnetic velocity pickup, besides development of vibration and temperature pick-ups, spectrum analysis, and computer-based diagnostic applications. Another important proposal was to use non-contacting vibration pick-ups: non-contact sensors for condition monitoring applications proposed during this period did not gain popularity due to technical limitations of sensors, cabling, and data acquisition electronics [5]. Partial discharge (PD) measurements have been widely used to detect insulation faults in machines above the low-voltage level. These have been correlated with types of insulation degradation for early warning of developing problems [6]. During the 1980s, Timperley showed that a combination of 'electromagnetic interference' measured at the neutral of an operating machine, together with wideband spectrum analysis, can provide significant diagnostic information with long lead time before complete failure [7].

The importance of monitoring systems that operate without shutdown of the machine (i.e., online) was further emphasized in [8], in which the benefits of daily monitoring of machines are shown through radio frequency (RF) measurements at the machine neutral using permanently installed RF sensors. The maturity of adjacent technologies during 2000 to 2010 in the area of sensing, digital data acquisition and integration of communication equipment has shown the increasing viability of continuous monitoring. The practitioners have made an effort to prove that the continuous diagnostics with real-time update of the plant-wide machine condition increases business profit through informed maintenance and service decisions [9,10]. The development of an integrated system prompted industries to use condition based maintenance through continuous assessment of equipment health across the plant. During the past three decades, there has been a steady growth in machine condition monitoring technologies owing to its importance concerning plant safety and mission criticality. Monitoring methods have been matured giving raise to the adaptation of reliable CM systems for fault prediction, detection, and fault location identification for various types of rotating machines in the plant's system critical chain.

Long service life, low cost, and useful fault indication are some of the most important aspects for the evolution of condition monitoring systems. The driving factors for this evolution are as given below:

- Efficient maintenance and reliable operation: accurate detection of incipient faults with sufficient lead time; thus, evolutions of health monitoring technologies are motivated by the increased productivity and/or reduced investments for maintenance.
- Maximized overall profitability: the monitoring technologies should have a low cost, to the point where further reduction would deteriorate the monitoring quality significantly.
- A sustainable business based on reliable monitoring systems whose service life equals that of the monitored equipment and at the same time offers lower total operating cost.

Figure 1 shows the condition monitoring technology trends that evolved over the years aligning to the plant life cycle functions. The plant life cycle can have five most important phases: (i) maintenance and service plans for developing the case for the use of condition monitoring in the plant; (ii) commissioning to deploy the monitoring applications; (iii) engineering to configure the system elements; (iv) plant operations for setting the human–machine interface (HMI) views, thresholds, and generating actions for maintenance, and (v) maintenance and service activities for keeping the main stream equipment in healthy conditions for its long and precise working [11].

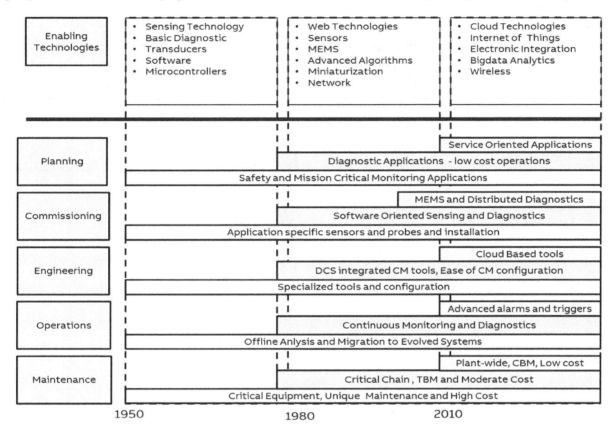

Figure 1. Technology trends, plant life cycle and monitoring system evolution.

Each phase has its technical challenges for employing condition monitoring solutions for rotating electrical machines. The planning phase deals with selection of the monitoring and the diagnostic applications based on the customer end use requirements. The rotating machine diagnostic applications are largely focused on detection of bearing and insulation failures using thermal, vibration, acoustic, magnetic, and partial discharge sensors. Identification of the correct sensor combination and corresponding diagnostic algorithms that can completely meet the end use application is one of the biggest challenges. The survey conducted addresses the steady growth that has been established in diagnostic algorithms and associated elements in areas of the rotating machine monitoring. The related challenges associated with these elements, such as sensor mounting and compensation, data acquisition, computation capability, etc., are assessed with regard to condition monitoring applications. For example, micro electro-mechanical sensors (MEMS) may bring down the costs, provided they exhibit long-term stability and reliable functioning throughout the service life of the equipment. During commissioning, it is vital to have a convenient setup for the hardware and software: this includes the installation of a measurement system that consists of sensors, sensor mount, data acquisition units, a communication unit, etc. The system configuration and architecture influence the monitoring cost per point. Hence, it is necessary to optimize the measurement, communication, and distribution of processing algorithm at various levels, in order to meet the monitoring application specifications [12,13]. The engineering phase is mainly related to configuring the system for its functionality; the greatest challenge is getting unified tools that can work along with the automation system native tools. Operation related activities are mainly related to maintaining the condition monitoring system, performing information acquisition, and running the diagnostics expert system for fault predictions. This system continuously updates the information related to early failures during the plant operations [14].

Condition based maintenance (CBM) in industrial automation has evolved to achieve reduction of maintenance effort and time. The challenges are the optimization of maintenance work flow, information system, timely maintenance, and equipment service. Various elements within the maintenance process are shown in Figure 2. The process optimization includes minimizing the activities associated with managing the measurement system, information system and maintenance work flow. It is also necessary to reduce time between a maintenance trigger and equipment service or repair or replacement, which is efficient enough to justify the condition monitoring investments [15].

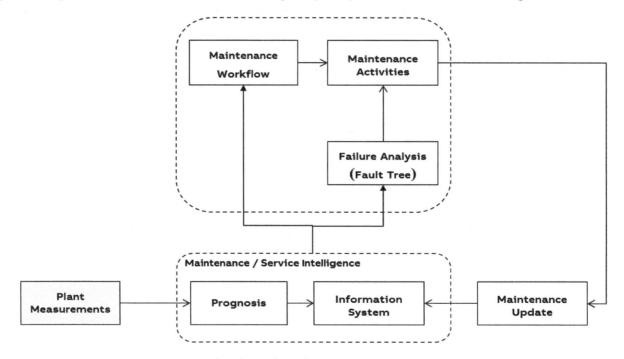

Figure 2. Condition based plant maintenance process.

Business models for condition monitoring evolved from a selective equipment monitoring to plant-wide diagnostics of all the dependent subsystems, i.e., a fleet of machines across the plant. The decision of using a condition monitoring system in the plant is mostly based on the trade-off between the risks associated with equipment failure and number of check points [16]. Recent trends show integration of monitoring functions with process control, leading to the development of suitable platforms, in particular smart sensors and related application software.

Figure 3 represents a typical process flow diagram for the plant-wide machine monitoring system. This type of deployment needs multi-point measurement, changing the economic scale of sensors required for machine condition monitoring. If the monitored device is in the critical chain, then the decision to employ health monitoring could be easy; if not, cost per monitoring point needs to be justified concerning business returns associated with machine maintenance, service revenue, product quality, and yield [17].

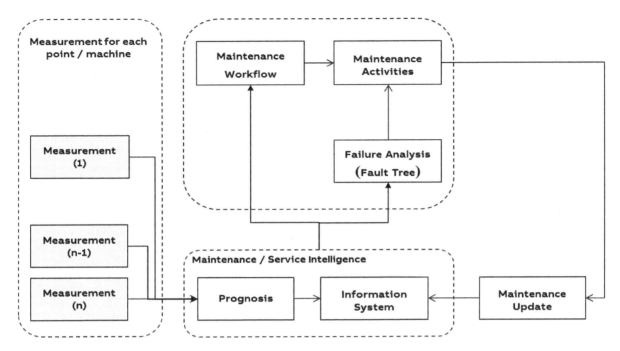

Figure 3. Condition based maintenance for plant-wide fleet monitoring.

3. Recent Trends in Rotating Machine Fault Diagnostics

Torque dynamics, torsional oscillations, shaft fatigue, and ageing of other mechanical parts, such as bearings and gearboxes, constitute anomalies internal to the machine, whereas extreme loading or sub synchronous resonance (SSR) mainly constitute anomalies external to the machine. The main areas of investigation in electrical machines condition diagnostics have been:

- Monitoring and distributed control system (DCS) integration,
- Diagnostic algorithms,
- Advanced diagnostic algorithms.

3.1. Monitoring and Distributed Control System (DCS) Integration

Integration of condition monitoring and the distributed control system is of crucial importance for better plant observability and operations. As shown in Figure 4, the integration approaches can be of three types: on-equipment, on-premise, and on-cloud. On-equipment integration provides a continuous monitoring of the equipment, and it usually has high update rates from the sensors for diagnostic algorithms.

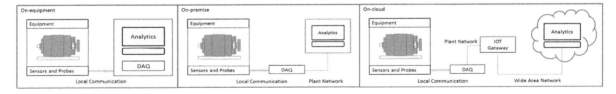

Figure 4. Monitoring system integration in industrial automation.

On-premise integration has a data acquisition unit to collect sensor data and send sensor information to diagnostics running as part of the plant operations. Typically, the data for the diagnostics is transferred using the plant network, and may have data rate limits. However, they offer sufficient computational resources for diagnostic algorithms. On-cloud diagnostics offer elasticity required for the data and computational resources. The cloud solution provides the historian for multi-dimensional data and sophisticated diagnostic algorithms for offline analysis and fault predictions. The equipment diagnostics may demand different resources for update rates, computation, notification mechanisms, and reinitialization. The evolutionary needs are improvement of diagnostics with an easy retrofit, augmentation of new methods for better fault diagnosis, error detection using newly invented techniques, and supporting systems. A general overview of diagnostic algorithms is given with an example of induction machine diagnostics. In addition, new diagnostic methods using digital signal processing and expert systems based on artificial intelligence are included. Still, the major challenge in automation system is scaling the system diagnostic abilities without losing the existing industrial practices. Automation system architecture uses a suitable integration method and it depends on the type of monitoring such as continuous versus offline, along with the diagnostic algorithms and maintenance trigger mechanisms.

3.2. Diagnostic Algorithms

In 2003, a literature survey was conducted and reported [18] to reflect the state-of-the-art development in this important area. The report points out the potential benefits through the utilization of advanced signal processing and artificial intelligence techniques in developing novel condition monitoring schemes. Rotating machines often run under a hostile operating environment, with possible uneven loading and overloading. Although precision processes govern the manufacturing of motors, any small but finite tolerance overrun leads to growing deterioration as time passes, resulting in eccentricities, torsional stress and other variants of stresses on the rotors, stators, gearbox, and other associated components that eventually will lead to their failures. Even the natural aging processes of the motor components get accelerated under unfavourable operating conditions. The magnetic coupling between the stator and rotor makes them work in coordination, and any anomaly to this coordinated work means that any malfunctioning of the rotors would continuously impact the working of the stator and vice versa. The most commonly used motors in the industry are induction motors, due to their robustness, ruggedness, reliability, efficiency, and low price. In this section, the induction motor is considered as an example to illustrate the recent developments in the machine diagnostics and monitoring applications. A similar approach is extended to other machines for their diagnostics on their specific parameters, such as torque ripples due to commutation in direct current (DC) machines, to manifest mechanical and electrical anomalies of the machines. Specific anomalies in rotating machines with industrial loads can be broadly categorized as listed below:

- Inter-turn short or open circuit in stator winding,
- Rotor eccentricities,
- Broken rotor bar or cracked rotor end rings,
- Static and/or dynamic air gap irregularities,
- Bent shaft (in the case of machine fleet),
- Bent shaft (fleet of turbine in case of utility),
- Shorted rotor field winding,
- Bearing and gearbox failures,
- Extremities of electrical loading and their dynamics (utilities),
- Extremities of mechanical loading and their dynamics (process plants),
- Interplay between electrical and mechanical counterparts (SSR).

Further refinement of the above list of anomalies leads to three broad categories of anomalies viz., the anomalies originating out of progressive mechanical defects, progressive electrical defects and lastly their coexistence and interplay. In an industrial application, such defects result in unscheduled maintenance leading to production loss and overall productivity reduction. Rotor failures seem to be the most common failures. Hence, development of rotor performance monitoring sensor, which works in a constrained environment is critical to the deployment of advanced techniques. Rotor failures may reflect in the motor performance by way of imbalance and modulation of stator current, heating, torque and speed ripple, increased losses, and decreased average torque leading to reduced efficiency. Stator current measurement, extraction of current signatures, and its analysis have evolved as a means to detect some rotor defects without rotor-based sensors. The most recent research work is aimed at electrical monitoring of the motor based on the use of electrical-signature-analysis techniques, with emphasis on inspecting its stator current [19–27]. Lower and upper sideband current assessment may not be a good solution, as it is hard to extract sidebands around the main frequency [28]. Therefore, it may be essential to develop different types of signal processing techniques [21,26,27,29,30], artificial intelligence (AI) [25,31,32] and parameter estimation [33] methods to be used in condition monitoring applications. Multiple voltage and current inputs enable computation and generation of derived data including space-vector current, air-gap torque, different instantaneous powers, etc., [23,34]. Similarly, signature analyses of the current Park's vector modulus [35,36], torque [34], single-phase power, partial, and total instantaneous active powers [37] have been proposed. These new techniques seem to be as effective as signature analysis when using only the line current [38]. The use of instantaneous reactive power has been shown for broken-rotor-bar diagnosis in three-phase induction motors [39].

Stator windings and their insulation system constitute the second potential source of failures since the various stresses that act on the motor cause gradual deterioration of stator insulation. Short-circuit and open-circuit of stator windings, and magnetic core failures are the failure modes associated with the stator. Coil-to-coil short circuit, phase-to-phase short circuit, and phase-to-ground short circuit variants are consequences of insulation failure, often due to a combination of temperature rise, mechanical and electric stress. On the other hand, open-circuit defects are comparatively rare and advantageously have a longer lead time to failures. Various surveys on motor reliability have been carried out over the years, and it is reported that the percentage of motor failures due to problems with the insulation is about 26% [40–42] or even 36% [43]. A turn-to-turn failure will eventually lead to complete failure of the winding, and the latency involved in this progression is unknown and hard to determine; therefore, online diagnostic techniques to detect turn-to-turn faults within a coil at an incipient stage may be needed [22,44].

The detection of stator winding faults has been reported as based on the analysis of vibration, axial leakage flux and stray flux, and these methods require the installation of sensors that may be infeasible or costly [39]. A non-invasive technique based on spectral analysis of motor current and winding temperature has been introduced to detect asymmetries of the machine. Currents may be transformed prior to analysis, e.g., by Park's transformation. Inter-turn fault detection through monitoring the presence of certain rotor-slot residual saturation and related terminal voltage harmonics is suggested. A spectral analysis based technique for diagnosis of both the rotor and stator winding faults is addressed. Specifically, signature based analysis of instantaneous active and reactive power obtained from the measured current and voltages has been carried out to detect stator winding faults. This method has been found to distinguish problems well from load torque oscillations and from other abnormalities that manifest identical behaviour. As far as the core failures, they are irreversible and they warrant the removal of motor from service [39].

3.3. Advanced Diagnostic Algorithms

The feasibility of digital signal processing (DSP) as part of analytics has triggered innovation in machine diagnostics. The signal processing has usually stringent requirements on the sensor data acquisition and update rates for accurate prediction of faults. Digital signal processing techniques for diagnostics are broadly classified as non-parametric, parametric, and high-resolution spectrum analysis methods.

Non-parametric are classical methods that begin by estimating the autocorrelation sequence from a given data, followed by the estimation of the power spectrum by employing a Fourier Transform. The Fast Fourier Transform (FFT) is computationally efficient and can give rise to a conceptually simple motor current signature analyzer (MCSA) [45]. On the other hand, in parametric methods, a process model is selected using sufficient prior knowledge and then model parameters are estimated from the process data. Finally, an estimation of the power spectrum is performed using the calculated parameters. The commonly used models are autoregressive (AR), moving average (MA), and autoregressive moving average (ARMA). Since the estimated parameters are small in number, it would be more efficient to transmit or store these parameters instead of signal values. The signal would be then reconstructed from the parameters. However, there are several applications such as steel rolling mills and cement industry processes where the operating point of the machine is not constant. This leads to the current, voltage, and power signals to be highly dynamic. Such non-stationary signals are handled by using short time Fourier transforms (STFTs). The use of STFTs and pattern-recognition techniques to detect faults in induction motors under varying operating conditions have been demonstrated [46]. The application of STFTs assumes stationary signals such as motor speed and load, which may be considered stationary in the analysis window. Accurate identification of frequency related faults can be accomplished by using a finite impulse response (FIR) filter bank combined with high-resolution spectrum analysis [47]. It corresponds to an eigenvalue analysis of the autocorrelation matrix of the motor current time series signals. The proposed method combines a FIR filter bank with high-resolution spectral analysis based on multiple signal classification for an accurate identification of the frequency-related fault. Results show the methodology potentiality as a deterministic detection technique that is suited for detecting multiple features of fault-related frequencies. For example, the bank of bandpass filters separates the original current and vibration signals into different fault-related bandwidths. Then, a high-resolution spectral analysis is applied to each bandwidth for an accurate identification of the frequency-related fault that identifies the presence of single or combined faults [47].

Another important consideration is wavelet analysis for fault diagnosis. STFTs are used to analyse non-stationary signals in a short signal window. In contrast, wavelet analysis non-stationary signals are simultaneously analysed at different resolutions: a larger window is used to get an estimate of stationary signal dynamics while smaller window is used for the transients. This multi-resolution or multi-scale view of the signal forms the foundation for wavelet analysis [48]. Non-stationary signals also originate out of starting current transients in induction motors. Wavelet analysis enables the detection of faults under no-load condition using the transient starting current of the motor [49,50].

Analytic wavelet transform (AWT) is another algorithm capable of detecting and tracking the fault frequencies. The analytical wavelet ridge detection [51] captures the small amplitude of the signals at the fault frequencies, whereas the phase information from the complex AWT facilitates tracking genuine faulty signals such as faulty stator current. Another example of use of AWT can be found in brushless DC (BLDC) Motor, where it is adapted to detect and track the fault frequencies of rotor faults such as dynamic eccentricity [47].

Another efficient algorithm is the combination of wavelet and power-spectral-density (PSD) techniques. This approach is popular in detecting faults similar to motor eccentricity and broken bars [52,53].

The typical diagnostic algorithms using wavelets, and fault signature in the frequency band interact with supply frequency. Time synchronous averaging (TSA) technique in association with Discrete Wavelet Transform can bring out more distinctive fault detection. Basically, TSA extracts the deterministic component from a signal. It is also shown that the reliability of the fault detection depends on the wavelet function [54].

3.4. Machine Diagnostics Using Artificial Intelligence

Artificial Intelligence diagnostic methods are based on learning algorithms: examples include artificial neural nets (ANN), fuzzy logic (FL), neuro-fuzzy techniques, and genetic algorithms (GA) constitute a second set of techniques termed as model independent techniques for condition monitoring. These models do not describe the dynamics and control of the monitored system but view the system as an input–output map.

Air-gap eccentricity and broken rotor bar faults can be detected using Artificial Neural Network (ANN) [32]. A three layered back propagation ANN, which has been trained using

Levenberg–Marquardt learning algorithm, uses a filtered vibration input signal. The trained NN is validated using known sets of training samples and the the residuals are generated using ANN output and the monitored data. These residuals trigger the fault indications depending on set residual thresholds. Similarly, the learning algorithms such as clustering techniques are used in combination with ANN. Moreover, it has been shown that ant behaviour based techniques implement clustering methods similar to K-means clustering for evaluation criterion [55]. It has been shown that both the electrical and the mechanical faults can be detected using Self Organizing Map (SOM) based Radial Basis Function (RBF) neural network along with its training algorithm. The SOM learning algorithm facilitates the design of the best possible network architecture according to the input data, and, by the very nature of the RBF neuron, it supports not only the fault detection but also its severity [56].

Fuzzy logic based techniques are another area of AI used mainly in condition based maintenance. Adaptive neuro–fuzzy inference systems (ANFISs), wavelet fuzzy logic wavelet packet transform, state vector machine (SVM), etc, are some of the intelligence building algorithms successfully demonstrated in machine fault predictions. ANFIS is trained using machine historical failure data, while the fault growth model is developed using trained ANFIS whose modeling noise manifests an higher order Markov model. A high-order particle filter uses this Markov model to predict the time evolution of the fault indicator in the form of a probability density function.

Another technique called Wavelet Fuzzy Logic uses a set of health condition states that constitute patterns. These patterns are mapped onto linguistic variables, which are, in turn, mapped to a fuzzy set by a set of membership functions. By means of the defined rule base, it is possible to map every set of health condition states onto one or more fuzzy logic based classification.

Wavelet Packet Transform (WPT) as an extension of the wavelet transform uses redundant basis functions and, hence, can provide an arbitrary time-frequency resolution over all frequency regions [57]. This enables the extraction of features from signals that combines stationary and non-stationary characteristics. The output of the WPT is used by the fuzzy logic functional block to generate the feature vector, which is then normalized and stored as a pattern from the experimental data. When the machine is in operation, the fault is classified comparing the feature vector extracted from real-time data with the stored patterns. As the WPT technique is an effective feature extractor, it can also be used along with support vector machine (SVM). This combination is proven for fault detection, severity assessment and detection of compound faults with high accuracy [58,59].

4. Plant-Wide Condition Monitoring

A typical deployment of a condition monitoring system consists of data acquisition (DAQ) unit, a local processing unit constantly communicating with the analytics [60]. Even though various advanced diagnostic methods like signature based techniques for the analysis of stator current and vibration signals are available in the literature, real life implementations of such methods are not numerous. The main reason is the overall commissioning and installation costs. Each DAQ unit is able to support a limited number of diagnostic algorithms and, therefore, in order to achieve a thorough plant condition monitoring, many units are required together with specialized sensors connected by means of proper cables and communication protocols [61]. Since DAQs are high performing and thus expensive, deployment cost rises significantly. The accuracy reached by these units in terms of condition monitoring is satisfactory, but, on the other hand, it is not sufficient to justify such high investment cost.

Therefore, the tendency is to monitor only a subset of the machines, and only very few physical parameters are tracked due to cost constraints. Needless to say, more economical solutions monitoring all machines' parameters are desirable. Even though cost reduction and plant-wide machine monitoring seem to go in opposite directions, with the use of latest low cost technology developments, it is possible to satisfy both of them [17]. In fact, low cost devices integrated with monitored machines can be built: they are miniaturized and more computationally powerful versions of the previous DAQ units, implementing various diagnostic algorithms at the same time and relying on board-integrated sensors. Clearly, these monitoring devices are significantly cheaper and can be mounted on the entire machine fleet without relevant economic impact. Unlike DAQs, each of them can track many machine parameters: this may result in reduced monitoring accuracy, since low cost technology may not be high performing, but the lower accuracy may be compensated by the overall plant overview given by machine fleet monitoring. The new integrated devices are able to perform first level analytics on retrieved data, but, in order to achieve deeper inference, constant communication with the remote service monitoring unit is required. Such communication needs to be standard-regulated, since various monitoring devices will be connected; even more importantly, it must meet the real-time requirements. Hence, it is necessary to adopt an integrated approach when we build the deployment architecture: distributed resource sharing, algorithms fusion and computational optimization are the key elements to consider along with the use of low-cost sensing, communication and computation devices.

The literature shows the implementation of intellectual property (IP) cores to enable hardware implementation of diagnostic algorithms [62], but they are not still practical yet. Smart sensor development and low cost continuous monitoring are ongoing research activities: the intended outcome is to offer low-cost, accurate and reliable monitoring sensors [63]. Some of the early work related to modelling and sensing are addressed [64–66]. The most recent development of plant-wide condition monitoring is the use of MEMS based smart sensors and low cost continuous monitoring system, for example: integrated sensing (vibration, temperature, acoustic, and magnetic sensors), communication, and a processing unit embedded with the motor. However, the remaining challenge is combining the state-of-the-art diagnostic algorithms in low cost computing platforms. The hardware platform using field programmable gate arrays (FPGA) seems to be promising, although not matured. A more elaborate research for the incipient detection of multiple faults along with an FPGA based implementation is presented in [67]. It employs the paradigm of information entropy and fuzzy logic. Again, these platforms have to be evaluated for their power consumption, footprint, etc., to make them usable for plant-wide monitoring.

Even though, to a good extent, the deployment of the CM system for individual machines might get duplicated in a meaningful manner, there are exceptions. The main areas of investigation for cost effective plant-wide condition monitoring are:

- Integration of sensing, computing and communication,
- Integrated monitoring, maintenance and service,
- Reliability and failure prediction performance,
- Shared and distributed platform, reduced overall cost per monitoring point.

In the process industry, a condition monitoring system is typically deployed along with the assessment of the risk associated with machine failure [68]. A common plant-wide automation system consists of hierarchically distributed monitoring modules, which are either integrated or embedded within the machines and are interconnected using wired or wireless industrial communication protocols. The information collected by diagnostic sensors is decomposed in order to be processed stepwise. Locally acquired data, such as vibration data from the device accelerometers, are first processed by the embedded device. Then, this processed information is sent to the plant controller or on-premise analytics operations. The information shared with other devices and/or controllers and/or analytics may get processed at each of these stages, enabling computation and communication load distribution. Afterwards, such processed information or objects are gradually aggregated and eventually used in decision support system or Service Intelligence Unit (SIU). The SIU, in turn, aggregates information or objects related to all the machines of the plant to perform overall predictive analysis.

The plant-wide deployment can be carried out using different architectures. Information flow and computation allocation are scheduled using a CM configuration engineering tool. Figure 5 shows a typical deployment using smart sensors. It usually includes an engineering tool, an on-premise operation server, a cloud enabled service or a remote service intelligence unit to support the embedded monitoring of plant-wide equipment. These monitoring devices are connected either to an input-output

(IO) network or to a control network belonging to automation hierarchy. Furthermore, the information flow is established using a suitable standard communication protocols. Engineering tools are used to configure the monitoring devices embedded in the monitored equipment. An instance of these smart sensors together with an association to the respective monitored machine is created [69]. Configuration information is shared among all the devices of the automation system, including controllers, operation servers, etc. This common configuration is further used in the analytics to develop the predictive diagnostics associated with these individually monitored machines. The SIU further integrates this information for generating the predictive diagnostics and maintenance report across the connected plants.

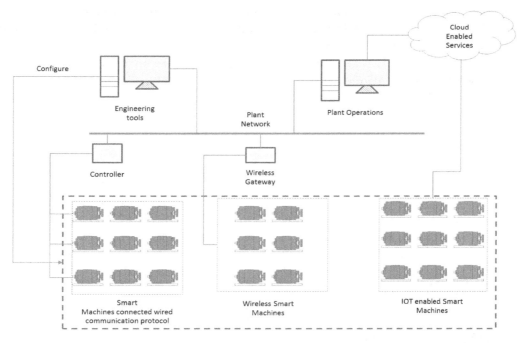

Figure 5. Plant-wide fleet condition monitoring.

The communication and processing requirements depend on the diagnostic application itself, its parameters and the number of machines being monitored in the plant. The recent advances in big data analytics help in handling the large amount of data and processing load. In general, a selected combination of application, sensors information, and maintenance activities are integrated in the service intelligence unit [70]. Several organizations have already matured smart condition monitoring for the entire fleet of machines in a plant, such as smart sensor solutions for plant-wide low voltage motor monitoring.

5. Conclusions

Condition monitoring of rotating electrical machines has evolved with the advent of sensing, machine diagnostic methods, data analytics platforms, communication and information management. Developments in these technologies now give a fresh opportunity to obtain automation and monitoring systems suited to diverse plant deployment scenarios that would not previously have been possible. The paper reviews the deployment methods in practice that allow integration of monitoring system on premise, on cloud, and even rogue internet of things (IOT) devices in industrial automation systems. The reviewed methods and techniques may need further analysis for adjacent technologies related to security. The security-enhanced, low-cost, and large-scale deployment is key to a future condition monitoring automation system for rotating electrical machines.

Acknowledgments: Authors would like to thank ABB., particularly Marco Sanguineti for the support and related discussions. Authors fully acknowledge the useful discussions with ABB colleagues Cajetan T.Pinto, Christopher Ganz, Stefan U. Svensson, and B S Nagabhushana from BMS college of Engineering, Bangalore.

Author Contributions: Mallikarjun Kande performed the literature review, wrote most of the text, and composed the diagrams. Alf J. Isaksson and Rajeev Thottappillil gave suggestions and guidance through project meetings, other discussion and proofreading. Nathaniel Taylor guided the work in regular meetings, and wrote or edited some parts of the text.

References

1. Kohler, J.L.; Sottile, J.; Trutt, F.C. Condition-based maintenance of electrical machines. In Proceedings of the Conference Record of the 1999 IEEE Industry Applications Conference, Thirty-Forth IAS Annual Meeting, Phoenix, AZ, USA, 3–7 October 1999; Volume 1, pp. 205–211.
2. Bram de Jonge, R.T.; Tinga, T. The influence of practical factors on the benefits of condition-based maintenance over time-based maintenance. *Reliab. Eng. Syst. Saf.* **2017**, *158*, 21–30.
3. Minou, C.A.; Olde Keizer, S.D.P.F.; Teunter, R.H. Condition-based maintenance policies for systems with multiple dependent components: A review. *Eur. J. Oper. Res.* **2017**, *261*, 405–420.
4. Walker, M. *The Diagnosing of Troubles in Electrical Machines*; Longmans Green and Co.: Harlow, UK, 1921; pp. 1–20.
5. Foster, G.B. Recent Developments in Machine Vibration Monitoring. *IEEE Trans. Ind. Gen. Appl.* **1967**, *IGA-3*, 149–158.
6. Kurtz, M.; Stone, G.C. Partial discharge testing of generator insulation. In Proceedings of the 1978 IEEE International Conference on Electrical Insulation, Philadelphia, PA, USA, 12–14 June 1978; pp. 73–77.
7. Timperley, J.E. Incipient Fault Identification Through Neutral RF Monitoring of Large Rotating Machines. *IEEE Trans. Power Appar. Syst.* **1983**, *PAS-102*, 693–698.
8. Harrold, R.T.; Emery, F.T. Radio Frequency Diagnostic Monitoring of Electrical Machines. *IEEE Electr. Insulation Mag.* **1986**, *2*, 18–24.
9. Kwan, C.; Ayhan, B.; Yin, J.; Liu, X.; Ballal, P.; Athamneh, A.; Ramani, A.; Lee, W.; Lewis, F. Real-time system condition monitoring using wireless sensors. In Proceedings of the 2009 IEEE Aerospace Conference, Big Sky, MT, USA, 7–14 March 2009; pp. 1–8.
10. Amaro, J.P.; Ferreira, F.J.T.E.; Cortesão, R.; Vinagre, N.; Bras, R.P. Low cost wireless sensor network for in-field operation monitoring of induction motors. In Proceedings of the 2010 IEEE International Conference on Industrial Technology, Vina del Mar, Chile, 14–17 March 2010; pp. 1044–1049.
11. Bratthall, L.G.; van der Geest, R.; Hofmann, H.; Jellum, E.; Korendo, Z.; Martinez, R.; Orkisz, M.; Zeidler, C.; Andersson, J.S. Integrating hundred's of products through one architecture—The industrial IT architecture. In Proceedings of the 24th International Conference on Software Engineering (ICSE 2002), Orlando, FL, USA, 19–25 May 2002; pp. 604–614.
12. Kolacinski, R.M.; Theeranaew, W.; Loparo, K.A. An information-theoretic architecture for advanced condition monitoring and control of power generating plants. In Proceedings of the 2012 Future of Instrumentation International Workshop (FIIW), Gatlinburg, TN, USA, 8–9 October 2012; pp. 1–4.
13. Vasel, J. One plant, one system: Benefits of integrating process and power automation. In Proceedings of the 65th Annual Conference for Protective Relay Engineers, College Station, TX, USA, 2–5 April 2012; pp. 215–250.
14. Loutfi, M.Y. Online condition monitoring network for critical equipment at Holcim's STE. genevieve plant. In Proceedings of the 2011 IEEE-IAS/PCA 53rd Cement Industry Technical Conference, St. Louis, MO, USA, 22–26 May 2011; pp. 1–11.
15. Simon, C.; Theilliol, D.; Sauter, D.; Orkisz, M. Remaining useful life assessment via residual generator approach—A SOH virtual sensor concept. In Proceedings of the 2014 IEEE Conference on Control Applications (CCA), Antibes/Nice, France, 8–10 October 2014; pp. 1202–1207.
16. Abu-Rub, H.; Bayhan, S.; Moinoddin, S.; Malinowski, M.; Guzinski, J. Medium-Voltage Drives: Challenges and existing technology. *IEEE Power Electron. Mag.* **2016**, *3*, 29–41.

17. Orkisz, M.; Wnek, M.; Kryczka, K.; Joerg, P. Variable frequency drive as a source of condition monitoring data. In Proceedings of the 2008 International Symposium on Power Electronics, Electrical Drives, Automation and Motion, Ischia, Italy, 11–13 June 2008; pp. 179–183.

18. Han, Y.; Song, Y.H. Condition monitoring techniques for electrical equipment—A literature survey. *IEEE Trans. Power Deliv.* **2003**, *18*, 4–13.

19. Bonnett, A.H.; Soukup, G.C. Rotor Failures in Squirrel Cage Induction Motors. *IEEE Trans. Ind. Appl.* **1986**, *IA-22*, 1165–1173.

20. Milimonfared, J.; Kelk, H.M.; Nandi, S.; Minassians, A.D.; Toliyat, H.A. A novel approach for broken-rotor-bar detection in cage induction motors. *IEEE Trans. Ind. Appl.* **1999**, *35*, 1000–1006.

21. Benbouzid, M.E.H.; Kliman, G.B. What stator current processing-based technique to use for induction motor rotor faults diagnosis? *IEEE Trans. Energy Convers.* **2003**, *18*, 238–244.

22. Nandi, S.; Toliyat, H.A.; Li, X. Condition monitoring and fault diagnosis of electrical motors—A review. *IEEE Trans. Energy Convers.* **2005**, *20*, 719–729.

23. Eltabach, M.; Charara, A.; Zein, I. A comparison of external and internal methods of signal spectral analysis for broken rotor bars detection in induction motors. *IEEE Trans. Ind. Electron.* **2004**, *51*, 107–121.

24. Kliman, G.B.; Koegl, R.A.; Stein, J.; Endicott, R.D.; Madden, M.W. Noninvasive detection of broken rotor bars in operating induction motors. *IEEE Trans. Energy Convers.* **1988**, *3*, 873–879.

25. Filippetti, F.; Franceschini, G.; Tassoni, C.; Vas, P. Recent developments of induction motor drives fault diagnosis using AI techniques. *IEEE Trans. Ind. Electron.* **2000**, *47*, 994–1004.

26. Kia, S.H.; Henao, H.; Capolino, G.A. A High-Resolution Frequency Estimation Method for Three-Phase Induction Machine Fault Detection. *IEEE Trans. Ind. Electron.* **2007**, *54*, 2305–2314.

27. Bellini, A.; Filippetti, F.; Franceschini, G.; Tassoni, C.; Kliman, G.B. Quantitative evaluation of induction motor broken bars by means of electrical signature analysis. *IEEE Trans. Ind. Appl.* **2001**, *37*, 1248–1255.

28. Arthur, N.; Penman, J. Induction machine condition monitoring with higher order spectra. *IEEE Trans. Ind. Electron.* **2000**, *47*, 1031–1041.

29. Jung, J.H.; Lee, J.J.; Kwon, B.H. Online Diagnosis of Induction Motors Using MCSA. *IEEE Trans. Ind. Electron.* **2006**, *53*, 1842–1852.

30. Bellini, A. Quad Demodulation: A Time-Domain Diagnostic Method for Induction Machines. *IEEE Trans. Ind. Appl.* **2009**, *45*, 712–719.

31. Ayhan, B.; Chow, M.Y.; Song, M.H. Multiple Discriminant Analysis and Neural-Network-Based Monolith and Partition Fault-Detection Schemes for Broken Rotor Bar in Induction Motors. *IEEE Trans. Ind. Electron.* **2006**, *53*, 1298–1308.

32. Su, H.; Chong, K.T. Induction Machine Condition Monitoring Using Neural Network Modeling. *IEEE Trans. Ind. Electron.* **2007**, *54*, 241–249.

33. Bazine, I.B.A.; Tnani, S.; Poinot, T.; Champenois, G.; Jelassi, K. On-line detection of stator and rotor faults occurring in induction machine diagnosis by parameters estimation. In Proceedings of the 8th IEEE Symposium on Diagnostics for Electrical Machines, Power Electronics Drives, Bologna, Italy, 5–8 September 2011; pp. 105–112.

34. Kral, C.; Wieser, R.S.; Pirker, F.; Schagginger, M. Sequences of field-oriented control for the detection of faulty rotor bars in induction machines-the Vienna Monitoring Method. *IEEE Trans. Ind. Electron.* **2000**, *47*, 1042–1050.

35. Cardoso, A.J.M.; Cruz, S.M.A.; Carvalho, J.F.S.; Saraiva, E.S. Rotor cage fault diagnosis in three-phase induction motors, by Park's vector approach. In Proceedings of the Industry Applications Conference, Thirtieth IAS Annual Meeting, IAS '95, Conference Record of the 1995 IEEE, Orlando, FL, USA, 8–12 October 1995; Volume 1, pp. 642–646.

36. Aboubou, A.; Sahraoui, M.; Zouzou, S.E.; Razik, H.; Rezzoug, A. Broken bars and/or end rings detection in three-phase induction motors by the extended Park's vector approach. In Proceedings of the 9th IEEE International Power Electronics Congress (CIEP 2004), Celaya, Gto., Mexico, 17–22 October 2004; pp. 128–133.

37. Zagirnyak, M.; Mamchur, D.; Kalinov, A. Induction motor diagnostic system based on spectra analysis of current and instantaneous power signals. In Proceedings of the IEEE SOUTHEASTCON 2014, Lexington, KY, USA, 13–16 March 2014; pp. 1–7.

38. Drif, M.; Cardoso, A.J.M. The instantaneous power factor approach for rotor cage faults diagnosis in three-phase induction motors. In Proceedings of the 2008 International Symposium on Power Electronics, Electrical Drives, Automation and Motion, Ischia, Italy, 11–13 June 2008; pp. 173–178.

39. Drif, M.; Cardoso, A.J.M. Stator Fault Diagnostics in Squirrel Cage Three-Phase Induction Motor Drives Using the Instantaneous Active and Reactive Power Signature Analyses. *IEEE Trans. Ind. Inform.* **2014**, *10*, 1348–1360.

40. Bell, R.N.; Mcwilliams, D.W.; O'Donnell, P.; Singh, C.; Wells, S.J. Report of Large Motor Reliability Survey of Industrial and Commercial Installations, Part I. *IEEE Trans. Ind. Appl.* **1985**, *IA-21*, 853–864.

41. Bell, R.N.; Heising, C.R.; O'Donnell, P.; Singh, C.; Wells, S.J. Report of Large Motor Reliability Survey of Industrial and Commercial Installations, Part II. *IEEE Trans. Ind. Appl.* **1985**, *IA-21*, 865–872.

42. O'Donnell, P.; Heising, C.; Singh, C.; Wells, S.J. Report of Large Motor Reliability Survey of Industrial and Commercial Installations: Part 3. *IEEE Trans. Ind. Appl.* **1987**, *IA-23*, 153–158.

43. Thorsen, O.V.; Dalva, M. A survey of faults on induction motors in offshore oil industry, petrochemical industry, gas terminals, and oil refineries. *IEEE Trans. Ind. Appl.* **1995**, *31*, 1186–1196.

44. Siddique, A.; Yadava, G.S.; Singh, B. A review of stator fault monitoring techniques of induction motors. *IEEE Trans. Energy Convers.* **2005**, *20*, 106–114.

45. Orman, M.; Orkisz, M.; Pinto, C.T. Slip estimation of a large induction machine based on MCSA. In Proceedings of the 8th IEEE Symposium on Diagnostics for Electrical Machines, Power Electronics Drives, Bologna, Italy, 5–8 September 2011; pp. 568–572.

46. Yazici, B.; Kliman, G.B. An adaptive statistical time-frequency method for detection of broken bars and bearing faults in motors using stator current. *IEEE Trans. Ind. Appl.* **1999**, *35*, 442–452.

47. Garcia-Perez, A.; de Jesus Romero-Troncoso, R.; Cabal-Yepez, E.; Osornio-Rios, R.A. The Application of High-Resolution Spectral Analysis for Identifying Multiple Combined Faults in Induction Motors. *IEEE Trans. Ind. Electron.* **2011**, *58*, 2002–2010.

48. Vetterli, M.; Herley, C. Wavelets and filter banks: Theory and design. *IEEE Trans. Signal Process.* **1992**, *40*, 2207–2232.

49. Burnett, R.; Watson, J.F.; Elder, S. The application of modern signal processing techniques to rotor fault detection and location within three phase induction motors. In Proceedings of the Integrating Intelligent Instrumentation and Control Instrumentation and Measurement Technology Conference (IMTC/95.), Waltham, MA, USA, 23–26 April 1995.

50. Eren, L.; Devaney, M.J. Motor bearing damage detection via wavelet analysis of the starting current transient. In Proceedings of the 18th IEEE Instrumentation and Measurement Technology Conference (IMTC 2001), Budapest, Hungary, 21–23 May 2001; Volume 3, pp. 1797–1800.

51. Rajagopalan, S.; Aller, J.M.; Restrepo, J.A.; Habetler, T.G.; Harley, R.G. Analytic-Wavelet-Ridge-Based Detection of Dynamic Eccentricity in Brushless Direct Current (BLDC) Motors Functioning Under Dynamic Operating Conditions. *IEEE Trans. Ind. Electron.* **2007**, *54*, 1410–1419.

52. CusidÓCusido, J.; Romeral, L.; Ortega, J.A.; Rosero, J.A.; Espinosa, A.G. Fault Detection in Induction Machines Using Power Spectral Density in Wavelet Decomposition. *IEEE Trans. Ind. Electron.* **2008**, *55*, 633–643.

53. Gu, F.; Wang, T.; Alwodai, A.; Tian, X.; Shao, Y.; Ball, A. A new method of accurate broken rotor bar diagnosis based on modulation signal bispectrum analysis of motor current signals. *Mech. Syst. Signal Process.* **2015**, *50*, 400–413.

54. Ngote, N.; Guedira, S.; Ouassaid, M.; Cherkaoui, M. Comparison of wavelet-functions for induction-motor rotor fault detection based on the hybrid Time Synchronous Averaging—Discrete Wavelet Transform approach. In Proceedings of the 2015 International Conference on Electrical and Information Technologies (ICEIT), Marrakech, Morocco, 25–27 March 2015; pp. 94–99.

55. Soualhi, A.; Clerc, G.; Razik, H. Detection and Diagnosis of Faults in Induction Motor Using an Improved Artificial Ant Clustering Technique. *IEEE Trans. Ind. Electron.* **2013**, *60*, 4053–4062.

56. Wu, S.; Chow, T.W.S. Induction machine fault detection using SOM-based RBF neural networks. *IEEE Trans. Ind. Electron.* **2004**, *51*, 183–194.

57. Escamilla-Ambrosio, P.J.; Liu, X.; Lieven, N.A.J.; Ramírez-Cortés, J.M. Wavelet-fuzzy logic approach to structural health monitoring. In Proceedings of the 2011 Annual Meeting of the North American Fuzzy Information Processing Society, El Paso, TX, USA, 18–20 March 2011; pp. 1–6.

58. Zhong, J.; Yang, Z.; Wong, S.F. Machine condition monitoring and fault diagnosis based on support vector machine. In Proceedings of the 2010 IEEE International Conference on Industrial Engineering and Engineering Management, Macao, China, 7–10 December 2010; pp. 2228–2233.

59. Sun, L.; Liu, M.; Qian, H.; Qiao, C. A New Method to Mechanical Fault Classification with Support Vector Machine. In Proceedings of the 2011 2nd International Conference on Digital Manufacturing Automation, Hunan, China, 5–7 August 2011; pp. 68–73.

60. Sheng, Z.; Liu, Z.; Wang, J.; Lu, Y. Development and application of condition monitoring system for plant production. In Proceedings of the 24th Chinese Control and Decision Conference (CCDC), Taiyuan, China, 23–25 May 2012; pp. 2490–2493.

61. Cardoso, F.J.A.; Faria, S.P.S.; Oliveira, J.E.G. A smart sensor for the condition monitoring of industrial rotating machinery. In Proceedings of the 2012 IEEE Sensors, Taipei, Taiwan, 28–31 October 2012; pp. 1–4.

62. Cabal-Yepez, E.; Osornio-Rios, R.A.; Romero-Troncoso, R.J.; Razo-Hernandez, J.R.; Lopez-Garcia, R. FPGA-Based Online Induction Motor Multiple-Fault Detection with Fused FFT and Wavelet Analysis. In Proceedings of the 2009 International Conference on Reconfigurable Computing and FPGAs, Quintana Roo, Mexico, 9–11 December 2009; pp. 101–106.

63. Wildermuth, S.; Ahrend, U.; Byner, C.; Rzeszucinski, P.; Lewandowski, D.; Orman, M. Condition monitoring of electric motors based on magnetometer measurements. In Proceedings of the 2015 IEEE 20th Conference on Emerging Technologies Factory Automation (ETFA), Luxembourg, 8–11 September 2015; pp. 1–4.

64. Stein, J.L.; Park, Y. Modeling and Sensing Issues for Machine Diagnostics. In Proceedings of the 1988 American Control Conference, Atlanta, GA, USA, 15–17 June 1988; pp. 1924–1930.

65. Su, C.Q. Smart condition monitoring. In Proceedings of the 2014 IEEE Electrical Insulation Conference (EIC), Philadelphia, PA, USA, 8–11 June 2014; pp. 138–141.

66. Zygmunt, M.; Budyn, M.; Orkisz, M.; Ottewill, J.; Jaramillo, V.; Nowak, A. Visual modeling of condition monitoring systems. In Proceedings of the 2012 IEEE 17th International Conference on Emerging Technologies Factory Automation (ETFA 2012), Krakow, Poland, 17–21 September 2012; pp. 1–4.

67. Romero-Troncoso, R.J.; Saucedo-Gallaga, R.; Cabal-Yepez, E.; Garcia-Perez, A.; Osornio-Rios, R.A.; Alvarez-Salas, R.; Miranda-Vidales, H.; Huber, N. FPGA-Based Online Detection of Multiple Combined Faults in Induction Motors Through Information Entropy and Fuzzy Inference. *IEEE Trans. Ind. Electron.* **2011**, *58*, 5263–5270.

68. Wasif, H.; Aboutalebi, A.; Brown, D.; Axel-Berg, L. Condition monitoring system for process industries a business approach. In Proceedings of the 2012 IEEE Symposium on Industrial Electronics and Applications, Bandung, Indonesia, 23–26 September 2012; pp. 251–256.

69. Josifovic, A.; Corney, J. Development of industrial process characterisation through data analysis. In Proceedings of the 2016 IEEE Symposium Series on Computational Intelligence (SSCI), Athens, Greece, 6–9 December 2016; pp. 1–7.

70. Xu, B.; Kumar, S.A. Big Data Analytics Framework for System Health Monitoring. In Proceedings of the 2015 IEEE International Congress on Big Data, New York, NY, USA, 27 June–2 July 2015; pp. 401–408.

3

Sensorless Control for IPMSM based on Adaptive Super-Twisting Sliding-Mode Observer and Improved Phase-Locked Loop

Shuo Chen [1], Xiao Zhang [1,*], Xiang Wu [1], Guojun Tan [1] and Xianchao Chen [2]

[1]   School of Electrical and Power Engineering, China University of Mining and Technology, Xuzhou 221116, China; ts17130047a3@cumt.edu.cn (S.C.); zb13060003@cumt.edu.cn (X.W.); gjtan@cumt.edu.cn (G.T.)

[2]   Xuzhou Yirui Construction Machinery Co. Ltd., Xuzhou 221000, China; xinchao623@163.com

*   Correspondence: zhangxiao@cumt.edu.cn.

Abstract: In traditional sensorless control of the interior permanent magnet synchronous motors (IPMSMs) for medium and high speed domains, a control strategy based on a sliding-mode observer (SMO) and phase-locked loop (PLL) is widely applied. A new strategy for IPMSM sensorless control based on an adaptive super-twisting sliding-mode observer and improved phase-locked loop is proposed in this paper. A super-twisting sliding-mode observer (STO) can eliminate the chattering problem without low-pass filters (LPFs), which is an effective method to obtain the estimated back electromotive forces (EMFs). However, the constant sliding-mode gains in STO may cause instability in the high speed domain and chattering in the low speed domain. The speed-related adaptive gains are proposed to achieve the accurate estimation of the observer in wide speed range and the corresponding stability is proved. When the speed of IPMSM is reversed, the traditional PLL will lose its accuracy, resulting in a position estimation error of $180°$. The improved PLL based on a simple strategy for signal reconstruction of back EMF is proposed to ensure that the motor can realize the direction switching of speed stably. The proposed strategy is verified by experimental testing with a 60-kW IPMSM sensorless drive.

Keywords: interior permanent magnet synchronous motor (IPMSM); sensorless control; adaptive algorithm; super-twisting sliding mode observer (STO); phase-locked loop (PLL)

1. Introduction

Recently, interior permanent magnet synchronous motors (IPMSMs) have been extensively utilized in the fields of electromechanical drives, electric vehicles, and numerical control servo systems due to their robustness, high efficiency, high power density, and compactness [1–4]. The usage of position sensors decreases the reliability and increases the cost and volume of IPMSM drives. In order to overcome these shortcomings caused by the use of mechanical position sensors, sensorless control technology has become one of the important research directions in related fields [5,6]. Generally, sensorless control strategies can be divided into two categories. The first one is called signal injection methods [7–9]. This method is based on the salient pole effect of the motor, which is mainly used in zero and low speed domains. The second one is called back EMF based methods [10–19], which utilizes the estimated back EMF signals to obtain the position information of the motor. Because the magnitude of back EMF is in proportion to the speed of the motor, the performance of back EMF based methods at ultra-low and zero speed is extremely poor [11]. Hence, back EMF based methods and signal injection methods are usually combined to achieve sensorless control for a whole speed range [12–14]. Back EMF based methods primarily includes the model adaptive method (MRAS) [16], the Kalman filtering method (EKF) [17], and the sliding mode observer (SMO) [2,18,19], etc. Compared with MRAS and EKF, SMO has simpler structure and stronger robustness. Hence, SMO is extensively applied in sensorless control strategy [19].

The signum function used in traditional SMO can introduce high frequency harmonics into the estimated signals, which eventually lead to the inevitable chattering phenomenon. Therefore, low-pass filters (LPFs) are commonly utilized to smooth the estimated signals. However, the LPFs in turn bring the disadvantages of phase delay of estimated signals. In [20], signum function is utilized to reduce the SMO chattering phenomenon caused by sigmoid function. In [21], an adaptive filter is proposed to reduce the negative effects of LPFs. However, these methods cannot completely avoid phase delay caused by LPFs. In [22,23], the super-twisting algorithm is proposed to eliminate the chattering phenomenon caused by signum function. The super-twisting sliding mode observer (STO) can effectively eliminate the sliding-mode chattering phenomenon without compromising robustness and avoid the use of LPFs. In [24], the stability of STO is further analyzed by using the Lyapunov function and the corresponding stability conditions are given. In [25], the sensorless control strategy based on STO and resistance identification is proposed for SPMSM. Resistance identification enhances the robustness of the super-twisting sliding mode observer. Although STO performs well in reducing chattering, there is still a problem to be solved. When the constant sliding-mode gains are adopted in this method, the sliding-mode gains should be big enough to meet the stability condition in the wide speed range. But the large sliding-mode gains will lead to a large chattering phenomenon, especially in a low speed domain [19].

Traditionally, the position information is obtained by the estimated back electromotive forces through arc-tangent method directly. However, the arc-tangent function makes position information susceptible to harmonics and noises. In order to improve estimation performance, the quadrature phase-locked loop algorithm is proposed in [6], which is called the traditional PLL in this paper. High-order harmonics can be filtered out due to the special structure of PLL. When the speed of IPMSM is reversed, the traditional PLL will lose its accuracy, resulting in a position estimation error of 180°. The reason for such drawback is that the sign of the back EMFs has an effect on the sign of the equivalent position error [26,27]. To solve the aforementioned problem, Refs. [26,27] proposed a kind of PLL, which constructs the equation of the equivalent position error based on tangent function. Such a scheme may overcome the problem, but it brings complexity to the algorithm and it is vulnerable to harmonics and noises due to the introduction of a tangent function.

In this paper, a new strategy based on adaptive super-twisting sliding mode observer and improved PLL for IPMSM sensorless control is proposed to overcome aforementioned limitations. Super-twisting sliding-mode observer is utilized to obtain the estimated back electromotive forces. Moreover, speed-related adaptive gains are proposed to achieve accurate estimation in a wide speed domain so that they widen the speed range of the super-twisting sliding-mode observer. On the basis of existing stability conditions in [24], the stability of the proposed adaptive STO is proved in this paper. To improve the shortcomings of the above-mentioned two kinds of PLL, a simple strategy for signal reconstruction of back EMF is proposed. Based on this strategy, the improved PLL can overcome the limitation of speed reversal existing in traditional PLL without the introduction of tangent function. Besides, the improved PLL has simple structure, great steady performance, and transient response. Finally, the proposed strategy based on adaptive STO and improved PLL is verified by experimental testing with a 60-kW IPMSM sensorless drive.

2. Adaptive Super-Twisting Sliding-Mode Observer

For the sake of convenience, magnetic saturation is neglected and it is assumed that the flux linkage distribution is perfectly sinusoidal. The model of IPMSM is shown in Figure 1. The ABC, $\alpha\beta$ and dq frames represent the natural, the stationary, and the rotating reference frames, respectively.

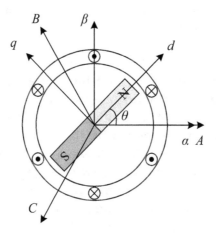

Figure 1. The model of interior permanent magnet synchronous motors (IPMSM).

The mathematic model of IPMSM in $\alpha\beta$ stationary reference frame is expressed as

$$u_\alpha = Ri_\alpha + L_d\frac{di_\alpha}{dt} + \omega_e(L_d - L_q)i_\beta + e_\alpha \tag{1}$$

$$u_\beta = Ri_\beta + L_q\frac{di_\beta}{dt} - \omega_e(L_d - L_q)i_\alpha + e_\beta \tag{2}$$

where u_α, u_β are stator voltages; i_α, i_β are stator currents; R is stator resistance; ω_e is electrical rotor speed; ψ_f is PM flux linkage; and L_d,L_q are stator inductances. e_α and e_β are the $\alpha\beta$-axis back EMFs of IPMSM, satisfying $e_\alpha = -E\sin\theta$ and $e_\beta = E\cos\theta$. θ is the rotor position and E is the amplitude of back EMF [28], satisfying

$$E = (L_d - L_q)\left(\omega_e i_d - \frac{di_q}{dt}\right) + \omega_e\psi_f \tag{3}$$

2.1. Super-Twisting Algorithm

A. Levant proposed the super-twisting algorithm to eliminate the chatter caused by the signum function in [23,29]. The fundamental form of this algorithm is written as follows:

$$\frac{d\hat{x}_1}{dt} = -k_1|\hat{x}_1 - x_1|\text{sign}(\hat{x}_1 - x_1) + \hat{x}_2 + \rho_1 \tag{4}$$

$$\frac{d\hat{x}_2}{dt} = -k_2\text{sign}(\hat{x}_1 - x_1) + \rho_2 \tag{5}$$

where x_i, $\hat{x}_i$, k_i, sign(), and ρ_i are state variables, estimation of state variables, sliding-mode gains, signum function, and perturbation terms, respectively. The corresponding conditions of the stability of the super-twisting algorithm have been educed in [24]. If ρ_1 and ρ_2 in Equations (6) and (7) satisfy the following conditions:

$$\rho_1 \leq \delta_1|x_1|^{\frac{1}{2}}, \rho_2 = 0 \tag{6}$$

where δ_1 is a positive constant and the sliding-mode gains k_1 and k_2 meet the condition:

$$k_1 > 2\delta_1, k_2 > k_1\frac{5\delta_1 k_1 + 4\delta_1^2}{2(k_1 - 2\delta_1)} \tag{7}$$

the stability of the system can be guaranteed.

2.2. Super-Twisting Sliding Mode Observer for IPMSM Sensorless Control

To estimate the back EMFs conveniently, the mathematic mode of IPMSM shown in Equations (3) and (4) is organized into the current model:

$$\frac{di_\alpha}{dt} = -\frac{R}{L_d}i_\alpha - \omega_e \frac{L_d - L_q}{L_d}i_\beta + \frac{u_\alpha}{L_d} - \frac{e_\alpha}{L_d} \tag{8}$$

$$\frac{di_\beta}{dt} = -\frac{R}{L_d}i_\beta + \omega_e \frac{L_d - L_q}{L_d}i_\alpha + \frac{u_\beta}{L_d} - \frac{e_\beta}{L_d} \tag{9}$$

The estimated currents are taken as state variables in Equations (4) and (5), then the STO for IPMSM sensorless control be represented as

$$\frac{d\hat{i}_\alpha}{dt} = -\frac{R}{L_d}\hat{i}_\alpha - \hat{\omega}_e \frac{L_d - L_q}{L_d}\hat{i}_\beta + \frac{u_\alpha}{L_d} - \frac{k_1}{L_d}|\bar{i}_\alpha|^{\frac{1}{2}}\mathrm{sign}(\bar{i}_\alpha) - \frac{1}{L_d}\int k_2\mathrm{sign}(\bar{i}_\alpha)dt \tag{10}$$

$$\frac{d\hat{i}_\beta}{dt} = -\frac{R}{L_d}\hat{i}_\beta + \hat{\omega}_e \frac{L_d - L_q}{L_d}\hat{i}_\alpha + \frac{u_\beta}{L_d} - \frac{k_1}{L_d}|\bar{i}_\beta|^{\frac{1}{2}}\mathrm{sign}(\bar{i}_\beta) - \frac{1}{L_d}\int k_2\mathrm{sign}(\bar{i}_\beta)dt \tag{11}$$

where $\bar{i}_\alpha = \hat{i}_\alpha - i_\alpha$, $\bar{i}_\beta = \hat{i}_\beta - i_\beta$ and $\hat{\ }$ represents the estimated variable. It should be noticed that, differently from the STO for SPMSM sensorless control in [26], the perturbation term ρ_1 in Equation (4) for IPMSM sensorless control is replaced by $-\frac{R}{L_d}\hat{i}_\alpha - \hat{\omega}_e \frac{L_d - L_q}{L_d}\hat{i}_\beta + \frac{u_\alpha}{L_d}$ and $-\frac{R}{L_d}\hat{i}_\beta + \hat{\omega}_e \frac{L_d - L_q}{L_d}\hat{i}_\alpha + \frac{u_\beta}{L_d}$, respectively.

By substituting the perturbation terms into Equation (6) and taking estimated currents as state variables, Equation (6) can be reformulated as

$$-\frac{R}{L_d}\hat{i}_\alpha - \hat{\omega}_e \frac{L_d - L_q}{L_d}\hat{i}_\beta + \frac{u_\alpha}{L_d} \leq \delta_1|\hat{i}_\alpha|^{\frac{1}{2}} \tag{12}$$

$$-\frac{R}{L_d}\hat{i}_\beta + \hat{\omega}_e \frac{L_d - L_q}{L_d}\hat{i}_\alpha + \frac{u_\beta}{L_d} \leq \delta_1|\hat{i}_\beta|^{\frac{1}{2}} \tag{13}$$

If δ_1 is large enough, the stable conditions can be guaranteed easily. By subtracting Equations (8) and (9) from Equations (10) and (11) respectively, the state equations of the current estimation errors can be obtained:

$$\frac{d\bar{i}_\alpha}{dt} = -\frac{R}{L_d}\bar{i}_\alpha - \frac{L_d - L_q}{L_d}(\hat{\omega}_e\hat{i}_\beta - \omega_e i_\beta) - \frac{k_1}{L_d}|\bar{i}_\alpha|^{\frac{1}{2}}\mathrm{sign}(\bar{i}_\alpha) - \frac{1}{L_d}\int k_2\mathrm{sign}(\bar{i}_\alpha)dt + \frac{e_\alpha}{L_d} \tag{14}$$

$$\frac{d\bar{i}_\beta}{dt} = -\frac{R}{L_d}\bar{i}_\beta + \frac{L_d - L_q}{L_d}(\hat{\omega}_e\hat{i}_\alpha - \omega_e i_\alpha) - \frac{k_1}{L_d}|\bar{i}_\beta|^{\frac{1}{2}}\mathrm{sign}(\bar{i}_\beta) - \frac{1}{L_d}\int k_2\mathrm{sign}(\bar{i}_\beta)dt + \frac{e_\beta}{L_d} \tag{15}$$

when STO reaches the sliding surface, it is approximately considered that the estimated value is equal to the actual value ($\hat{\omega}_e \approx \omega_e$, $\hat{i}_\alpha \approx i_\alpha$ and $\hat{i}_\beta \approx i_\beta$). Then the equivalent control law of the back EMFs is expressed as

$$\hat{e}_\alpha = k_1|\bar{i}_\alpha|^{\frac{1}{2}}\mathrm{sign}(\bar{i}_\alpha) + \int k_2\mathrm{sign}(\bar{i}_\alpha)dt \tag{16}$$

$$\hat{e}_\beta = k_1|\bar{i}_\beta|^{\frac{1}{2}}\mathrm{sign}(\bar{i}_\beta) + \int k_2\mathrm{sign}(\bar{i}_\beta)dt \tag{17}$$

The linear term $k_1|\bar{i}_\alpha|^{\frac{1}{2}}\mathrm{sign}(\bar{i}_\alpha)$ determines the convergence rate of the STO and the integral term $\int k_2\mathrm{sign}(\bar{i}_\alpha)dt$ is related to the suppression of chattering phenomena. Hence, k_2 usually has a large value.

2.3. Adaptive Super-Twisting Sliding Mode Observer for IPMSM Sensorless Control

Although STO performs well in reducing chattering, there is still a problem to be solved. When the constant sliding-mode gains are adopted in this method, the sliding-mode gains should be large enough to meet the stable conditions when the IPMSM runs at high speed. However, due to the excessive sliding mode gains, the performance of the STO in the low speed domain will be seriously deteriorated [19]. In order to extract accurate rotor position in wide speed range, the STO for IPMSM with speed-related adaptive gains is proposed in this paper. The speed-related adaptive gains k_1 and k_2 are adopted as

$$k_1 = l_1 \omega_e^*, \ k_2 = l_2 \omega_e^{*2} \tag{18}$$

$$\omega_e^* = \begin{cases} \omega_{emin} & 0 \le \hat{\omega}_e < \omega_{emin} \\ \text{LPF}(\hat{\omega}_e) & \omega_{emin} \le \hat{\omega}_e \le \omega_{emax} \\ \omega_{emax} & \hat{\omega}_e > \omega_{emax} \end{cases} \tag{19}$$

where l_1 and l_2 are adaptive coefficients, ω_{emax} is the maximum electrical rotor speed of motor, ω_{emin} is the minimum electrical rotor speed allowed by the STO for back EMFs observation. The first-order LPF in the STO is utilized to smooth the gain variations and improve the robustness of the observer in the transient process. Its cut-off frequency is determined according to ω_{emax} and switching frequency. The stability of adaptive STO is proved as follows:

In Equations (12) and (13), compared with $\frac{u_\alpha}{L_d}$ and $\frac{u_\beta}{L_d}$, $\frac{R}{L_d}\hat{i}_\alpha$, $\hat{\omega}_e \frac{L_d - L_q}{L_d}\hat{i}_\beta$, $\frac{R}{L_d}\hat{i}_\beta$ and $\hat{\omega}_e \frac{L_d - L_q}{L_d}\hat{i}_\alpha$ can be neglected. Then, the perturbation terms can be simplified as

$$\rho_1(i_\alpha) \approx \frac{u_\alpha}{L_d}, \ \rho_1(i_\beta) \approx \frac{u_\beta}{L_d} \tag{20}$$

then, Equation (6) can be rewritten as

$$|\rho_1(i_\alpha)| \approx \left| \frac{u_\alpha}{L_d} \right| \approx \frac{\omega_e \psi_f}{L_d} \le \delta_1 |\hat{i}_\alpha|^{\frac{1}{2}} \tag{21}$$

when STO reaches the sliding surface, $|\hat{i}_\alpha|^{\frac{1}{2}}$ is in a certain range and $\omega_e^* \approx \omega_e$. δ_1 is replaced by $\lambda \omega_e$ in Equation (21), Equation (21) can be rewritten as

$$|\rho_1(i_\alpha)| \approx \left| \frac{u_\alpha}{L_d} \right| \approx \frac{\omega_e \psi_f}{L_d} \le \lambda |\hat{i}_\alpha|^{\frac{1}{2}} \omega_e \tag{22}$$

This formula can be satisfied by choosing a large λ. Substituting $\delta_1 = \lambda \omega_e$, $k_1 = l_1 \omega_e$ and $k_2 = l_2 \omega_e^2$ into Equation (7), Equation (7) can be rewritten as

$$k_1 = l_1 \omega_e > 2\delta_1 = 2\lambda \omega_e \tag{23}$$

$$k_2 = l_2 \omega_e^2 > k_1 \frac{5\delta_1 k_1 + 4\delta_1^2}{2(k_1 - 2\delta_1)} = l_1 \frac{5\lambda l_1 + 4\lambda^2}{2l_1 - 4\lambda} \omega_e^2 \tag{24}$$

It is obvious that when the adaptive coefficients l_1 and l_2 satisfy the condition $l_1 > 2\lambda$ and $l_2 > l_1 \frac{5\lambda l_1 + 4\lambda^2}{2l_1 - 4\lambda}$, the stability conditions of adaptive STO can be satisfied. The black diagram of adaptive STO for IPMSM sensorless control is shown in Figure 2.

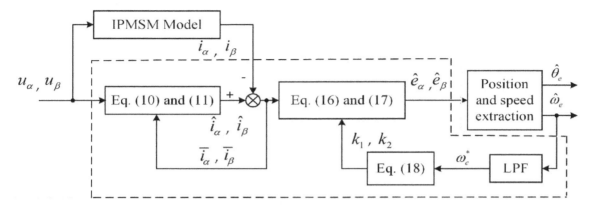

Figure 2. The black diagram of adaptive super-twisting sliding-mode observer (STO) for IPMSM sensorless control.

3. Acquisition of Position Information

Traditionally, the position information is obtained by the estimated back electromotive forces through arc-tangent method directly.

$$\hat{\theta}_e = -\arctan \cdot \left(\frac{\hat{e}_\alpha}{\hat{e}_\beta} \right) \tag{25}$$

The electrical rotor speed can be calculated by $\hat{\omega}_e = \frac{d\hat{\theta}_e}{dt}$. However, the estimated position and speed is susceptible to noise and harmonics because of the usage of arc-tangent method. Especially when $\hat{e}_\beta$ crosses zero, the obvious estimation errors may be produced. Ref. [6] proposed the quadrature phase-locked loop algorithm to mitigate the adverse effect. In this paper, this algorithm is called the traditional PLL.

3.1. Traditional PLL

The transfer function of the traditional PLL can be written as

$$G(s) = \frac{\hat{\theta}_e}{\theta_e} = \frac{EK_p s + EK_i}{s^2 + EK_p s + EK_i} \tag{26}$$

where K_p is the proportional gain, K_i is the integral gain. The structure of the traditional PLL is represented in Figure 3.

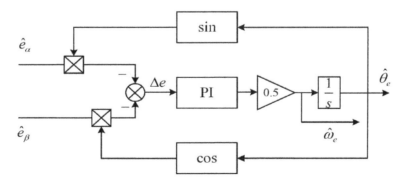

Figure 3. The structure of the traditional phase-locked loop (PLL).

The bode diagram of Equation (26) with different E is shown in Figure 4. As shown in Figure 4, E varies with the rotor speed, so the bandwidth of the PLL is influenced by the operating frequency of motor. This could make the design of system parameters more difficult and deteriorate the accuracy of the position estimation. Therefore, the normalization of the back EMFs is necessary.

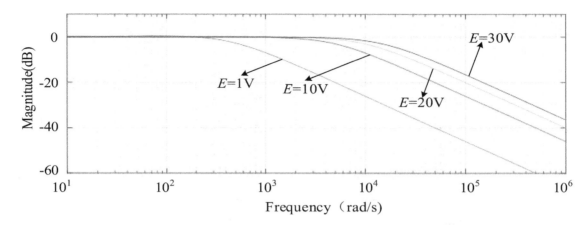

Figure 4. Bode diagram of the traditional PLL transfer function with different E.

By normalizing the estimated back EMF, the equivalent position error Δe can be written as

$$
\begin{aligned}
\Delta e &= \frac{1}{\sqrt{\hat{e}_\alpha^2 + \hat{e}_\beta^2}} \left[-\hat{e}_\alpha \cos(\hat{\theta}_e) - \hat{e}_\beta \sin(\hat{\theta}_e) \right] \\
&= -\hat{e}_{\alpha n} \cos(\hat{\theta}_e) - \hat{e}_{\beta n} \sin(\hat{\theta}_e) \\
&= \sin(\theta_e) \cos(\hat{\theta}_e) - \sin(\hat{\theta}_e) \cos(\theta_e) \\
&= \sin(\theta_e - \hat{\theta}_e) \approx \theta_e - \hat{\theta}_e
\end{aligned}
\tag{27}
$$

where $\hat{e}_{\alpha n}$ and $\hat{e}_{\beta n}$ are the normalized back EMFs, and the closed-loop transfer function of the traditional PLL with back EMF normalization can be obtained by

$$
G(s) = \frac{\hat{\theta}_e}{\theta_e} = \frac{K_p s + K_i}{s^2 + K_p s + K_i}
\tag{28}
$$

The traditional PLL has the characteristics of LPF. High-order harmonics can be filtered out due to the special structure of phase-locked loop. However, when the speed of IPMSM is reversed, the traditional PLL will lose its accuracy, resulting in a position estimation error of 180°. When the parameters of PLL are set for one direction of rotation, the estimation of rotor position is correct for this direction only and an error of 180° will be produced in the other direction. Such a drawback makes the traditional PLL not suitable for applications where the motor needs to switch the direction of rotation. The theoretical analysis of the above problem is shown in Section 3.3.

3.2. Tangent-Based PLL

To solve the aforementioned problem, Refs. [26,27] proposed a kind of PLL scheme, which constructs the equivalent position error equation based on tangent function.

$$
\Delta e = \frac{\frac{\hat{e}_\alpha}{\hat{e}_\beta} - \frac{\sin\left(\frac{\hat{\theta}_e}{2}\right)}{\cos\left(\frac{\hat{\theta}_e}{2}\right)}}{1 + \frac{\hat{e}_\alpha}{\hat{e}_\beta} \cdot \frac{\sin\left(\frac{\hat{\theta}_e}{2}\right)}{\cos\left(\frac{\hat{\theta}_e}{2}\right)}} = \frac{\tan(\theta_e) - \tan\left(\frac{\hat{\theta}_e}{2}\right)}{1 + \tan(\theta_e) \cdot \tan\left(\frac{\hat{\theta}_e}{2}\right)}
$$

$$
= \tan\left(\theta_e - \frac{\hat{\theta}_e}{2}\right)
\tag{29}
$$

The structure of the tangent-based PLL is shown in Figure 5. When the system achieves the steady point, rotor position can be calculated as

$$
\theta_e = \frac{\hat{\theta}_e}{2}
\tag{30}
$$

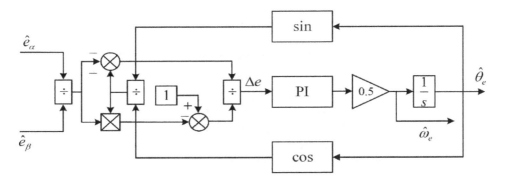

Figure 5. The structure of the tangent-based PLL.

This kind of PLL can solve the reversal problem. However, it increases the complexity of the algorithm. And it is vulnerable to harmonic and noise interference due to the introduction of tangent function. Especially, during $\hat{e}_\beta$ crosses zero and the rotor position crosses $\pm\frac{\pi}{2}$, the obvious estimation error may occur.

3.3. Improved PLL

The improved PLL is based on a simple EMF signals reconstruction strategy. The structure of the improved PLL is depicted in Figure 6 and the equation of the equivalent position error in the proposed scheme can be expressed as

$$
\begin{aligned}
\Delta e &= -\hat{e}_{\alpha n}\hat{e}_{\beta n}\cos(2\hat{\theta}_e) + \frac{(\hat{e}_{\alpha n}{}^2 - \hat{e}_{\beta n}{}^2)}{2}\sin(2\hat{\theta}_e) \\
&= \tfrac{1}{2}[\sin(2\theta_e)\cos(2\hat{\theta}_e) - \sin(2\hat{\theta}_e)\cos(2\theta_e)] \\
&= \tfrac{1}{2}\sin(2(\theta_e - \hat{\theta}_e))
\end{aligned}
\tag{31}
$$

when the system reaches the stable point, Δe can be derived as

$$
\begin{aligned}
\Delta e &= \tfrac{1}{2}\sin(2(\theta_e - \hat{\theta}_e)) \\
&\approx \theta_e - \hat{\theta}_e
\end{aligned}
\tag{32}
$$

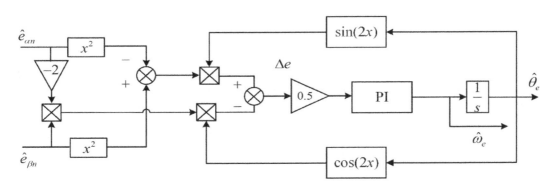

Figure 6. The structure of the improved PLL.

In the positive speed range of the motor,

$$
\hat{e}_{\alpha n} = -\sin(\theta_e), \quad \hat{e}_{\beta n} = \cos(\theta_e)
\tag{33}
$$

and the dynamic equations of the traditional PLL with back EMF normalization are represented as

$$
\frac{de_\theta}{dt} = e_\omega
\tag{34}
$$

$$\frac{de_\omega}{dt} = -K_p \cos(e_\theta)e_\omega - K_I \sin(e_\theta) \tag{35}$$

where $e_\theta = \theta_e - \hat{\theta}_e$, $e_\omega = \omega_e - \hat{\omega}_e$. The phase trajectory of the traditional PLL for positive speed is shown in Figure 7a. As shown in Figure 7a, there are three equilibrium points in the system, which are $(0,0)$, $(\pi,0)$ and $(-\pi,0)$. Among the three equilibrium points, only $(0,0)$ is stable point. The others are saddle points. That means the trajectories in the phase trajectory of traditional PLL for positive speed will move to the origin. In other words, e_θ and e_ω can converge to $(0,0)$ in limited time, which meets the requirements of estimation performance.

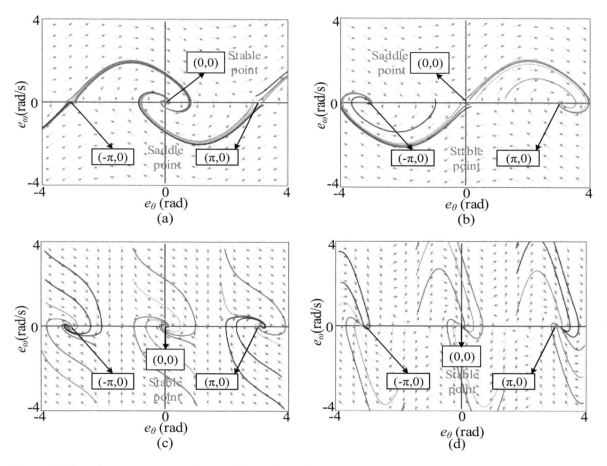

Figure 7. The phase trajectory of (**a**) traditional PLL for positive speed. (**b**) traditional PLL for negative speed. (**c**) tangent-based PLL for both positive and negative speed. (**d**) improved PLL for both positive and negative speed.

But when the direction of rotation is reversed, the symbols of the back EMF change and the same symbolic change can be detected on the equivalent position error signal Δe:

$$\hat{e}_{\alpha n} = \sin(\theta_e), \ \hat{e}_{\beta n} = -\cos(\theta_e) \tag{36}$$

$$\begin{aligned}
\Delta e &= -\hat{e}_{\alpha n} \cos(\hat{\theta}_e) - \hat{e}_{\beta n} \sin(\hat{\theta}_e) \\
&= -\sin(\theta_e)\cos(\hat{\theta}_e) + \sin(\hat{\theta}_e)\cos(\theta_e) \\
&= \sin(\hat{\theta}_e - \theta_e) \approx -(\theta_e - \hat{\theta}_e)
\end{aligned} \tag{37}$$

And the dynamic equations of the traditional PLL are rewritten as

$$\frac{de_\theta}{dt} = e_\omega \tag{38}$$

$$\frac{de_\omega}{dt} = K_p \cos(e_\theta)e_\omega + K_I \sin(e_\theta) \tag{39}$$

The phase trajectory of the traditional PLL for negative speed is given in Figure 7b. The system has the same three equilibrium points, which are $(0,0)$, $(\pi,0)$ and $(-\pi,0)$. However, $(0,0)$ changes into saddle point and $(\pm\pi,0)$ become stable points. The trajectories in the nonlinear system depart from $(0,0)$ to reach the stable points $(\pm\pi,0)$ so that the system produce a position estimation error of $180°$. Although this problem can be solved by resetting the gains of the PI controller, it is difficult to implement in real-time control system. Therefore, the traditional PLL cannot meet the requirements of applications where the motor needs to switch the direction of rotation.

The phase trajectory of the tangent-based PLL for both positive and negative speed is shown in Figure 7c. More details can be found in [26,27]. In this kind of PLL system, $(0,0)$, $(\pi,0)$, and $(-\pi,0)$ are three stable points. By setting the proper parameters of PI regulator, e_θ and e_ω can converge to $(0,0)$. That means the tangent-based PLL can solve the reversal problem. But due to the introduction of tangent function, it is vulnerable to harmonic and noise interference. Especially when $\hat{e}_\beta$ crosses zero and the position crosses $\pm\frac{\pi}{2}$, the obvious estimation errors will be produced. This algorithm is difficult to adopt in practice.

Compared with the traditional PLL and the tangent-based PLL, the improved PLL makes the speed reversal of motor not cause the symbolic change of the equivalent position error Δe by using a simple back EMF signals reconstruction strategy without tangent function. The dynamic equations are the same for both positive and negative speed and can be represented as

$$\frac{de_\theta}{dt} = e_\omega \tag{40}$$

$$\frac{de_\omega}{dt} = \frac{1}{2}\left[-K_p \cos(2e_\theta)2e_\omega - K_I \sin(2e_\theta)\right] \tag{41}$$

There are five equilibrium points in the system, which are $(0,0)$, $(\pm\pi,0)$ and $(\pm\frac{\pi}{2},0)$. In order to confirm the properties of equilibrium points in the system conveniently, the nonlinear equation of state is linearized. The Jacobian matrix $J(e_\theta, e_\omega)$ for (40) and (41) is represented as

$$J(e_\theta, e_\omega) = \begin{bmatrix} 0 & 1 \\ 2K_p \sin(2e_\theta)e_\omega - K_I \cos(2e_\theta) & -K_p \cos(2e_\theta) \end{bmatrix} \tag{42}$$

Substituting $(e_\theta, e_\omega) = (0,0)$ and $(e_\theta, e_\omega) = (\pm\pi,0)$ into (42) respectively, the expression is the same at these points:

$$J(e_\theta, e_\omega)_{(e_\theta,e_\omega)=(0,0),(\pm\pi,0)} = \begin{bmatrix} 0 & 1 \\ -K_I & -K_p \end{bmatrix} \tag{43}$$

The eigenvalues of (43) can be expressed as

$$\lambda_1 = \frac{-K_p + \sqrt{K_p^2 - 4K_I}}{2}, \quad \lambda_2 = \frac{-K_p - \sqrt{K_p^2 - 4K_I}}{2} \tag{44}$$

Because $K_p > 0$ and $K_I > 0$, λ_1 and λ_2 have negative real parts. That means $(0,0)$ and $(\pm\pi,0)$ are stable points.

Substituting $(e_\theta, e_\omega) = (\pm\frac{\pi}{2},0)$ into Equation (42) respectively, the expression is the same at these points:

$$J(e_\theta, e_\omega)_{(e_\theta,e_\omega)=(\pm\frac{\pi}{2},0)} = \begin{bmatrix} 0 & 1 \\ K_I & K_p \end{bmatrix} \tag{45}$$

The eigenvalues of Equation (45) can be expressed as

$$\lambda_1 = \frac{K_p + \sqrt{K_p^2 + 4K_I}}{2} > 0, \lambda_2 = \frac{K_p - \sqrt{K_p^2 + 4K_I}}{2} < 0 \tag{46}$$

Because $\lambda_1 > 0$ and $\lambda_2 < 0$, $(\pm\frac{\pi}{2},0)$ are saddle points in the system. In summary, among the five equilibrium points, (0,0) and $(\pm\pi,0)$ are stable points and $(\pm\frac{\pi}{2},0)$ are saddle points. The phase trajectory of the improved PLL for both positive and negative speed is shown in Figure 7d. Similar to the tangent-based PLL, each of these stable points is a focal point that the neighborhood phase trajectories will be attracted to. Moreover, because there is no introduction of the arctangent function, this method has better robustness than the tangent-based PLL. By selecting the appropriate gains of the PI regulator, e_θ and e_ω will converge to the origin. That means the motor can switch the speed direction steadily by adopting the proposed PLL.

4. Experimental Results

The control diagram of proposed sensorless control strategy for IPMSM based on adaptive STO and improved PLL is shown in Figure 8. The double closed-loop vector control is adopted. The details of the adaptive STO and the improved PLL are shown in Figures 2 and 6, respectively.

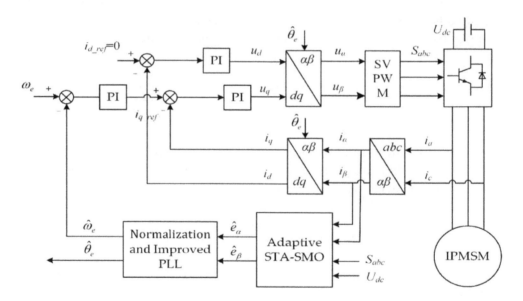

Figure 8. The control diagram of proposed sensorless control strategy for IPMSM.

An experimental prototype is shown in Figure 9 and the corresponding experimental platform was established as shown in Figure 10. The platform is mainly composed of two water-cooled IPMSMs, one rectifier, two inverters, and three controllers. The motor 1 is connected with inverter 1, and the proposed strategy is implemented by the controller 1. The motor 2 is a load motor which is controlled by the inverter 2, which is controlled by controller 2. Table 1 lists the parameters of the IPMSM. A 540 V dc-link voltage is obtained by the PWM rectifier for testing and verifying the performance of the proposed strategy. The rectifier is controlled by controller 3. In the experiment, TMS320F2812 DSP is adopted to carry out the new sensorless control strategy. All signals are converted by a digital-to-analog chip (TLV5610) and displayed on a digital oscilloscope. The traditional two-level inverter topology is adopted [30]. Switching frequency of the inverter and sampling frequency of the control system are set to 10 kHz. A rotary decoder (PGA411-Q1) is employed to obtain the actual position and speed of the motor, which are used for comparing and verifying the performance of the proposed strategy.

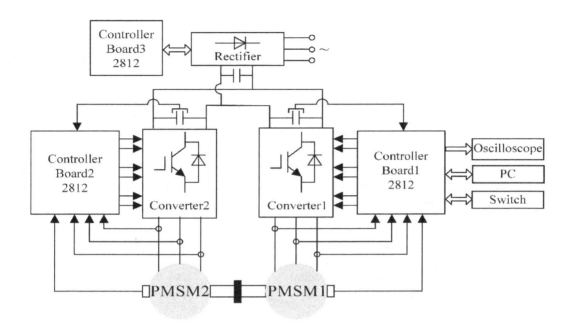

Figure 9. The experimental prototype.

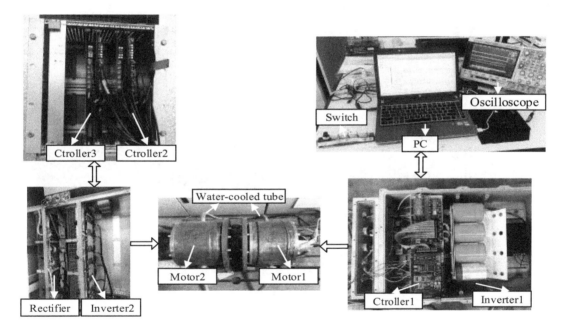

Figure 10. The experimental platform.

Table 1. Parameters of the IPMSM.

Parameter	Value
Flux linkage	0.225 Wb
d/q-axis inductor	0.95/2.05 mH
Resistance	0.1 Ω
Pole pairs	4
Rated power	60 kW
Rated speed	3000 rpm

4.1. Experimental Results of Adaptive Super-Twisting Sliding Mode Observer

The performances of the STO with constant sliding-mode gains in different speed ranges are presented in Figures 11 and 12. The parameters of the STO are $k_1 = 15$ and $k_2 = 60,000$ and the parameters of the PI regulator in the PLL are $K_p = 250$ and $K_i = 20,000$. Since the STO is based on the back electromotive forces model, the performance of STO is unreliable in ultra-low and zero speed domains. In this paper, IF control is adopted to ensure the start-up for IPMSM sensorless control. The threshold of speed that transiting from IF control to sensorless control is set to 300 r/min. The Figure 11 shows the performance of STO with no load from 0 to 1000 rpm.lo The IPMSM starts up in open-loop by using IF control at 1 s and switches to sensorless control at 2 s. Obviously, the estimation errors are large in the process of start-up and it takes about 1 s for the observer to get accurate rotor position information. When the IPMSM operates at 1000 r/min under sensorless control, the speed estimation error is within ±8 r/min and the position estimation error is between $1.08°$ and $7.2°$. The estimated back EMFs have good sinusoidal properties. This means the STO with $k_1 = 15$ and $k_2 = 60,000$ can operate perfectly at 1000 r/min.

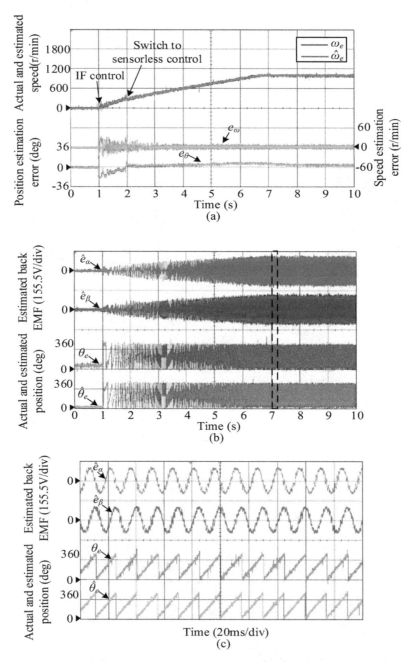

Figure 11. The performance of STO with no load from 0 to 1000 rpm. (**a**) Actual and estimated speed, speed estimation error, and position estimation error. (**b**) Estimated back electromotive forces (EMFs) and Actual and estimated position. The waveforms in (**b**) at 1000 r/min are zoomed in (**c**).

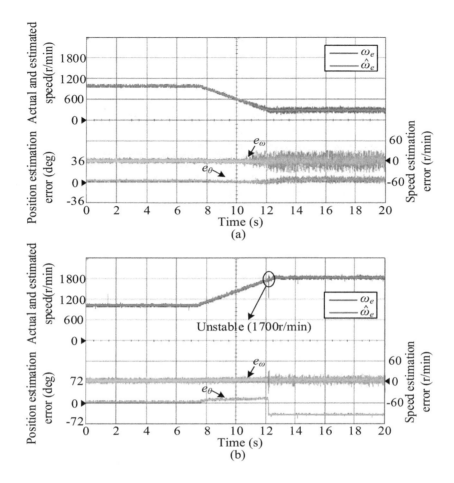

Figure 12. The performance of STO in wide speed range. (**a**) The performance of STO with no load from 1000 r/min to 300 r/min in closed-loop. (**b**) The performance of STO with no load from 1000 r/min to 1800 r/min in open-loop.

The performances of the STO with $k_1 = 15$ and $k_2 = 60,000$ from 1000 r/min to 300 r/min in closed-loop and from 1000 r/min to 1800 r/min in open-loop are shown in the Figure 12. In the process of motor speed decreasing from 1000 r/min to 300 r/min, the error of speed and position estimation increases significantly. That is because excessive sliding-mode gains lead to the large chattering of the estimated signals, resulting in severe chattering of the motor. It is dangerous to test the STO for the IPMSM in high speed range and closed-loop, so the speed is raised from 1000 r/min to 1800 r/min in open-loop. The corresponding performance is given in Figure 12b. The STO becomes unreliable at about 1700 r/min. At about 1700 r/min, the position estimation error jumps abruptly from 10.8° to $-40°$ and the estimated speed has a large flutter. This means the IPMSM cannot operate at high speed over 1700 r/min in closed-loop. That is because the sliding-mode gains are too small to meet the stability conditions of STO. Experimental results presented in Figure 12 illustrate that the performance of STO in low and high speed range is limited by the constant sliding-mode gains and it is necessary to adopt speed-related adaptive sliding-mode gains.

The adaptive coefficients of the observer can be calculated by $l_1 = \frac{k_1}{\omega_e}$ and $l_2 = \frac{k_2}{\omega_e^2}$. The STO with $k_1 = 15$ and $k_2 = 60,000$ can operate perfectly at 1000 r/min ($\omega_e \approx 418.9$ rad/s). So in this paper, $l_1 = \frac{15}{418.9} \approx 0.036$ and $l_2 = \frac{60,000}{418.9^2} \approx 0.342$. After applying the proposed adaptive STO, the IPMSM works well in wide speed range and closed-loop as shown in Figure 13. Throughout the operation, the speed estimation error is within ± 10 r/min and the position estimation error is less than 10.8°. It is obvious that the position and speed estimation errors are significantly lower than the observer with constant sliding-mode gains, when the IPMSM runs in low and high speed range.

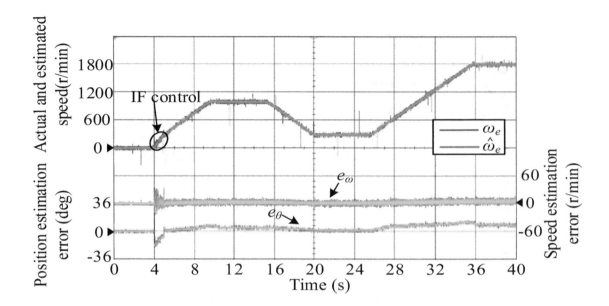

Figure 13. The performance of adaptive STO with no load in closed-loop under variable speed: raises from 0 r/min to 1000 r/min, drops to 300 r/min, and raises to 1800 r/min.

The dynamic performance of adaptive STO at 1800 r/min is shown in Figure 14. A 40 N·m load is enabled at 3 s and disabled at 6.2 s. The estimated speed can track the actual speed accurately and the estimated position error is less than 10.8° in the course of operation. The DC error of the position estimation increases by about 5° after loading and this is due to the mismatch of parameters caused by the increase of current after loading [12,31]. Hence, the performance of the adaptive STO could be verified.

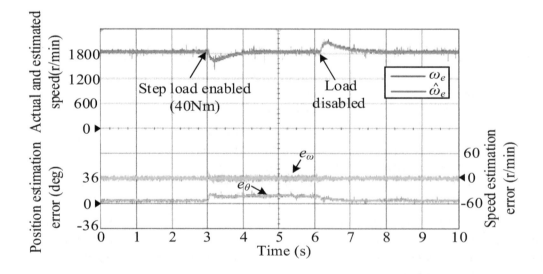

Figure 14. The dynamic performance of adaptive STO at 1800 r/min.

4.2. Experimental Results of the Proposed Improved PLL

The performances of traditional PLL, tangent-based PLL, and proposed improved PLL when the IPMSM turns from positive speed to reverse speed in open-loop are shown in Figure 15. For comparative purposes, three kinds of PLL operate under the same conditions: $K_p = 250$ and $K_i = 20,000$. The speed command is turned from 600 r/min to −600 r/min at 0.6 s.

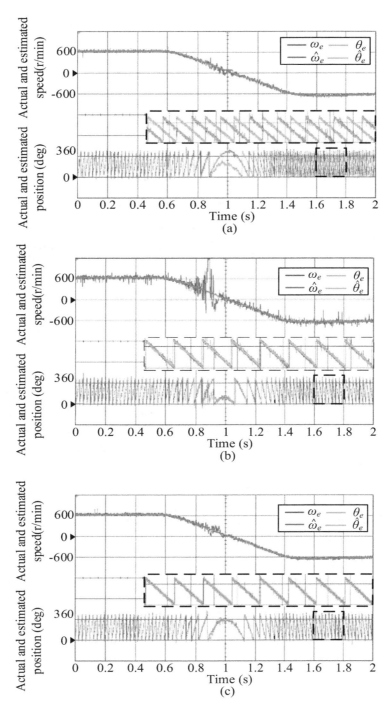

Figure 15. The performance of (**a**) traditional PLL, (**b**) tangent-based PLL, (**c**) improved PLL in open-loop from 600 r/min to −600 r/min.

As shown in Figure 15a, The estimated speed follows the actual speed accurately, when the rotation direction of the motor is positive. But when the speed of IPMSM is reversed, the conventional PLL loses its accuracy and produces a large position estimation error (180°). This prevents the motor from turning from positive speed to reverse speed in closed-loop. The performance of tangent-based PLL is shown in Figure 15b. Although tangent-based PLL can solve the speed reversal problem, the introduction of division and tangent functions increases the complexity of the algorithm and makes the tangent-based PLL vulnerable to harmonic and noise, especially when the back EMF crosses zero and the position crosses $\pm\frac{\pi}{2}$ where an obvious estimation error may occur. Excessive speed and position chattering shown in Figure 15b means the algorithm cannot be adopted in practice. The performances of the proposed improved PLL in open-loop and closed-loop are shown in Figures 15c and 16, respectively. It is clearly that the improved PLL has great performance when the IPMSM turns from positive speed to reverse speed. Thus, the effectiveness of the proposed improved PLL can be verified.

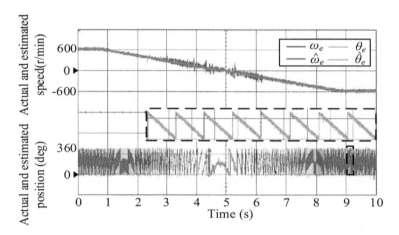

Figure 16. The performance of proposed improved PLL in closed-loop from 600 r/min to −600 r/min.

5. Conclusions

A new strategy for IPMSM sensorless control based on adaptive STO and improved PLL is proposed in this paper. STO is utilized to obtain the estimated back electromotive forces and the speed-related adaptive gains are proposed to achieve the accurate estimation of the observer in wide speed range. Moreover, the improved PLL based on a simple strategy for signal reconstruction of back EMF is proposed to overcome the limitation of speed reversal existing in traditional PLL without the introduction of tangent function. The experimental results show that the speed range of the super-twisting sliding-mode observer can be widened by adopting the proposed adaptive algorithm and the improved PLL has great performance so that IPMSM can realize the direction switching of speed stably.

Author Contributions: S.C. proposed the new sensorless control strategy. S.C. and X.W. performed the experiments and analyzed the data. S.C. wrote the first draft and X.W., X.Z., G.T. and X.C. guided and revised the manuscript.

References

1.   Li, S.; Zhou, X. Sensorless Energy Conservation Control for Permanent Magnet Synchronous Motors Based on a Novel Hybrid Observer Applied in Coal Conveyer Systems. *Energies* **2018**, *11*, 2554. [CrossRef]
2.   Wang, Y.; Wang, X.; Xie, W.; Dou, M. Full-Speed Range Encoderless Control for Salient-Pole PMSM with a Novel Full-Order SMO. *Energies* **2018**, *11*, 2423. [CrossRef]
3.   Wang, G.; Ding, L.; Li, Z.; Lin, Z.; Xu, J.; Zhang, G.; Zhan, H.; Zhan, H.; Ni, R.; Xu, D. Enhanced Position Observer Using Second-Order Generalized Integrator for Sensorless Interior Permanent Magnet Synchronous Motor Drives. *IEEE Trans. Energy Convers.* **2014**, *29*, 486–495.
4.   Wang, M.-S.; Tsai, T.-M. Sliding Mode and Neural Network Control of Sensorless PMSM Controlled System for Power Consumption and Performance Improvement. *Energies* **2017**, *10*, 1780. [CrossRef]
5.   Joo, K.J.; Park, J.S.; Lee, J. Study on Reduced Cost of Non-Salient Machine System Using MTPA Angle Pre-Compensation Method Based on EEMF Sensorless Control. *Energies* **2018**, *11*, 1425. [CrossRef]
6.   Wang, G.; Li, Z.; Zhang, G.; Yu, Y.; Xu, D. Quadrature PLL-Based High-Order Sliding-Mode Observer for IPMSM Sensorless Control with Online MTPA Control Strategy. *IEEE Trans. Energy Convers.* **2013**, *28*, 214–224. [CrossRef]

7. Tian, L.; Zhao, J.; Sun, J. Sensorless Control of Interior Permanent Magnet Synchronous Motor in Low-Speed Region Using Novel Adaptive Filter. *Energies* **2016**, *9*, 1084. [CrossRef]

8. Liu, J.; Zhu, Z. Novel Sensorless Control Strategy with Injection of High-Frequency Pulsating Carrier Signal into Stationary Reference Frame. *IEEE Trans. Ind. Appl.* **2014**, *50*, 2574–2583. [CrossRef]

9. Yoon, Y.; Sul, S.; Morimoto, S.; Ide, K. High-Bandwidth Sensorless Algorithm for AC Machines Based on Square-Wave-Type Voltage Injection. *IEEE Trans. Ind. Appl.* **2011**, *47*, 1361–1370. [CrossRef]

10. Jung, T.-U.; Jang, J.-H.; Park, C.-S. A Back-EMF Estimation Error Compensation Method for Accurate Rotor Position Estimation of Surface Mounted Permanent Magnet Synchronous Motors. *Energies* **2017**, *10*, 1160. [CrossRef]

11. Cho, Y. Improved Sensorless Control of Interior Permanent Magnet Sensorless Motors Using an Active Damping Control Strategy. *Energies* **2016**, *9*, 135. [CrossRef]

12. Tuovinen, T.; Hinkkanen, M. Adaptive Full-Order Observer with High-Frequency Signal Injection for Synchronous Reluctance Motor Drives. *IEEE J. Emerg. Sel. Top. Power Electron.* **2014**, *2*, 181–189. [CrossRef]

13. Yousefi-Talouki, A.; Pescetto, P.; Pellegrino, G.; Ion, B. Combined Active Flux and High Frequency Injection Methods for Sensorless Direct Flux Vector Control of Synchronous Reluctance Machines. *IEEE Trans. Power Electron.* **2018**, *33*, 2447–2457. [CrossRef]

14. Yousefi-Talouki, A.; Pescetto, P.; Pellegrino, G. Sensorless Direct Flux Vector Control of Synchronous Reluctance Motors Including Standstill, MTPA and Flux Weakening. *IEEE Trans. Ind. Appl.* **2017**, *53*, 3598–3608. [CrossRef]

15. Wang, G.; Yang, R.; Xu, D. DSP-Based Control of Sensorless IPMSM Drives for Wide-Speed-Range Operation. *IEEE Trans. Ind. Electron.* **2013**, *60*, 720–727. [CrossRef]

16. Zhao, Y.; Qiao, W.; Wu, L. Improved Rotor Position and Speed Estimators for Sensorless Control of Interior Permanent-Magnet Synchronous Machines. *IEEE J. Emerg. Sel. Top. Power Electron.* **2014**, *2*, 627–639. [CrossRef]

17. Park, J.B.; Wang, X. Sensorless Direct Torque Control of Surface-Mounted Permanent Magnet Synchronous Motors with Nonlinear Kalman Filtering. *Energies* **2018**, *11*, 969. [CrossRef]

18. Qiao, Z.; Shi, T.; Wang, Y.; Yan, Y.; Xia, C.; He, X. New Sliding-Mode Observer for Position Sensorless Control of Permanent-Magnet Synchronous Motor. *IEEE Trans. Ind. Electron.* **2013**, *60*, 710–719. [CrossRef]

19. Lin, S.; Zhang, W. An Adaptive Sliding-Mode Observer with a Tangent function-based PLL Structure for Position Sensorless PMSM Drives. *Int. J. Electr. Power Energy Syst.* **2017**, *88*, 63–74. [CrossRef]

20. Kim, H.; Son, J.; Lee, J. A High-Speed Sliding-Mode Observer for the Sensorless Speed Control of a PMSM. *IEEE Trans. Ind. Electron.* **2011**, *58*, 4069–4077.

21. Cascella, G.L.; Salvatore, N.; Salvatore, L. Adaptive Sliding-Mode Observer for Field Oriented Sensorless Control of SPMSM. In Proceedings of the 2003 IEEE International Symposium on Industrial Electronics (Cat. No. 03TH8692), Rio de Janeiro, Brazil, 9–11 June 2003; Volume 2, pp. 1137–1143.

22. Levant, A. Principles of 2-sliding Mode Design. *Automatica* **2007**, *43*, 576–586. [CrossRef]

23. Levant, A. Sliding Order and Sliding Accuracy in Sliding Mode Control. *Int. J. Control* **1993**, *58*, 1247–1263. [CrossRef]

24. Moreno, J.A.; Osorio, M. A Lyapunov Approach to Second-Order Sliding Mode Controllers and Observers. In Proceedings of the 47th IEEE Conference on Decision and Control, Cancun, Mexico, 9–11 December 2008; pp. 2856–2861.

25. Liang, D.; Li, J.; Qu, R. Sensorless Control of Permanent Magnet Synchronous Machine Based on Second-Order Sliding-Mode Observer with Online Resistance Estimation. *IEEE Trans. Ind. Appl.* **2017**, *53*, 3672–3682. [CrossRef]

26. Olivieri, C.; Tursini, M. A Novel PLL Scheme for a Sensorless PMSM Drive Overcoming Common Speed Reversal Problems. In Proceedings of the IEEE International Symposium on Power Electronics, Electrical Drives, Automation and Motion, Sorrento, Italy, 20–22 June 2012.

27. Olivieri, C.; Parasiliti, F.; Tursini, M. A Full-Sensorless Permanent Magnet Synchronous Motor Drive with an Enhanced Phase-Locked Loop Scheme. In Proceedings of the IEEE International Conference on Electrical Machines, Marseille, France, 2–5 September 2012; pp. 2202–2208.

28. Chen, Z.; Tomita, M.; Doki, S.; Okuma, S. An Extended Electromotive Force Model for Sensorless Control of Interior Permanent-Magnet Synchronous Motors. *IEEE Trans. Ind. Electron.* **2007**, *43*, 576–586.

29. Levant, A. Robust Exact Differentiation via Sliding Mode Technique. *Automatica* **1998**, *34*, 379–384. [CrossRef]

30. Wu, X.; Tan, G.; Ye, Z.; Liu, Y.; Xu, S. Optimized Common-Mode Voltage Reduction PWM for Three-Phase Voltage Source Inverters. *IEEE Trans. Power Electron.* **2016**, *31*, 2959–2969. [CrossRef]

31. Li, Y.; Zhu, Z.; Howe, D.; Bingham, C. Improved Rotor Position Estimation in Extended Back-EMF Based Sensorless PM Brushless AC Drives with Magnetic Saliency. In Proceedings of the IEEE International Electric Machines & Drives Conference, Antalya, Turkey, 3–5 May 2007; pp. 214–229.

4

Online Current Loop Tuning for Permanent Magnet Synchronous Servo Motor Drives with Deadbeat Current Control

Zih-Cing You, Cheng-Hong Huang and Sheng-Ming Yang *

Electrical Engineering, National Taipei University of Technology, Taipei 10608, Taiwan;
carefree60024@gmail.com (Z.-C.Y.); yyu124p@gmail.com.tw (C.-H.H.)
* Correspondence: smyang@ntut.edu.tw

Abstract: High bandwidths and accurate current controls are essential in high-performance permanent magnet synchronous (PMSM) servo drives. Compared with conventional proportional–integral control, deadbeat current control can considerably enhance the current control loop bandwidth. However, because the deadbeat current control performance is strongly affected by the variations in the electrical parameters, tuning the controller gains to achieve a satisfactory current response is crucial. Because of the prompt current response provided by the deadbeat controller, the gains must be tuned within a few control periods. Therefore, a fast online current loop tuning scheme is proposed in this paper. This scheme can accurately identify the controller gain in one current control period because the scheme is directly derived from the discrete-time motor model. Subsequently, the current loop is tuned by updating the deadbeat controller with the identified gains within eight current control periods or a speed control period. The experimental results prove that in the proposed scheme, the motor current can simultaneously have a critical-damped response equal to its reference in two current control periods. Furthermore, satisfactory current response is persistently guaranteed because of an accurate and short time delay required for the current loop tuning.

Keywords: deadbeat current control; PMSM servo motor drives; auto tuning; parameter identification

1. Introduction

A modern servo motor drive usually includes current, speed, and position control loops. In general, the current loop bandwidth is considerably higher than the bandwidth of the speed and position loops. Therefore, a current loop with a high bandwidth can fundamentally enhance the performance of the servo motor drive.

When the current loop is implemented with a digital signal processor (DSP), because of the limited computation capability, the calculated voltage command requires one control period delay for the pulse width modulation (PWM) module to output voltage to the motor. This time delay causes an underdamped or unstable current response when a proportional–integral (PI) controller is used for motor current regulation [1–4]. The discretized PI controller directly designed in the z-domain has been proposed in [3,4]; however, limited improvement in the current loop bandwidth was achieved and current overshoots were persistent. To eliminate the influences of the time delay, schemes based on the predictive current control [5–11] and deadbeat current control [12–17] have been proposed. The predictive current controller generates the optimal voltage vector by minimizing a specific cost function. This voltage vector allows the motor current to reach its reference value as fast as possible with minimum overshoot. Deadbeat current control is well-known for its zero overshoot, zero steady-state error, and minimum rise time characteristics. Consequently, the motor current can reach its reference value with minimum control periods without overshoot. Compared with predictive current control, deadbeat control is simple to implement and requires less computation. However, its performance is parameter-dependent, as reported in [15–17]. In particular, inductance is sensitive to the current level. Online controller gain tuning is an effective method to mitigate the effects of parameter variations.

Numerous online electrical parameter identification strategies have been proposed. The observer-based methods in [18–21] identify the parameters by converging the error between the sampled and estimated current to zero. In [18], the identified inductance was used for the predictive current controller to improve the robustness of the current loop. Observer-based methods often require long execution times because of the delay of the observer and may encounter stability problems. The authors in [22–24] performed the recursive least-square (RLS) algorithm to identify electrical parameters. The motor model was used to develop the RLS algorithm. Then, the parameters were identified by minimizing the discrepancy between the sampled and calculated current. Although the latency caused by the observer does not exist in RLS-based methods, accurately identifying the parameters in a few control periods is still difficult. In addition, the electrical parameters are generally identified instead of the controller gains in these methods. However, the effect caused by the parameter mismatch can be treated as a disturbance to the current controller. To compensate for this disturbance, the compensation voltage, which was obtained through the disturbance observer in [12,17] and through adaptive control in [25], is added to the current loop. Despite their effectiveness, the schemes in [17,25] involve a complex design procedure to achieve satisfactory performance.

In this study, a deadbeat current controller was designed to enhance the current loop bandwidth for its simple implementation. A novel online current loop tuning strategy is proposed to reduce the effect of parameter variations. The proposed method is simple and effective because the method is directly derived from the discrete-time motor model. In addition, the proposed method directly identifies the gains of the deadbeat controller instead of the electrical parameters. After the controller gains are identified, the gains are averaged to further improve accuracy. Then, the current loop is tuned by updating the deadbeat controller with the average gains.

2. Discrete-Time Motor Model

The stator voltage of a PMSM in the rotor reference frame can be expressed as follows:

$$\begin{bmatrix} v_{qs}^r \\ v_{ds}^r \end{bmatrix} = \begin{bmatrix} r_s + sL_{qs} & \omega_r L_{ds} \\ -\omega_r L_{qs} & r_s + sL_{ds} \end{bmatrix} \begin{bmatrix} i_{qs}^r \\ i_{ds}^r \end{bmatrix} + \begin{bmatrix} \omega_r \lambda_m \\ 0 \end{bmatrix} \tag{1}$$

where v_{qs}^r, v_{ds}^r, i_{qs}^r, and i_{ds}^r are the q- and d-axis voltages and currents, respectively; L_{qs} and L_{ds} are the q- and d-axis inductance, respectively; r_s, ω_r, and λ_m are the phase resistance, rotor electrical speed, and magnet flux, respectively; and s denotes the Laplace operator. When the current loop and PWM function of the PMSM are implemented digitally, a time delay is inevitably introduced. Figure 1 displays the time sequence of the current sampling, voltage command calculation, and voltage command output, where T_s is the sampling period of the motor currents and the control period of the current loop. As depicted in Figure 1, the voltage command is calculated at t_0 and outputs to the PWM module at $t_0 + T_s$. Then, the voltage command is activated by the PWM module and held for one sampling period during $t_0 + T_s$ to $t_0 + 2T_s$. The motor current induced by the corresponding voltage command is then sampled at $t_0 + 2T_s$. Therefore, a time delay of two sampling periods is generated in the current loop. The PWM function and calculation delay can be modeled together by using a zero-order hold involving one sampling period delay. Accordingly, the stator voltage in Equation (1) can be discretized as follows:

$$G_q(z) = \frac{i_{qs}^r(z)}{v_{qs}^r(z)} = Z\left\{ e^{-sT_s} \cdot ZOH\left(\frac{1}{L_{qs}s + r_s} \right) \right\} = \frac{B_{mq}z^{-2}}{1 - A_{mq}z^{-1}} \tag{2}$$

$$G_d(z) = \frac{i_{ds}^r(z)}{v_{ds}^r(z)} = Z\left\{ e^{-sT_s} \cdot ZOH\left(\frac{1}{L_{ds}s + r_s} \right) \right\} = \frac{B_{md}z^{-2}}{1 - A_{md}z^{-1}} \tag{3}$$

where Z{} is the Z-transform; the model gains A_{mq} and B_{mq} are $e^{-T_s r_s/L_{qs}}$ and $(1 - A_{mq})/r_s$, respectively; and the model gains A_{md} and B_{md} are $e^{-T_s r_s/L_{ds}}$ and $(1 - A_{md})/r_s$, respectively. Because the back-EMF and cross-coupling voltages are approximately constant within one control period, these voltages are assumed to be decoupled from the current controller and are not represented in Equations (2) and (3). The q- and d-axis decoupling voltages, namely v_{qff} and v_{dff}, respectively, are derived from Equation (1) by using the estimated electrical parameters and rotor speed in the following expression:

$$v_{qff} = \omega_r \hat{L}_{ds} i_{ds}^r + \omega_r \lambda_m$$
$$v_{dff} = -\omega_r \hat{L}_{qs} i_{qs}^r$$

(4)

where "^" denotes the estimated quantity.

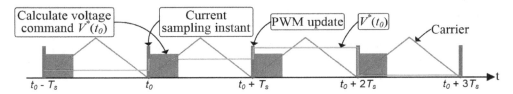

Figure 1. Time sequence for current sampling, voltage command calculation, and output, where PWM is pulse-width modulation.

3. Overall Control System

3.1. Servo Control System

Figure 2 illustrates the overall servo control system, where v_{an}, v_{bn}, and v_{cn} are the phase voltages; θ_m, θ_r, and ω_m denote the mechanical position, electrical angle, and speed, respectively; and "*" denotes the command value. The motor current is regulated using a deadbeat current controller. The classical proportional position with proportional-plus-integral velocity (P-PI) control is implemented to regulate the motor speed and position [26]. The bandwidths of the speed loop and position loop are set as 100 and 10 Hz, respectively. The proposed online current loop tuning algorithm continuously tunes the gains in the current loop to achieve a satisfactory current response. The variables associated with the online tuning are defined in the following text.

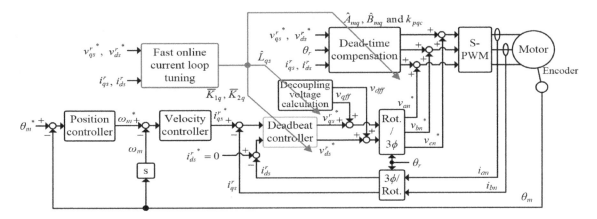

Figure 2. Block diagram of servo control system with the proposed online current loop tuning strategy.

3.2. Dead-Time Compensation

High-performance servo motor drives generally include dead-time compensation. The voltage error caused by the dead time can be measured through the steady-state voltage command and current feedback [27]. The voltage error at different phase currents for the inverter used in this study is depicted in Figure 3. The configuration of the motor drive is listed in Table A1. The voltage error saturates when the magnitude of the phase current is higher than 1 A. After the voltage error is calculated with the phase current feedback, dead-time compensation is performed by adding the voltage error to the current loop, as depicted in Figure 2.

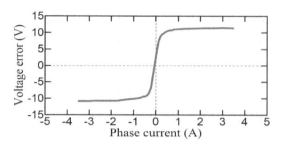

Figure 3. The voltage error caused by the dead-time at different phase current.

The time delay in the current loop can degrade the effectiveness of dead-time compensation. To mitigate the influence of this time delay, the estimated phase current is used to calculate the voltage error. Figure 4 illustrates the proposed current estimator. A PI controller is used to reduce the error between the sampled and estimated current. By ignoring the PI controller, the estimated current can be expressed as follows:

$$\hat{i}_{qs}^r(k) = \hat{B}_{mq} \cdot v_{qs}^{r\,*}(k) + \hat{A}_{mq} \cdot \hat{i}_{qs}^r(k-1) \tag{5}$$

$$\hat{i}_{ds}^r(k) = \hat{B}_{md} \cdot v_{ds}^{r\,*}(k) + \hat{A}_{md} \cdot \hat{i}_{ds}^r(k-1) \tag{6}$$

When the parameters are correct, the estimated current approximates the current sampled at the $(k + 2)$th sampling instant, which is induced by $v_{qs}^{r*}(k)$ and $v_{ds}^{r*}(k)$. An operator z^{-2} is added in the feedback path of the estimator because the estimated current is from two sampling periods before the present sampled current. Then, the estimated phase current can be calculated as follows:

$$\begin{bmatrix} \hat{i}_{an} \\ \hat{i}_{bn} \\ \hat{i}_{cn} \end{bmatrix} = \begin{bmatrix} 1 & 0 \\ -1/2 & -\sqrt{3}/2 \\ -1/2 & \sqrt{3}/2 \end{bmatrix} \begin{bmatrix} \cos\theta_r & \sin\theta_r \\ -\sin\theta_r & \cos\theta_r \end{bmatrix} \begin{bmatrix} \hat{i}_{qs}^r \\ \hat{i}_{ds}^r \end{bmatrix} \tag{7}$$

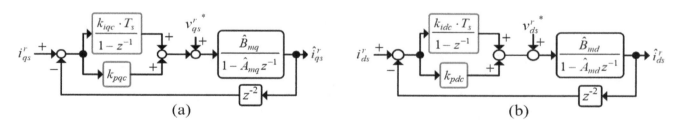

Figure 4. The proposed (**a**) q-axis and (**b**) d-axis current estimator.

The pole-zero cancelation technique is used to design the PI controller of the current estimators. By canceling the plant pole with the controller zero, the proportional gain k_{pqc} and k_{pdc} are calculated as follows:

$$k_{pqc} = k_{iqc} \cdot \hat{A}_{mq} \cdot T_s / \left(1 - \hat{A}_{mq}\right) \tag{8}$$

$$k_{pdc} = k_{idc} \cdot \hat{A}_{md} \cdot T_s / \left(1 - \hat{A}_{md}\right) \tag{9}$$

where k_{iqc} and k_{idc} are the integral gains. Figure 5 shows the damping ratio and bandwidth of the current estimator for various k_{iqc} values. The bandwidth of the current estimator is determined from the integral gain, and the proportional gains are then calculated using Equations (8) and (9).

In this study, the bandwidth of the current estimator was set to 900 Hz because this can simultaneously ensure that the estimated current strictly follows the sampled feedback current and the estimator predicts the sampled current accurately. In addition, the current estimator has a critical-damped response.

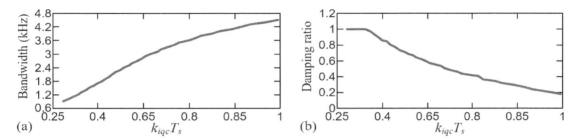

Figure 5. The (**a**) bandwidth and (**b**) damping ratio of the current estimator at different integral gain.

4. Deadbeat Current Controller

When a conventional PI controller is used to regulate the motor current, the time delay in the control loop can cause stability problems because of the degraded phase margin. Consequently, the current loop bandwidth is limited to maintaining an acceptable overshoot on the motor current. To enhance the bandwidth of the current loop to its theoretical maximum, a deadbeat current controller was developed in this study.

Deadbeat Controller Design

Figure 6 depicts the schematics of the q- and d-axis current control loops with the deadbeat controller, where $C_q(z)$ and $C_d(z)$ are the deadbeat controllers for the q- and d-axes, respectively. Deadbeat controller design is conducted entirely in the z-domain. All the closed-loop poles are placed at the origin in the z-domain. The q-axis transfer function is expressed as follows:

$$\frac{i_{qs}^r}{i_{qs}^{r*}} = \frac{C_q(z) \cdot G_q(z)}{1 + C_q(z) \cdot G_q(z)} = \frac{h(z)}{z^n} \tag{10}$$

where the numerator $h(z)$ provides an additional degree of freedom for the controller design and n is the number of poles. The q-axis current should strictly follow the command value without a steady-state error. By applying the finite-value theorem to Equation (10), the following result is obtained:

$$\lim_{z=1}\left[(z-1) \cdot \left(\frac{i_{qs}^r}{i_{qs}^{r*}} \cdot \frac{z}{z-1}\right)\right] = \lim_{z=1}\frac{h(z)}{z^n} = 1 \tag{11}$$

For convenience, $h(z)$ is set as 1. The difference form of Equation (10) can be derived as follows:

$$i_{qs}^r(k) = i_{qs}^{r*}(k-n) \tag{12}$$

The results indicate that the q-axis current lags the command value by n control periods when $h(z) = 1$. Except for the zero steady-state error, the q-axis current can also reach the command value without an overshoot.

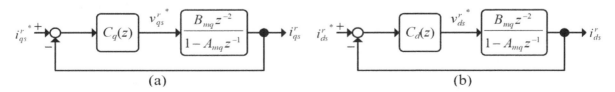

Figure 6. (**a**) q-axis and (**b**) d-axis current loop with the deadbeat controller when the motor is at standstill.

According to Equation (10) and the relation $h(z) = 1$, the deadbeat controller $C_q(z)$ can be derived using the estimated electrical parameters. The deadbeat controller $C_q(z)$ is expressed as follows:

$$C_q(z) = \frac{z^2 - \hat{A}_{mq} \cdot z}{\hat{B}_{mq} \cdot (z^n - 1)} \tag{13}$$

The transfer function of the q-axis voltage command is given as follows:

$$\frac{v_{qs}^{r\,*}}{i_{qs}^{r\,*}} = \frac{z^2 - \hat{A}_{mq} z}{\hat{B}_{mq} z^n} \tag{14}$$

To satisfy the causality, the following inequality must be satisfied:

$$n \geq \deg\{z^2 - \hat{A}_{mq} z\} = 2 \tag{15}$$

where $\deg\{\}$ denotes the highest order of the polynomial. When n is selected to be 2 and the parameters are perfectly matched, the q-axis current can attain the steady-state and equal the command value in two control periods. In addition, the voltage command can attain the steady state in two control periods after the current command changes. Therefore, $C_q(z)$ is modified as follows:

$$C_q(z) = \frac{z^2 - \hat{A}_{mq} \cdot z}{\hat{B}_{mq} \cdot (z^2 - 1)} \tag{16}$$

The q-axis voltage command at the kth control instant can be derived from (16) as follows:

$$v_{qs}^{r\,*}(k) = v_{qs}^{r\,*}(k - 2) + \hat{K}_{1q} \cdot \left(i_{qs}^{r\,*}(k) - i_{qs}^{r}(k)\right) - \hat{K}_{2q} \cdot \left(i_{qs}^{r\,*}(k - 1) - i_{qs}^{r}(k - 1)\right) \tag{17}$$

where $\hat{K}_{1q} = 1/\hat{B}_{mq}$, and $\hat{K}_{2q} = \hat{A}_{mq}/\hat{B}_{mq}$. Similarly, the deadbeat controller $C_d(z)$ can be derived as follows:

$$C_d(z) = \frac{z^2 - \hat{A}_{md} \cdot z}{\hat{B}_{md} \cdot (z^2 - 1)} \tag{18}$$

The d-axis voltage command at the kth sampling instant is expressed as follows:

$$v_{ds}^{r\,*}(k) = v_{ds}^{r\,*}(k - 2) + \hat{K}_{1d} \cdot \left(i_{ds}^{r\,*}(k) - i_{ds}^{r}(k)\right) - \hat{K}_{2d} \cdot \left(i_{ds}^{r\,*}(k - 1) - i_{ds}^{r}(k - 1)\right) \tag{19}$$

where $\hat{K}_{1d} = 1/\hat{B}_{md}$, and $\hat{K}_{2d} = \hat{A}_{md}/\hat{B}_{md}$. Figure 7 depicts the detailed schematics of $C_q(z)$ and $C_d(z)$ with the decoupling voltage and voltage limitation.

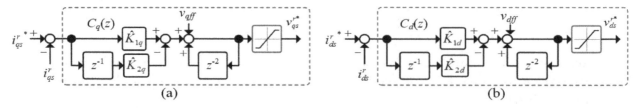

(a)   (b)

Figure 7. Block diagram of the deadbeat controller with the decoupling voltage and voltage limitation block, (a) $C_q(z)$ and (b) $C_d(z)$.

5. Simulation Results

A 400-W servo motor was used in the simulation. The motor parameters are listed in Table A2. The drive losses were ignored in the simulation. The voltage command was limited to half of the DC voltage because sinusoidal PWM was implemented, as illustrated in Figure 7. Because the d-axis current is expected to have a similar response as the q-axis current, only the q-axis current simulation results are presented.

5.1. Results with Correct Motor Parameters

Figures 8 and 9 illustrate the q-axis current, current command, and voltage command when the current steps from 0 to 1 A and from 0 to 4 A, respectively, when the motor is at standstill. As depicted in Figure 8, the q-axis current does not exhibit overshoot and is exactly equal to the command value in two control periods after the current command changes. The voltage command is generated immediately after the current command changes. The voltage commands at the kT_s and $(k + 1)T_s$ control periods can be calculated as follows:

$$v_{qs}^{r}{}^{*}(kT_s) \approx L_{qs} \frac{i_{qs}^{r}((k+2)T_s) - i_{qs}^{r}((k+1)T_s)}{T_s} = 84.14 \text{ V} \tag{20}$$

$$v_{qs}^{r}{}^{*}((k+1)T_s) \approx r_s \cdot i_{qs}^{r}{}^{*}((k+2)T_s) = 2.1 \text{ V} \tag{21}$$

Note that these command values are less than half of the DC supply.

Conversely, as depicted in Figure 9, the q-axis current increases slowly and requires approximately six control periods to attain the command value because the voltage required for the current to increase from 0 to 4 A in one control period exceeds the command limit. The voltage command saturates several times before the q-axis current reaches its command value, which is in agreement with the control law presented in Equation (17). The actual rise time is dependent on the motor inductance.

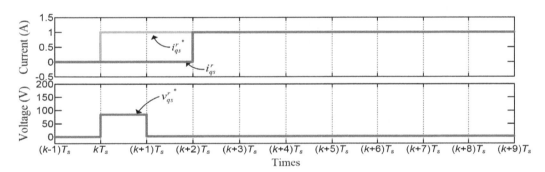

Figure 8. The simulated q-axis current, current command, and voltage command when the current command steps from 0 A to 1 A.

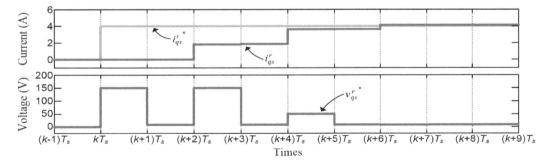

Figure 9. The simulated q-axis current, current command, and voltage command when the current command steps from 0 A to 4 A.

Figure 10 illustrates the frequency response of the deadbeat controller without voltage limitations. The current loop gain is 0 dB at low frequencies and is flat until the Nyquist frequency. This indicates that the current can follow its command without overshoot and steady-state error. However, the phase lag increases with frequency. The phase margin decreases to 0 at 4.575 kHz. Therefore, the maximum theoretical bandwidth of the proposed deadbeat controller is one-fourth of the control frequency.

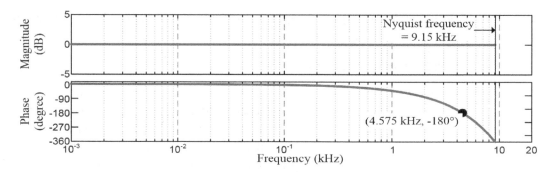

Figure 10. The frequency response of the deadbeat current controller without voltage limitation.

5.2. Results with Parameter Mismatch

The phase resistance and q- and d-axis inductances are required to design a deadbeat controller. The deadbeat controller performance is dependent on the accuracy of the estimated electrical parameters. Figures 11 and 12 display the dominant poles of the q-axis current loop in the z-domain and the corresponding q-axis current response when the current command steps from 0 to 1 A and the phase resistance varies 50% and 150% from its nominal value, respectively. The figures indicate that the poles remain near the origin regardless of the variations in the phase resistance. The q-axis current can still reach its command value in two control periods; however, marginal overshoot is observed. This implies that the influence of the resistance mismatch to the deadbeat controller is trivial. However, the q-axis current depicted in Figure 11 is marginally lower than its command value at the steady state because the resistance is smaller than its nominal value. Conversely, the q-axis current illustrated in Figure 12 is marginally higher than its command value at the steady state because the resistance is larger than its nominal value.

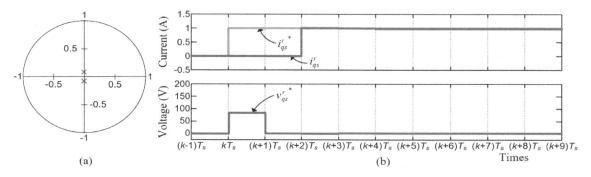

Figure 11. (a) The dominant poles and (b) simulated q-axis current and voltage command with $\hat{r}_s = 0.5 r_s$ when the current command steps from 0 A to 1 A. The motor is at standstill.

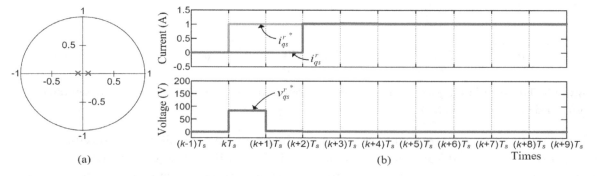

Figure 12. (a) The dominant poles and (b) simulated q-axis current and voltage command with $\hat{r}_s = 1.5 r_s$ when the current command steps from 0 A to 1 A. The motor is at standstill.

Figures 13 and 14 illustrate the dominant poles of the q-axis current loop and the corresponding q-axis current response when the current command steps from 0 to 1 A and the q-axis inductance varies 50% and 120% from its nominal value, respectively. In contrast to the results depicted in Figures 11 and 12, the variations in the inductance considerably deteriorate the system performance. As depicted in Figure 13, the current response becomes overdamped when the inductance is smaller than its nominal value because the poles mitigate toward the unit circle along the real axis. However, in Figure 14, the current response becomes underdamped when the inductance is larger than its nominal value because the poles mitigate toward the unit circle along the imaginary axis. Although no steady-state error is observed, the transient response of the q-axis current is considerably affected. In addition, the current loop can become unstable if the poles mitigate outside the unit circle because of the mismatched inductance.

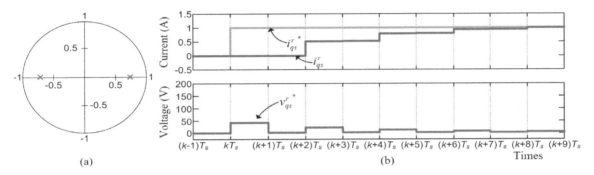

Figure 13. (a) The dominant poles and (b) simulated q-axis current and voltage command response with $\hat{L}_{qs} = 0.5 L_{qs}$ when the current command steps from 0 A to 1 A, the motor is at standstill.

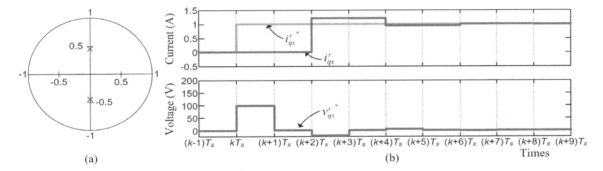

Figure 14. (a) The dominant poles and (b) simulated q-axis current and voltage command response with $\hat{L}_{qs} = 1.2 L_{qs}$ when the current command steps from 0 A to 1 A, the motor is at standstill.

6. Online Current Loop Tuning

The deadbeat controller performance is considerably affected by parameter mismatch because the voltage command is directly related to the voltage drop on the inductance and resistance. In this study, a novel online current loop tuning strategy was developed to preserve the deadbeat controller performance. Only the q-axis current loop is discussed because similar results can be obtained for the d-axis current loop.

6.1. Controller Gain Identification

From Equation (2), the q-axis current sampled at the kth and $(k-1)$th control instants can be expressed as follows:

$$\begin{bmatrix} i_{qs}^r(k) \\ i_{qs}^r(k-1) \end{bmatrix} = \begin{bmatrix} B_{mq} \cdot v_{qs}^r(k-2) + A_{mq} \cdot i_{qs}^r(k-1) \\ B_{mq} \cdot v_{qs}^r(k-3) + A_{mq} \cdot i_{qs}^r(k-2) \end{bmatrix} \tag{22}$$

Equation (22) can be rearranged as follows:

$$\begin{bmatrix} i_{qs}^r(k) & -i_{qs}^r(k-1) \\ i_{qs}^r(k-1) & -i_{qs}^r(k-2) \end{bmatrix} \begin{bmatrix} K_{1q} \\ K_{2q} \end{bmatrix} = \begin{bmatrix} v_{qs}^r(k-2) \\ v_{qs}^r(k-3) \end{bmatrix} \tag{23}$$

where K_{1q} and K_{2q} are defined as $K_{1q} = 1/B_{mq}$ and $K_{2q} = A_{mq}/B_{mq}$, respectively. Because the voltage error caused by the dead-time is satisfactorily compensated, the controller gains K_{1q} and K_{2q} can be reasonably estimated using the command values, which are expressed as follows:

$$\begin{bmatrix} \hat{K}_{1q} \\ \hat{K}_{2q} \end{bmatrix} = \begin{bmatrix} i_{qs}^r(k) & -i_{qs}^r(k-1) \\ i_{qs}^r(k-1) & -i_{qs}^r(k-2) \end{bmatrix}^{-1} \begin{bmatrix} v_{qs}^{r*}(k-2) \\ v_{qs}^{r*}(k-3) \end{bmatrix} \tag{24}$$

To solve Equation (24), the determinant of the inverse matrix must be a nonzero value. This condition is expressed as follows:

$$\det\left(i_{qs}^r\right) = i_{qs}^r(k-1)^2 - i_{qs}^r(k) \cdot i_{qs}^r(k-2) \neq 0 \tag{25}$$

As presented in Equation (24), the controller gains can be estimated using the sampled currents and voltage commands.

6.2. Estimation Accuracy Improvement

The controller gains cannot be identified in the steady state because $\det\left(i_{qs}^r\right)$ is 0. In addition, although the current ripples caused by the speed and position controller or current sensor noise yield nonzero $\det\left(i_{qs}^r\right)$, these currents cannot be used to identify controller gains because they have a low correlation with the motor parameters and consequently a low signal-to-noise-ratio (SNR). Figure 15a depicts a steady-state q-axis current with a current ripple. Although the current ripple is unpredictable in practice, the ripple is modeled as a square wave with an amplitude of Δi for convenience of analysis. Then, $\det\left(i_{qs}^r\right)$ with the current ripple is calculated as follows:

$$\det\left(i_{qs}^r\right)\Big|_{SS} = \left(i_{qs}^{r*} + \Delta i\right)^2 - i_{qs}^{r*} \cdot \left(i_{qs}^{r*} - \Delta i\right) = 3i_{qs}^{r*} \cdot \Delta i + \Delta i^2 \tag{26}$$

Figure 15b depicts a plot of $\det\left(i_{qs}^r\right)\Big|_{SS}$ versus the q-axis current command when Δi is set as 10% of the command value. It can be seen that $\det\left(i_{qs}^r\right)\Big|_{SS}$ increases with the current level. Therefore, a threshold for $\det\left(i_{qs}^r\right)$ must be set to avoid identification error in the steady state. Accordingly, controller gain identification is performed only when the following condition is satisfied:

$$\left|\det\left(i_{qs}^r\right)\right| > \det_{thres}\left(i_{qs}^r\right) \tag{27}$$

where $\det_{thres}\left(i_{qs}^r\right)$ is the threshold value. In general, the threshold value can be tuned through experiments.

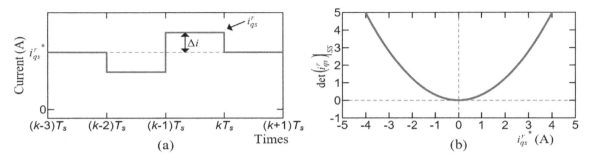

Figure 15. (a) Steady-state q-axis current with current ripple, and (b) $\det\left(i_{qs}^r\right)\Big|_{SS}$ versus current command.

The identification accuracy can be further improved by averaging the controller gains calculated in the last m control periods. In addition, each identified gain is weighted using its $\det\left(i_{qs}^r\right)$. The averaged controller gains are calculated as follows:

$$\overline{K}_{1q} = \sum_{y=1}^{m} \hat{K}_{1q,y} \cdot \left|\det_y\left(i_{qs}^r\right)\right| / \sum_{y=1}^{m} \left|\det_y\left(i_{qs}^r\right)\right| \tag{28}$$

$$\overline{K}_{2q} = \sum_{y=1}^{m} \hat{K}_{2q,y} \cdot \left|\det_y\left(i_{qs}^r\right)\right| / \sum_{y=1}^{m} \left|\det_y\left(i_{qs}^r\right)\right| \tag{29}$$

Because the average controller gain is dominated by the identified gain with higher $\det\left(i_{qs}^r\right)$, the identification accuracy improves.

After the average controller gains are calculated, the model gains can be determined as follows:

$$\hat{A}_{mq} = \overline{K}_{2q} / \overline{K}_{1q} \tag{30}$$

$$\hat{B}_{mq} = 1 / \overline{K}_{2q} \tag{31}$$

Because the effect of resistance variation on the current response is trivial, the estimated q-axis inductance can be approximated as follows:

$$\hat{L}_{qs} \approx -T_s r_s / \ln\left(\hat{A}_{mq}\right) \tag{32}$$

The model gains $\hat{A}_{md}$ and $\hat{B}_{md}$ as well as the estimated d-axis inductance can be obtained similarly. As depicted in Figure 1, the estimated inductances and model gains are used for the decoupling voltage calculation and dead-time compensation, respectively.

6.3. Identification When Voltage Command Is Limited

As illustrated in Figure 7, the stator voltage saturates to a maximum voltage V_{max} as follows:

$$\sqrt{{v_{qs}^r}^2 + {v_{ds}^r}^2} \leq V_{max} \tag{33}$$

V_{max} depends on the DC voltage and the dead-time of the inverter. The motor used in this study has almost identical q- and d-axis inductances. Thus, the d-axes current is controlled to 0 to generate the required torque with a minimum stator current. Consequently, the d-axis voltage approximates to the decoupling voltage and the steady-state q-axis voltage can be calculated using Equation (34) when the stator voltage saturates to V_{max}.

$$v_{qs}^r = \pm\sqrt{V_{max}^2 - {v_{ds}^r}^2} = \pm\sqrt{V_{max}^2 - v_{dff}^2} \tag{34}$$

Subsequently, the voltage command used to identify the controller gains when the voltage is limited is obtained as follows:

$$v_{qs}^{r\,*} = v_{qs}^r - v_{qff}$$

(35)

6.4. Gain Update Method

Figure 16 illustrates the timing for identifying and updating the gains in a deadbeat controller and current estimator, where the green bar denotes the execution of the current control and the blue bar denotes the execution of the speed and position control. Because the d-axis current is controlled to 0, only the gains in the q-axis current loop are identified. However, because $L_{qs} \approx L_{ds}$, the gains in $C_d(z)$ and the d-axis current estimator are set equal to the corresponding q-axis values. The controller gains are identified when the current control loop is executed. Because the current control executes eight times faster than the speed and position control, at most eight controller gains are identified before the next speed control is executed. Then, $\overline{K}_{1q}$, $\overline{K}_{2q}$, $\hat{A}_{mq}$, $\hat{B}_{mq}$, $\hat{L}_{qs}$, and k_{pqc} are calculated using Equations (8) and (28)–(32) when the speed control is executed. Subsequently, the gains in the deadbeat controller and the parameters in the current estimator are updated in the next execution of the current control because the motor current generally reaches steady state at this instant. The PI controller in the current estimator is updated two control periods after the model gain is updated because the sampled current is two control periods behind the estimated current.

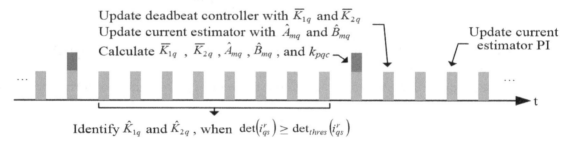

Figure 16. Time sequence of the gain identification, calculation and updating.

7. Experimental Results

A 400-W servo motor was used for experimental verifications. The parameters of the motor are provided in Table A1. Figure 17 illustrates the experimental system. The proposed online current loop tuning scheme is implemented using a Texas Instruments TMS320F28335 DSP. The detailed configuration of the drive is detailed in Table A2. In this study, V_{max} was set to 139 V to account for the losses caused by the dead time. The motor position and speed were measured using an encoder with a resolution of 2500 pulse/rev.

Figure 17. Experimental system.

The experimental results shown in Figures 18–21 were obtained without the online current loop tuning algorithm. Figure 18 illustrates the q-axis current and voltage command response when the current increased from 0 to 1 A and from 0 to 4 A, respectively, when the motor was at standstill. The voltage command was less than the limit for the 1 A step but exceeded the limit for the 4 A step. As depicted in Figure 18a, the q-axis current was exactly two current control periods behind the command value. Furthermore, overshoot and steady state error were not observed for the current. However, the q-axis current presented in Figure 18b required approximately seven control periods to reach the command value because the voltage was limited to 139 V. These results highly concur with the simulation results described in Section 5. Thus, the effectiveness of the deadbeat current controller was verified.

Figure 19a demonstrates the q-axis current, voltage, and speed response when the motor had rotation speeds between −3000 and 3000 rpm. The q-axis current followed the command value closely regardless of the motor speed. Figure 19b,c depicts the amplified views of the situation when the current increased from −4 to 4 A and decreased from 4 to −4 A, respectively. The deadbeat controller produced pulse-wise voltage because the stator voltage was limited. Although the voltage command had an opposite polarity to that of the decoupling voltage, the q-axis current required seven control periods to reach the command value. According to the aforementioned results, the performance of the deadbeat controller was independent of the motor speed. However, a marginal current overshoot is observed in Figures 18b and 19b,c because of the magnetic saturation.

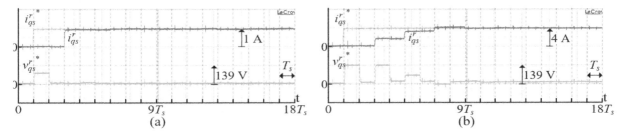

Figure 18. q-axis current and voltage when the motor is at standstill and the current command steps from (**a**) 0 A to 1 A, (**b**) 0 A to 4 A.

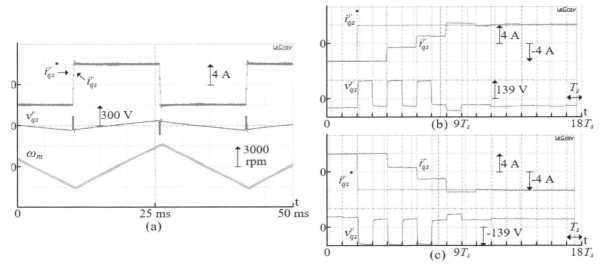

Figure 19. q-axis current, voltage, and speed when the motor cycles between −3000 rpm and 3000 rpm with step current command, (**a**) complete waveform, (**b**) amplified view when current command steps from −4 A to 4 A, (**c**) amplified view when current command steps from 4 A to −4 A.

Figures 20a and 21a illustrate the q-axis current and voltage response when the current increased from 0 to 1 A as the estimated inductance was set as 50% and 120% of its nominal value, respectively. The experiments were performed when the motor was at standstill. As depicted in Figure 20a, when $\hat{L}_{qs} = 0.5L_{qs}$, the q-axis current became overdamped and required seven control periods to reach the command value. However, as depicted in Figure 21a, when $\hat{L}_{qs} = 1.2L_{qs}$, the q-axis current became underdamped and had an observable overshoot. The experimental results are similar to the simulation results presented in Section 5.

Figure 20b,c displays the calculated $\det(i_{qs}^r)$ and controller gains for the transient response depicted in Figure 20a, respectively. Similarly, Figure 21b,c displays the calculated $\det(i_{qs}^r)$ and controller gains for the transient response presented in Figure 21a, respectively. For convenience of observation, the controller gains were normalized by their nominal values. Moreover, only the gains within ±200% of their nominal value are displayed. As depicted in the aforementioned figures, a large current difference resulted in a high $\det(i_{qs}^r)$ magnitude. Consequently, highly accurate gains were obtained because of a superior SNR. In general, the controller gain could be accurately identified for $\left|\det(i_{qs}^r)\right| \geq 0.1$. The maximum error between the identified controller gains and their nominal values were within 16%. Moreover, the proposed identification method could identify the controller gains in one current control period provided $\left|\det(i_{qs}^r)\right|$ was sufficiently large.

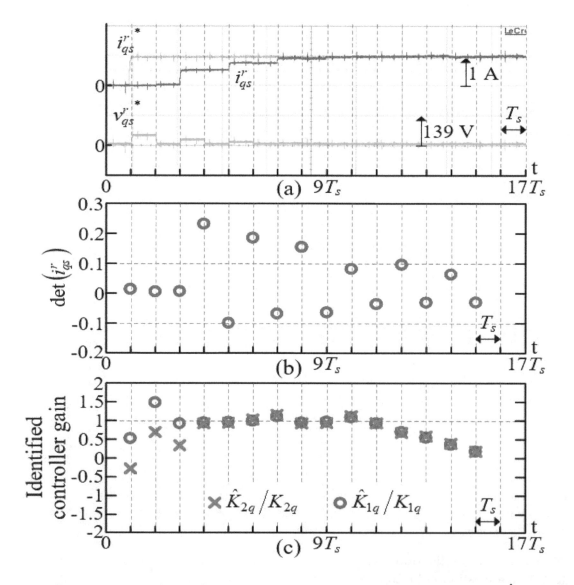

Figure 20. Current command steps from 0 A to 1 A when the motor is at standstill and $\hat{L}_{qs} = 0.5L_{qs}$, (**a**) q-axis current and voltage command, (**b**) $\det(i_{qs}^r)$, (**c**) normalized identified controller gains.

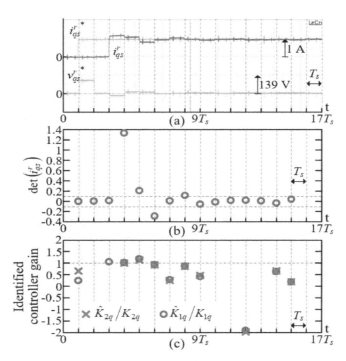

Figure 21. Current command steps from 0 A to 1 A when the motor is at standstill and $\hat{L}_{qs} = 1.2 L_{qs}$, (a) q-axis current and voltage command, (b) $\det(i_{qs}^r)$, (c) normalized identified controller gains.

Figure 22 displays the speed, position, and current responses when the motor was controlled in the positioning mode. The motor moved forward to 11π and then back to 0. The maximum speed was 3000 rpm, which is the rated speed of the motor. Furthermore, the motor was accelerating and decelerating with its rated current. An observable position error $\theta_m^* - \theta_m$ was obtained only when the motor was accelerating and decelerating. In the following experiments, the waveforms in the acceleration region were amplified to examine the effectiveness of the online current loop tuning scheme. In addition, the lowest bound of $\det_{min}(i_{qs}^r)$ for the controller gain calculation was set as 0.2 to ensure sufficient identification accuracy.

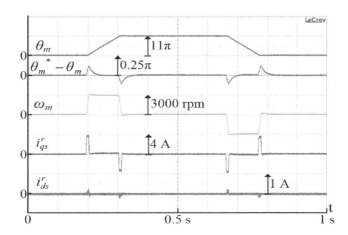

Figure 22. Speed, position, and current waveforms when the motor is controlled in the positioning mode.

Figure 23a, Figure 24a, and Figure 25a depict the current response with $\hat{L}_{qs} = L_{qs}$, $\hat{L}_{qs} = 0.5 L_{qs}$, and $\hat{L}_{qs} = 1.2 L_{qs}$ respectively, when online current loop tuning was deactivated. Conversely, Figure 23b, Figure 24b, and Figure 25b display the same waveforms but with online current loop tuning activated. The average controller gains were normalized by their nominal value for a clear observation. Figure 23a indicates that even with the correct inductance, overshoot and undershoot were observed for the

q-axis current at high current levels because of the magnetic saturation. By contrast, as indicated in Figure 23b, no apparent overshoot was observed after online current loop tuning was activated.

As depicted in Figure 24a, because the estimated inductance was set to half of the nominal value, the q-axis current response became overdamped. In addition, the d-axis current had a marginal steady-state error. By contrast, as illustrated in Figure 24b, the q-axis current was tuned to reach its reference without overshoot within a speed control period and the d-axis current had no steady state error after online current loop tuning was activated. The q-axis current in Figure 25a exhibits considerable overshoot despite the current level because the q-axis inductance is 20% higher than its nominal value. This caused additional ripples to appear on the d-axis current. However, as depicted in Figure 25b, the overshoot was eliminated within a speed control period after online current loop tuning was activated. It can be observed in Figures 24b and 25b that after the deadbeat controller is tuned by the proposed method, the required sampling period for current to reach its command value is reduced from nine to two sampling periods, and the overshoot on the current is reduced from 0.4 A to 0.09 A. These experimental results verify that the proposed method is effective and can greatly reduce the sensitivity of the deadbeat controller to the variations in inductance.

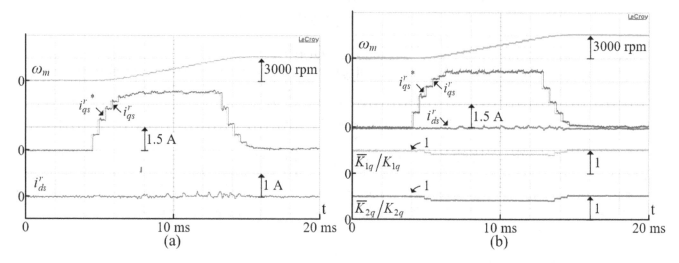

Figure 23. The amplified current response in the acceleration region of Figure 23 with $\hat{L}_{qs} = L_{qs}$ when the online current loop tuning is (**a**) de-activated and (**b**) activated.

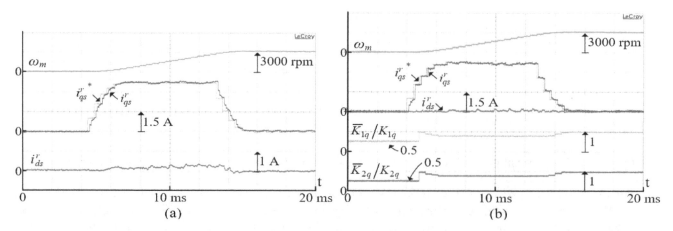

Figure 24. The amplified current response in the acceleration region of Figure 23 with $\hat{L}_{qs} = 0.5L_{qs}$ when the online current loop tuning is (**a**) de-activated and (**b**) activated.

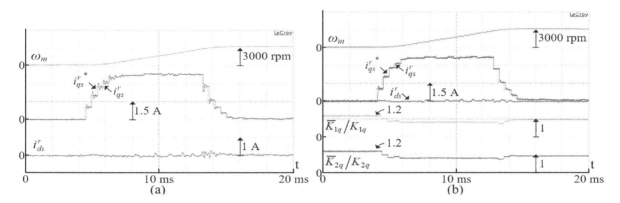

Figure 25. The amplified current response in the acceleration region of Figure 23 with $\hat{L}_{qs} = 1.2 L_{qs}$ when the online current loop tuning is (**a**) de-activated and (**b**) activated.

Figure 26 displays the measured and calculated frequency response of the q-axis deadbeat current controller. In the measurements, voltage was within the limit and the motor was at standstill. It can be seen that the current amplitude did not vary with frequency. However, the phase delay gradually increased with frequency. This is because the deadbeat controller was designed to reach its reference in two control periods, and the phase delay for two time periods was small at low frequencies but large at high frequencies.

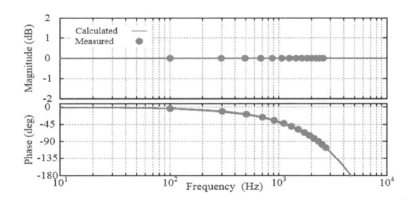

Figure 26. The measured and the calculated frequency response of the deadbeat current controller.

8. Conclusions

In this study, we present an online controller gain tuning scheme for deadbeat current control. The experimental results verify that the motor current can reach its reference value without overshoot in two current control periods with the deadbeat controller and correct parameters. However, the current response can easily become overdamped or underdamped when the controller gains are calculated using incorrectly estimated inductances. The proposed online controller gain tuning scheme is derived on the basis of the discrete-time motor model. The experimental results indicate that the correct controller gains can be identified in one current control period, and the control loop is tuned in a speed control period. Consequently, the deadbeat controller can persistently control the motor current to its reference value in two sampling periods without overshoot irrespective of the inductance variations. Furthermore, the proposed scheme is easy to implement and requires limited computations.

Author Contributions: Conceptualization, Z.-C.Y., C.-H.H., and S.-M.Y.; methodology, Z.-C.Y. and C.-H.H.; software, C.-H.H.; validation, Z.-C.Y., C.-H.H., and S.-M.Y.; formal analysis, Z.-C.Y. and C.-H.H.; investigation, Z.-C.Y. and C.-H.H.; resources, S.-M.Y.; data curation, C.-H.H.; writing—original draft preparation, Z.-C.Y.; writing—review and editing, S.-M.Y.; visualization, Z.-C.Y.; supervision, S.-M.Y.; project administration, S.-M.Y.

Appendix A

Table A1. Main drive parameters.

	Value	Unit
DC voltage	300	V
Sampling period for current loop (T_s)	55	µs
Sampling period for speed and position loop	440	µs
Dead-time	2	µs

Table A2. Main motor parameters.

	Value	Unit
Rated speed/pole pairs	3000/5	rpm
Rated current	4	A
Magnet flux (λ_m)	0.042	Wb-turns
Stator resistance (r_s)	1.4	Ω
d-axis inductance (L_{ds})	4.46	mH
q-axis inductance (L_{qs})	4.54	mH

References

1.   Bocker, J.; Beineke, S.; Bahr, A. On the Control Bandwidth of Servo Drives. In Proceedings of the 13th European Conference Power Electronics Applications, Barcelona, Spain, 8–10 September 2009; pp. 1–10.
2.   Huh, K.K.; Lorenz, R.D. Discrete-Time Domain Modeling and Design for AC Machine Current Regulation. In Proceedings of the Conference Record 42th IEEE/IAS Annual Meeting, New Orleans, LA, USA, 23–27 September 2007; pp. 2066–2073.
3.   Kim, H.; Degner, M.W.; Guerrero, J.M.; Briz, F.; Lorenz, R.D. Discrete-Time Current Regulator Design for AC Machine Drives. *IEEE Trans. Ind. Appl.* **2011**, *46*, 1425–1435.
4.   Yepes, A.G.; Vidal, A.; Malvar, J.; Lopez, O.; Jesus, D.G. Tuning Method Aimed at Optimized Setting Time and Overshoot for Synchronous Proportional-Integral Current Control in Electric Machines. *IEEE Trans. Power Electron.* **2014**, *29*, 3041–3054. [CrossRef]
5.   Andersson, A.; Thiringer, T. Assessment of an Improved Finite Control Set Model Predictive Current Controller for Automotive Propulsion Applications. *IEEE Trans. Ind. Electron.* **2019**, *67*, 91–100. [CrossRef]
6.   Cortes, P.; Kazmierkowski, M.P.; Kennel, R.M.; Quevedo, D.E.; Rodriguez, J. Predictive Control in Power Electronics and Drives. *IEEE Trans. Ind. Electron.* **2008**, *55*, 4312–4324. [CrossRef]
7.   Zhang, Y.; Xu, D.; Liu, J.; Gao, S.; Xu, W. Performance Improvement of Model-Predictive Current Control of Permanent Magnet Synchronous Motor Drives. *IEEE Trans. Ind. Appl.* **2017**, *53*, 3683–3695. [CrossRef]
8.   Lin, C.K.; Liu, T.H.; Yu, J.T.; Fu, L.C.; Hsiao, C.F. Model-Free Predictive Current Control for Interior Permanent-Magnet Synchronous Motor Based on Current Difference Detection Technique. *IEEE Trans. Ind. Electron.* **2014**, *61*, 667–681. [CrossRef]
9.   Lin, C.K.; Yu, J.T.; Lai, Y.S.; Yu, H.C. Improved Model-Free Predictive Current Control for Synchronous Reluctance Motor Drives. *IEEE Trans. Ind. Electron.* **2016**, *63*, 3942–3953. [CrossRef]
10.   Young, H.A.; Perez, M.A.; Rodriguez, J. Analysis of Finite-Control-Set Model Predictive Current Control with Model Parameter Mismatch in a Three-Phase Inverter. *IEEE Trans. Ind. Electron.* **2016**, *63*, 3100–3107. [CrossRef]
11.   Ahmed, A.A.; Koh, B.K.; Lee, Y.I. A Comparison of Finite Control Set and Continuous Control Set Model Predictive Control Schemes for Speed Control of Induction Motors. *IEEE Trans. Ind. Informat.* **2018**, *14*, 1334–1346. [CrossRef]
12.   Yang, H.; Zhang, Y.; Liang, J.; Xia, B.; Walker, P.D.; Zhang, N. Deadbeat Control Based on a Multipurpose Disturbance Observer for Permanent Magnet Synchronous Motors. *IET Electr. Pwer Appl.* **2018**, *12*, 708–716. [CrossRef]
13.   Isermann, R. *Digital Control System*; Springer: Berlin, Germany, 1981.

14. Moon, H.T.; Kim, H.S.; Youn, M.J. A Discrete-Time Predictive Current Control for PMSM. *IEEE Trans. Power Electron.* **2003**, *18*, 464–472. [CrossRef]

15. Walz, S.; Lazar, R.; Buticchi, G.; Liserre, M. Dahlin-Based Fast and Robust Current Control of a PMSM in Case of Low Carrier Ratio. *IEEE Access* **2019**, *7*, 102199–102208. [CrossRef]

16. Yang, S.M.; Lee, C.H. A Deadbeat Current Controller for Field Oriented Induction Motor Drives. *IEEE Trans. Power Electron.* **2002**, *17*, 772–778. [CrossRef]

17. Zhang, X.; Hou, B.; Mei, Y. Deadbeat Predictive Current Control of Permanent-Magnet Synchronous Motors with Stator Current and Disturbance Observer. *IEEE Trans. Power Electron.* **2017**, *32*, 3818–3834. [CrossRef]

18. Zhang, X.; Zhang, L.; Zhang, Y. Model Predictive Current Control for PMSM Drives with Parameter Robustness Improvement. *IEEE Trans. Power Electron.* **2019**, *34*, 1645–1657. [CrossRef]

19. Boileau, T.; Leboeuf, N.; Babak, N.M.; Farid, M.T. Online Identification of PMSM Parameters: Parameter Identifiability and Estimator Comparative Study. *IEEE Trans. Ind. Appl.* **2011**, *47*, 1944–1957. [CrossRef]

20. Liu, K.; Zhu, Z.Q.; Zhang, Q.; Zhang, J. Influence of Nonideal Voltage Measurement on Parameter Estimation in Permanent-Magnet Synchronous Machine. *IEEE Trans. Ind. Electron.* **2012**, *59*, 2438–2447. [CrossRef]

21. Hamida, M.A.; Leon, J.D.; Glumineau, A.; Boisliveau, R. An Adaptive Interconnected Observer for Sensorless Control of PM Synchronous Motors with Online Parameter Identification. *IEEE Trans. Ind. Electron.* **2013**, *60*, 739–748. [CrossRef]

22. Ichikawa, S.; Timita, M.; Doki, S.; Okuma, S. Sensorless Control of Permanent-Magnet Synchronous Motors Using Online Parameter Identification Based on System Identification Theory. *IEEE Trans. Ind. Electron.* **2006**, *53*, 363–372. [CrossRef]

23. Inoue, Y.; Yamada, K.; Morimoto, S.; Sanada, M. Effectiveness of Voltage Error Compensation and Parameter Identification for Model-Based Sensorless Control. *IEEE Trans. Ind. Appl.* **2009**, *45*, 213–221. [CrossRef]

24. Feng, G.; Lai, C.; Mukherjee, K.; Kar, N.C. Current Injection-Based Online Parameter and VSI Nonlinearity Estimation for PMSM Drives Using Current and Voltage DC Components. *IEEE Trans. Transp. Electrific.* **2016**, *2*, 119–128. [CrossRef]

25. Mohamed, Y.A.I.; Saadany, E.F.E. Robust High Bandwidth Discrete-Time Predictive Current Control with Predictive Internal Model–A Unified Approach for Voltage-Source PWM Converters. *IEEE Trans. Power Electron.* **2008**, *23*, 126–136. [CrossRef]

26. Yang, S.M.; Lin, K.W. Automatic Control Loop Tuning for Permanent-Magnet AC Servo Motor Drives. *IEEE Trans. Ind. Electron.* **2016**, *63*, 1499–1506. [CrossRef]

27. Shen, G.; Yao, W.; Chen, B.; Wang, K.; Lee, K.; Lu, Z. Automeasurement of the Inverter Output Voltage Delay Curve to Compensate for Inverter Nonlinearity in Sensorless Motor Drives. *IEEE Trans. Power. Electron.* **2014**, *29*, 5542–5553. [CrossRef]

5

Robust Speed Controller Design using H_infinity Theory for High-Performance Sensorless Induction Motor Drives

Ahmed A. Zaki Diab [1,*], Abou-Hashema M. El-Sayed [1], Hossam Hefnawy Abbas [1] and Montaser Abd El Sattar [2]

[1] Electrical Engineering Department, Faculty of Engineering, Minia University, Minia 61111, Egypt;
    dr_mostafa555@yahoo.com (A.-H.M.E.-S.); hosamhe@yahoo.com (H.H.A.)
[2] El-Minia High Institute of Engineering and Technology, Minia 61111, Egypt; mymn2013@yahoo.com
*   Correspondence: a.diab@mu.edu.eg.

Abstract: In this paper, a robust speed control scheme for high dynamic performance sensorless induction motor drives based on the H_infinity (H_∞) theory has been presented and analyzed. The proposed controller is robust against system parameter variations and achieves good dynamic performance. In addition, it rejects disturbances well and can minimize system noise. The H_∞ controller design has a standard form that emphasizes the selection of the weighting functions that achieve the robustness and performance goals of motor drives in a wide range of operating conditions. Moreover, for eliminating the speed encoder—which increases the cost and decreases the overall system reliability—a motor speed estimation using a Model Reference Adaptive System (MRAS) is included. The estimated speed of the motor is used as a control signal in a sensor-free field-oriented control mechanism for induction motor drives. To explore the effectiveness of the suggested robust control scheme, the performance of the control scheme with the proposed controllers at different operating conditions such as a sudden change of the speed command/load torque disturbance is compared with that when using a classical controller. Experimental and simulation results demonstrate that the presented control scheme with the H_∞ controller and MRAS speed estimator has a reasonable estimated motor speed accuracy and a good dynamic performance.

Keywords: Sensorless; induction motors; H_infinity; drives; vector control; experimental implementation

1. Introduction

The development of effective induction motor drives for various applications in industry has received intensive effort for many researchers. Many methods have been developed to control induction motor drives such as scalar control, field-oriented control and direct torque control, among which field-oriented control [1–5] is one of the most successful and effective methods. In field orientation, with respect to using the two-axis synchronously rotating frame, the phase current of the stator is represented by two component parts: the field current part and the torque-producing current part. When the component of the field current is adjusted constantly, the electromagnetic torque of the controlled motor is linearly proportional to the torque-producing components, which is comparable to the control of a separately excited DC motor. The torque and flux are considered as input commands for a field-oriented controlled induction motor drive, while the three-phase stator reference currents after a coordinated transformation of the two-axis currents are considered as the output commands. To achieve the decoupling control between the torque and flux currents components, the three-phase currents of the induction motor are controlled so that they follow their reference current commands through the use of current-regulated pulse-width-modulated (CRPWM) inverters [2–8]. Moreover, the controls of the rotor magnetic flux level and the electromagnetic torque are entirely decoupled using an additional outer feedback speed loop. Therefore, the control scheme has two loops; the inner loop of the decoupling the currents components of flux and torque, and the outer control loop which controls the rotor speed and produces the reference electromagnetic torque. Based on that, the control of an induction motor drive can be considered as a multi feedback-loop control problem consisting of current control and speed-control loops. The classical Proportional-Integral (PI) controller is frequently used in a speed-control loop due to its simplicity and stability. The parameters of the PI controller are designed through trial-and-error [3–5]. However, PI controllers often yield poor dynamic responses to changes in the load torque and moment of inertia. To overcome this problem of classical PI controllers and to improve the dynamic performance, various approaches have been proposed in References [6] and [7]. The classical two-degree-of-freedom controller (phase lead compensator and PI controller) [6] was used for indirect vector control of an induction motor drive. However, the parameters of this controller are still obtained through trial-and-error to reach a satisfactory performance level.

In Reference [8], the authors presented a control scheme for the induction motors drives based on fuzzy logic. The proposed control scheme has been applied to improve the overall performance of an induction motor drive system. This controller does not require a system model and it is insensitive to external load torque disturbances and information error. On the other hand, the presented control scheme suffers from drawbacks such as large oscillations in transient operation. Moreover, the control system requires an optical encoder to measure the motor speed [8].

A linear quadratic Gaussian controller was applied in References [9] and [10] to regulate motor speed and improve the motor's dynamic performance. The merits of this controller are as follows: fast response, robustness and the ability to operate with available noise data. However, this controller's drawbacks are that it needs an accurate system model, does not guarantee a stability margin and requires more computation.

Recently, H_infinity (H_∞) control theory has been widely implemented for its robustness against model uncertainty perturbations, external disturbances and noise. Some applications of this technique in different systems, such as permanent magnet DC motors [11], switching converters [12] and synchronous motors [13], have been reported. Moreover, researchers have worked to apply the H_∞ controllers in the induction machines drives. In Reference [14], a control scheme based on the H_∞ is presented for control the speed of the induction motors. However, the control system is validated only through the simulation results. Additionally, the speed sensor which used to measure the rotor speed is reduced the control system reliability and also its cost. In Reference [15], a vector control scheme for the induction machines based on H_∞ has been designed and experimentally validated. However, the authors used a sensor to measure the rotor speed. Additionally, a comparison between the performances of the sensor vector-control scheme of induction motor based on the PI controller and is presented. The results show the priority of the H_∞ control scheme rather than the PI controller. The main drawbacks of the control system of Ref. [15] were that speed sensor data simulation verification results were not included. Another control scheme based on the H_∞ has been presented in Reference [16]. The introduced control scheme in Ref. [16] has many drawbacks such as the need for a speed encoder, and the control law which is based on the linear parameter varying (LPV) should be updated online which increases the cost of implementation. However, validation of the control scheme has been carried out based on only a simulation using the MATLAB/Simulink (2014a, MathWorks, Natick, MA, USA) package. The authors of this paper recommended future work to eliminate the speed sensor and to minimization the implementation time. An interesting research work about the application of induction machines drives has been presented in Reference [17]. The control scheme is applied for Electric Trains application. The control system suffers from reliability reduction because of presence of the speed sensor and also the increasing of the implementation time because of the time which is needed to reach the solution of the Riccati equation. From the previous discussion, further research work is required to enhance the dynamic performance of the induction motor drives with the application of the H_∞ control theory. Moreover, the application of sensorless algorithms with the H∞ based induction machine drives is an essential research point. Furthermore, the experimental implementation of the induction machines drives based on H∞ is required for greater validation of the control scheme. Additionally, the major aspect of an H∞ control is to synthesize a feedback law that forces the closed-loop system to satisfy a prescribed H∞ norm constraint. This aspect achieves the desired stability and tracking requirements.

In this paper, a robust speed controller design for high-performance sensor-free induction motor drives based on the H∞ theory is proposed. The proposed speed controller is used to achieve both robust stability and good dynamic performance even under system parameter variations. It can withstand disturbances well and ignores system noise. Moreover, it is simple to implement and has a low computational cost. Additionally, this paper formulates the design problem of an H∞ controller in a standard form with an emphasis on the selection of the weighting functions that reflect the robustness and performance goals of motor drives. The motor speed is estimated based on a presented model-reference adaptive system (MRAS). The estimated motor speed is used as a control signal in a sensor-free field-oriented control mechanism for induction motor drives. To demonstrate the effectiveness of the proposed controller, the motor speed response following a step-change in speed command and load torque disturbance is compared with that when using a classical controller. The presented experimental and simulation results demonstrate that the proposed control system achieves reasonable estimated motor speed accuracy and good dynamic performance.

2. Mathematical Model of an Induction Motor

The induction motor can be modeled in the following mathematical differential equations represented in the rotating reference frame [8,9]:

$$\begin{bmatrix} v_{ds} \\ v_{qs} \\ 0 \\ 0 \end{bmatrix} = \begin{bmatrix} R_s + pL_s & -\omega_s L_s & pL_m & -\omega_s L_m \\ \omega_s L_s & R_s + pL_s & \omega_s L_m & pL_m \\ pL_m & -s\omega_s L_m & R_r pL_r & -s\omega_s L_r \\ s\omega_s L_m & pL_m & s\omega_s L_r & R_r + pL_r \end{bmatrix} \begin{bmatrix} i_{ds} \\ i_{qs} \\ i_{dr} \\ i_{qr} \end{bmatrix} \tag{1}$$

where, p is the differential operator, d/dt, L^n_m is the equivalent magnetizing inductance and s represents the difference between the synchronous speed and the rotor speed and it refers to slip. The self-inductances of the motor can be represented as the following:

$$L^n_s = L^n_m + L_{ls}$$
$$L^n_r = L^n_m + L_{lr} \quad '$$

where L_{ls} and L_{lr} are the stator and rotor leakage reactances, respectively.

The torque equation, in this case, is expressed as

$$T_e = \frac{3}{2}P\frac{L^n_m}{L^n_r}(i_{qs}\psi_{dr} - i_{ds}\psi_{qr}) \tag{2}$$

The previous Equation (2) is to calculate the electromagnetic torque of the motor as a function of the stator currents components, rotor flux components, pole pairs P and rotor and magnetizing inductances. Moreover, the rotor flux linkage can be written as the following equations [10]:

$$\psi_{dr} = L^n_r i_{dr} + L^n_m i_{ds}$$

$$\psi_{qr} = L^n{}_r i_{qr} + L^n{}_m i_{qs}.$$

The equation of motion is

$$J_m \frac{d\omega_r}{dt} + f_d \omega_r + T_l = T_e \tag{3}$$

where J_m is the moment of inertia, ω_r is the angular speed the rotor shaft, f_d is to express the damping coefficient, T_e indicates the electromagnetic torque of the induction motor and T_l indicates the load torque.

For achieving the finest decoupling between the ds- and qs-axis currents components, the two components of the rotor flux can be written as the following:

$$\psi_{qr} = 0 \text{ and } \psi_{dr} = \psi_r \tag{4}$$

Based on operational requirements, when the rotor flux is set to a constant, the equation of the electromagnetic torque Equation (2) will be as follows:

$$T_e = K_T i_{qs} \tag{5}$$

where

$$K_T = \frac{3P}{2} \frac{L^n{}_m}{L^n{}_r} \psi_r.$$

Equation (3) can be rewritten in the s-domain as follows:

$$\omega_r = G_p(s)(T_e(s) - T_l(s)) \tag{6}$$

where

$$G_p(s) = \frac{1/J_m}{s + f_d/J_m} \tag{7}$$

A block diagram representing an indirect vector-controlled induction motor drive is shown in Figure 1. The diagram consists mainly of three sub-models; a model for an induction motor under load when considering the core-loss, a hysteresis current-controlled pulse-width-modulated (PWM) inverter, and vector-control technique followed by a coordinate transformation and an outer speed-control loop.

In the vector-control scheme of Figure 1, the currents i^*_{ds} and i^*_{qs} are, respectively, the magnetizing and torque current components commands.

Where $I^*_s = \sqrt{i^{*2}_{ds} + i^{*2}_{qs}}$, $\theta^*_t = \tan^{-1}(i^*_{qs}/i^*_{ds})$, and $\omega^*_s = \omega^*_{sl} + \omega_r$ (the * refers to the command value). The stator current commands of phase "a", which is the reference current command for the CRPWM inverter, is presented in References [3–6].

$$i^*_{as} = I^*_s \cos(\omega^*_s t + \theta^*_t) \tag{8}$$

The commands for the other two stator phases are defined below. Referring to Figure 1, the slip speed command is calculated by

$$\omega^*_{sl} = \frac{1}{T_r} \frac{i^*_{qs}}{i^*_{ds}} \tag{9}$$

The torque current component command, i^*_{qs}, is obtained from the error of the speed, which applied to a speed controller provided that i^*_{ds} remains constant according to the operational requirements.

According to the above-mentioned analysis, the dynamic performance of the entire drive system, described in Figure 1, can be represented by the control system block diagram in Figure 2. This block diagram calls for accurate K_T parameters and the transfer function blocks of $G_p(s)$. In this paper, the speed controller $K(s)$ is designed using H_∞ theory to eliminate the problems inherent to classical controllers.

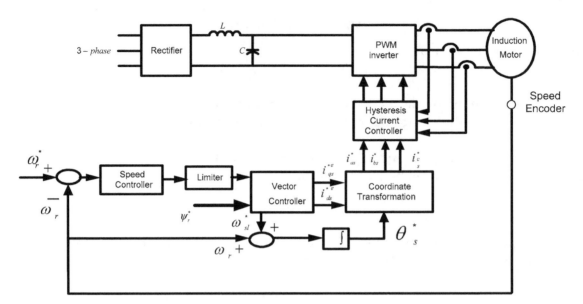

Figure 1. Block diagram of an indirect field-oriented (IFO)-controlled induction motor drive.

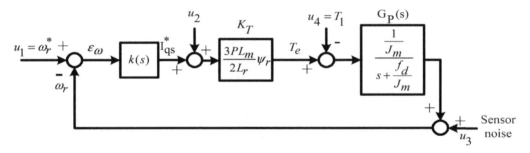

Figure 2. Block diagram of the speed-control system of an IFO-controlled induction motor drive.

3. Design of the Proposed Controller Based on H∞ Theory

The proposed controller is designed to achieve the following objectives:

(1)   Minimum effect of the measurement noise at high frequency

(2)   Maximum bounds on closed-loop signals to prevent saturation

(3)   Minimum effect of load disturbance rejection, reducing the maximum speed dip

(4)   Asymptotic and good tracking for sudden changes in command signals, in addition to a rapid and excellent damping response

(5)   Survivability against system parameter variations.

The H∞ theory offers a reliable procedure for synthesizing a controller that optimally verifies singular value loop-shaping specifications [11–17]. The standard setup of the H∞ control problem consists of finding a static or dynamic feedback controller such that the H∞ norm (a standard quantitative measure for the size of the system uncertainty) of the closed-loop transfer function below a given positive number under the constraint that the closed-loop system is internally stable.

H∞ synthesis is performed in two stages:

i.   Formulation: the first stage is to select the optimal weighting functions. The proper selections of the weighting functions give the ability to improve the robustness of the system at different operation condition and varying the model parameters. Moreover, this to reject the disturbance and noises besides the parameter uncertainties.

ii.   Solution: The transfer function of the weights has been updated to reach the optimal configuration. In this paper, the MATLAB optimization toolbox in the Simulink is used to determine the best weighting functions.

Figure 3 illustrates the block diagram of the H_∞ design problem, where G(s) is the transfer function of the supplemented plant (nominal plant Gp(s)) plus the weighting functions that represent the design features and objectives. u_1 is the control signal and w is the exogenous input vector, which generally comprises the command signals, perturbation, disturbance, noise and measurement interference; and y_1 is the controller inputs such as commands, measured output to be controlled, and measured disturbance; its components are typically tracking errors and filtered actuator signal; and z is the exogenous outputs; "error" signals to be minimized.

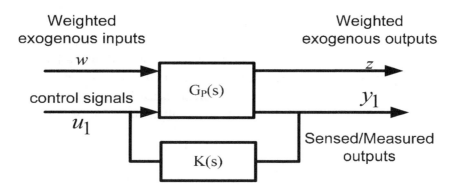

Figure 3. General setup of the H^∞ design problem.

The objective of this problem is to design a controller $K(s)$ for the augmented plant G(s) to have desirable characteristics for the input/output transfer based on the information of $y1$ (inputs to the controller $k(s)$) to generate the control signal $u1$. Therefore, the design and selection of the $K(s)$ should counteract the influence of w and z. As a conclusion, the H_∞ design problem can be subedited as detecting an equiponderating feedback control law $u1$ (s) = K(s) y_1(s) to neutralizes the effect of w and z and so to minimize the closed loop norm from w to z.

In the proposed control system that includes the H_∞ controller, one feedback loop is designed to adjust the speed of the motor, as given in Figure 4. The nominal system Gp(s) is augmented with the weighting transfer functions $W_1(s)$, $W_2(s)$ and $W_3(s)$, which penalize the error signals, control signals, and output signals, respectively. The selection of the appropriate weighting functions is the quintessence of the H_∞ control. The wrong weighting function may cause the system to suffer from poor dynamic performance and instability characteristics.

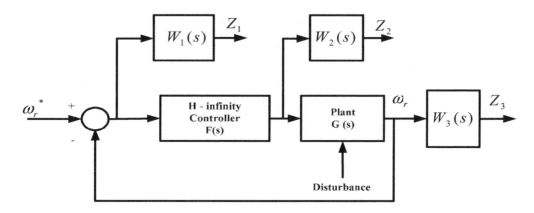

Figure 4. Simplified block diagram of the augmented plant, including the H_∞ controller.

Consider the augmented system shown in Figure 3. The following set of weighting transfer functions are selected to represent the required robustness and operation objectives:

A good choice for $W_1(s)$ is helpful for achieving good input reference tracking and good disturbance rejection. The matrix of the weighted error transfer function Z_1, which is needed for regulation, can be driven as follows:

$$Z_1 = W_1(s) \left[\omega_{ref} - \omega_r \right].$$

A proper selection for the second weight $W_2(s)$ will assist in excluding actuator saturation and provide robustness to plant supplemented disturbances. The matrix of the weighted control function Z_2 can be expressed as:

$$Z_2 = W_2(s) \cdot u(s),$$

where $u(s)$ is the transfer function matrix of the control signal output of the H$_\infty$ controller.

Additionally, a proper selection for the third weight $W_3(s)$ will restrict the bandwidth of the closed loop and achieve robustness to plant output multiplicative perturbations and sensor noise attenuation at high frequencies. The weighted output variable can be provided as:

$$Z_3 = \omega_r W_3(s).$$

In summary, the transfer functions of interest that determine the behavior of the voltage and power closed-loop systems are:

(a) Sensitivity function: $S = [I + G(s) \cdot K(s)]^{-1}$,

where G(s) and F(s) are the transfer functions of the nominal plant and the H$_\infty$ controller, respectively, while I is the identity matrix. Therefore, when S is minimized at low frequencies, it will secure perfect tracking and disturbance rejection.

(b) Control function: $C = K(s) [I + G(s) \cdot K(s)]^{-1}$.

Minimizing C will preclude saturation of the actuator and acquire robustness to plant additional disturbances.

(c) Complementary function: $T = I - S$.

Minimizing T at high frequencies will ensure robustness to plant output multiplicative perturbations and achieve noise attenuation.

4. Robust Speed Estimation Based on MRAS Techniques for an IFO Control

The using of speed encoder in induction machines drives spoils the ruggedness and simplicity of the induction motor. Moreover, the speed sensor increases the cost of the induction motor drives. To eliminate the speed sensor, the calculation of the speed may be based on the coupled circuit equations of the motor [18–34]. The following explanation and analysis of the stability of the Model Reference Adaptive System (MRAS) speed estimator. In this work, the stability estimator is proven based on Popov's criterion. The measured stator voltages and currents have been used in a stationary reference frame to describe the stator and rotor models of the induction motor. The voltage model (stator model) and the current model (rotor equation) can be written as the following in the stationary reference frame $\alpha - \beta$ [18–29]:

The voltage model (stator equation):

$$p \begin{bmatrix} \psi_{\alpha r} \\ \psi_{\beta r} \end{bmatrix} = \frac{L_r}{L_m} \left(\begin{bmatrix} V_{\alpha s} \\ V_{\beta s} \end{bmatrix} - \begin{bmatrix} (R_s + \sigma L_s p) & 0 \\ 0 & (R_s + \sigma L_s p) \end{bmatrix} \begin{bmatrix} i_{\alpha s} \\ i_{\beta s} \end{bmatrix} \right) \tag{10}$$

The current model (rotor equation):

$$p \begin{bmatrix} \psi_{\alpha r} \\ \psi_{\beta r} \end{bmatrix} = \begin{bmatrix} (-1/T_r) & -\omega_r \\ \omega_r & (-1/T_r) \end{bmatrix} \begin{bmatrix} \psi_{\alpha r} \\ \psi_{\beta r} \end{bmatrix} + \frac{L_m}{T_r} \begin{bmatrix} i_{\alpha s} \\ i_{\beta s} \end{bmatrix}$$

(11)

Figure 5 illustrates an alternative way of observing the rotor speed using MRAS. Two independent rotor flux observers are constructed to estimate the components of the rotor flux vector: one based on Equation (10) and the other based on Equation (11). Because Equation (10) does not involve the quantity ω_r, this observer may be regarded as a reference model of the induction motor, while Equation (11), which does involve ω_r, may be regarded as an adjustable model. The states of the two models are compared and the error between them is applied to a suitable adaptation mechanism that produces the observed $\hat{\omega}_r$ for the adjustable model until the estimated motor speed tracks well against the actual speed.

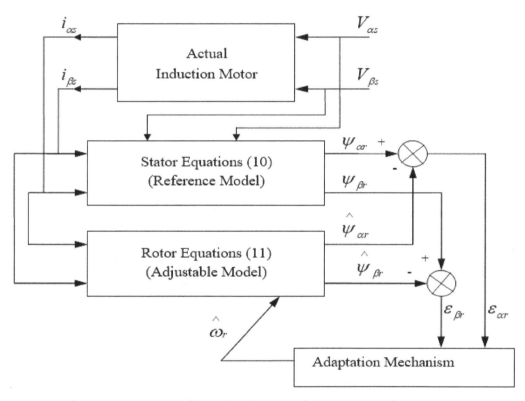

Figure 5. Structure of the MRAS system for motor speed estimation.

When eliciting an adaptation mechanism, it is adequate to initially act as a constant parameter of the reference model. By subtracting Equation (11) for the adjustable model from the corresponding equations belonged to the reference model Equation (10) for the rotor equations, the following equations for the state error can be obtained:

$$p \begin{bmatrix} \varepsilon_{\alpha r} \\ \varepsilon_{\beta r} \end{bmatrix} = \begin{bmatrix} (-1/T_r) & -\omega_r \\ \omega_r & (-1/T_r) \end{bmatrix} \begin{bmatrix} \varepsilon_{\alpha r} \\ \varepsilon_{\beta r} \end{bmatrix} + \begin{bmatrix} -\hat{\psi}_{\beta r} \\ \hat{\psi}_{\alpha r} \end{bmatrix} (\omega_r - \hat{\omega}_r)$$

(12)

that is,

$$p [\varepsilon] = [A_r] [\varepsilon] - [W]$$

(13)

Because $\hat{\omega}_r$ is a function of the state error, Equations (12) and (13) represent a non-linear feedback system, as shown in Figure 6.

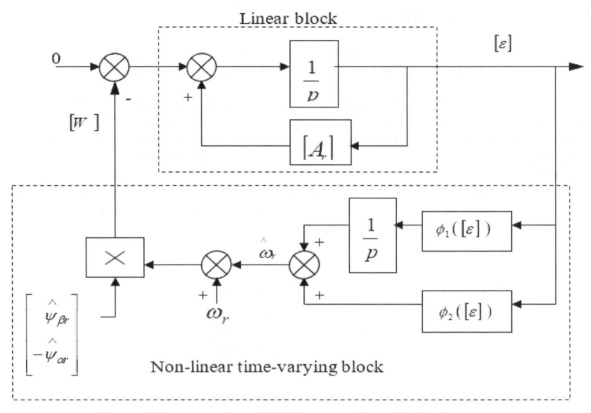

Figure 6. Representation of MRAS as a non-linear feedback system.

According to Landau, hyperstability is confirmed as long as the linear time-invariant forward-path transfer matrix is precisely positive real and that the non-linear feedback (which includes the adaptation mechanism) comply with Popov's hyperstability criterion. Popov's criterion demands a bounded negative limit on the input/output inner product of the non-linear feedback system. Assuring this criterion leads to the following candidate adaptation mechanism [31–34]:

Let

$$\hat{\omega}_r = \phi_2([\varepsilon]) + \int_0^t \phi_1([\varepsilon])d\tau \qquad (14)$$

then, Popov's criterion presupposes that

$$\int_0^{t_1} [\varepsilon]^T [W] \, dt \geq -\gamma_0^2 \quad \text{for all } t_1 \geq 0 \qquad (15)$$

where γ_0^2 is a positive constant. Substituting for $[\varepsilon]$ and $[W]$ in this inequality using the definition of $\hat{\omega}_r$, Popov's criterion for the system under study will be

$$\int_0^{t_1} \left\{ [\varepsilon_{\alpha r} \, \psi_{\beta r} - \varepsilon_{\beta r} \, \psi_{\alpha r}] \left[\omega_r - \phi_2([\varepsilon]) - \int_0^t \phi_1([\varepsilon]) \, d\tau \right] \right\} dt \geq -\gamma_0^2 \qquad (16)$$

A proper solution to this inequality can be realized via the following well-known formula:

$$\int_0^{t_1} k(p \, f(t)) \, f(t) \, dt \geq -\frac{1}{2} k \, f(0)^2, \quad k \succ 0 \qquad (17)$$

Using this expression, Popov's inequality is satisfied by the following functions:

$$\phi_1 = K_P\left(\varepsilon_{\beta r}\hat{\psi}_{\alpha r} - \varepsilon_{\alpha r}\hat{\psi}_{\beta r}\right) = K_P\left(\psi_{\beta r}\hat{\psi}_{\alpha r} - \psi_{\alpha r}\hat{\psi}_{\beta r}\right).$$

$$\phi_2 = K_I\left(\varepsilon_{\beta r}\hat{\psi}_{\alpha r} - \varepsilon_{\alpha r}\hat{\psi}_{\beta r}\right) = K_I\left(\psi_{\beta r}\hat{\psi}_{\alpha r} - \psi_{\alpha r}\hat{\psi}_{\beta r}\right).$$

Figure 7 illustrates the block diagram of the MRAS. The outputs of the two models the rotor flux components. Moreover, the measured stator voltages and currents are in the stationary reference frame have been applied to be as the inputs of MRAS.

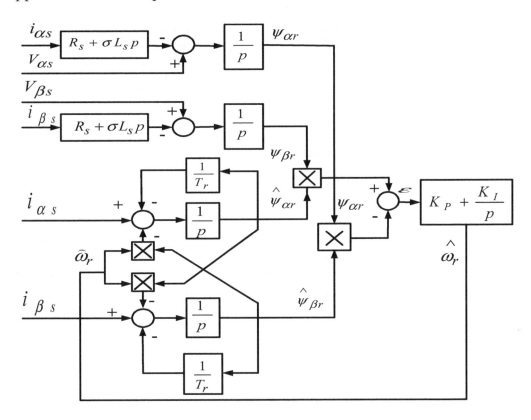

Figure 7. Block diagram of the Model Reference Adaptive System (MRAS) speed estimation system.

5. Proposed Sensor-Free Induction Motor Drive for High-Performance Applications

Figure 8 shows a block diagram of the proposed controller sensor-free induction motor drive system. The control system is composed of a robust controller based on H∞ theory, hysteresis controllers, a vector rotator, a digital pulse with modulation (PWM) scheme for a transistor bridge voltage source inverter (VSI) and a motor speed estimator based on MRAS. The speed controller makes speed corrections by assessing the error between the command and estimated motor speed. The speed controller is used to generate the command q-component of the stator current i^*_{qs}. The vector rotator and phase transform in Figure 8 are used to transform the stator current components command to the three-phase stator current commands (i^*_{as}, i^*_{bs} and i^*_{cs}) using the field angle position $\hat{\theta}_r$. The field angle is obtained by integrating the summation of the estimated speed and slip speed. The hysteresis current control compares the stator current commands to the actual currents of the machine and switches the inverter transistors in such a way as to obtain the desired command currents. Moreover, the MRAS rotor speed estimator can observe the rotor speed $\hat{\omega}_r$ based its inputs of measured stator voltages and currents. This estimated speed is fed back to the speed controller. Additionally, the estimated speed is added to the slip speed, and the sum is integrated to obtain the field angle θ_s.

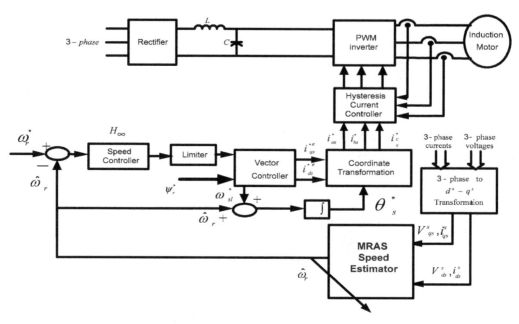

Figure 8. The block diagram of the proposed sensorless induction motor drive based on H∞ theory.

6. Simulation and Experimental Results

6.1. Simulation Results

The complete block diagram of the field-oriented induction motor drive with the proposed H∞ and MRAS speed estimator has been presented in Figure 8. The proposed system has been simulated using MATLAB/Simulink under different operating conditions. The parameters and data specifications of the entire system used in the simulation are given in Table 1. The following set of weighting functions are obtained based on the application of the optimization toolbox in MATLAB/Simulink to achieve the proposed robustness and operation objectives:

$$W_1 = \gamma_1 \frac{s + 1e - 4}{s + 0.001}, \ W_2 = \gamma_2 \frac{s + 1e - 4}{s + 10}, \ W_3 = \gamma_3,$$

where $\gamma_1 = 12$, $\gamma_2 = 0.00001$, $\gamma_3 = 0$.

Table 1. Parameters and data specifications of the induction motor.

Rated Power (W)	180	Rated Voltage (V)	220
Rated current (A)	1.3–1.4	Rated frequency (Hz)	60
R_s (Ω)	11.29	R_r (Ω)	6.11
L_s (H)	0.021	L_r (H)	0.021
L_m (H)	0.29	Rated rotor flux, (wb)	0.3
J (kg.m^2)	0.00940	Rated speed (rpm)	1750

The transient behavior of the proposed sensorless control system is evaluated by applying and removing the motor-rated torque (1 N.m), as shown in Figure 9. Figure 9a shows the performance of the proposed control scheme with H∞ controller. While Figure 9b illustrates the performance of sensorless induction motor (IM) drive based on the conventional PI controller for the comparison purpose. Figure 9 shows that the motor speed can be effectively estimated and accurately tracks the actual speed when using the proposed sensorless scheme. Moreover, the figure shows that the performance of the two controllers of the PI controller and the proposed H∞ controller have acceptable dynamic performance. Furthermore, the figure also indicates that a fast and precise transient response to motor torque is achieved with the H∞ controller. Additionally, the stator phase current matches the value of the application and removal of the motor-rated torque.

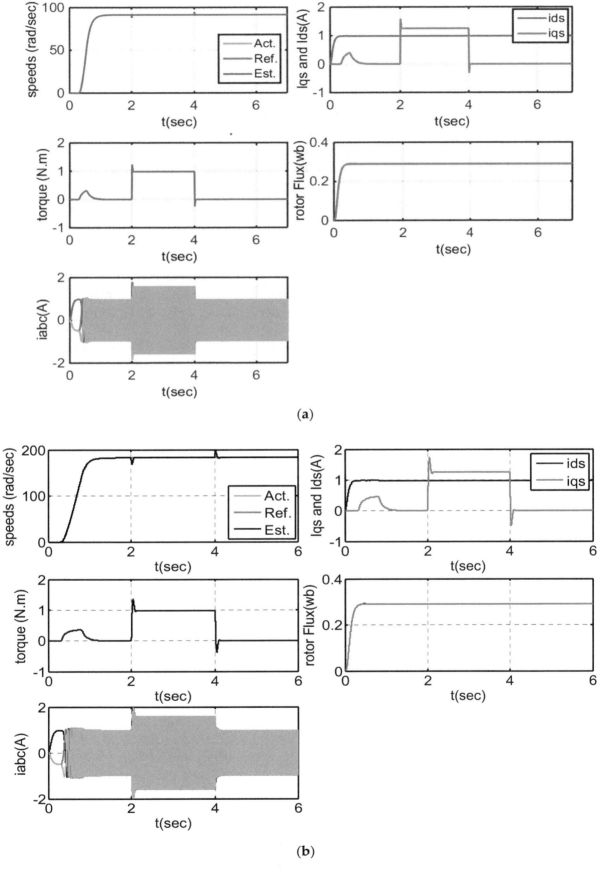

Figure 9. Simulation results of speed transient with load step changes of rated values using the proposed sensorless drive system; (**a**) with H_∞ controller and (**b**) with PI controller.

For more validating of the H_∞ controller, Figure 10 describes the dynamic response of the control scheme based on the proposed H_∞ controller versus the PI controller when the system is subjected to a step change in the load torque. From the figure, the results show an acceptable dynamic performance of the control scheme with the PI controllers and the H_∞ against the step change in the load torque. However, the application of the H_∞ controller causes improvements in the dynamic response of the rotor speed. Clearly, the PI controller requires a long rise time compared to the H_∞ controller and the speed response has a larger overshoot with respect to the H_∞ controller. The reasons for these results of the priority of the H_∞ based induction motor drive may be because the H_∞ controller has been designed taking the weighting function for the disturbance. The discussion of these results can be more discussed as follows: The fixed parameter controllers such as PI controllers are developed at nominal operating points. However, it may not be suitable under various operating conditions. However, the real problem in the robust nonlinear feedback control system is to synthesize a control law which maintains system response and error signals to within prespecified tolerances despite the effects of uncertainty on the system. Uncertainty may take many forms but among the most significant are noise, disturbance signals, and modeling errors. Another source of uncertainty is unmodeled nonlinear distortion. Consequently, researchers have adopted a standard quantitative measure the size of the uncertainty, called the H_∞. Therefore, the dynamic performance of the control scheme with the H_∞ controller is improved rather than the acceptable performance with the PI controller.

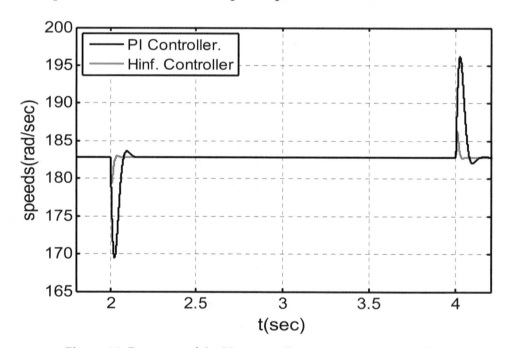

Figure 10. Response of the H_∞ controller versus the PI controller.

Figure 11 shows the actual and estimated speed transient, motor torque and stator phase current during acceleration and deceleration at different speeds. The estimated speed agreed satisfactorily with the actual speed. Additionally, a small deviation occurs from the actual speed before reaching steady-state and subsequently tracking quickly towards the command value. The motor torque response exhibits good dynamic performance. Figure 11 also shows the stator phase current during acceleration. From these simulation results, the proposed sensorless drive system is capable of operating at high speed, as illustrated in Figure 11.

In the last case of the study, the transient performance of the sensorless induction machine drive is examined by reversing the rotor speed. Figure 12 shows that induction motor drive based on the proposed robust controller has high dynamic performance at the reversing the rotor speed from 150 rad/s. Moreover, the speed estimator MRAS can accurately observe the rotor speed.

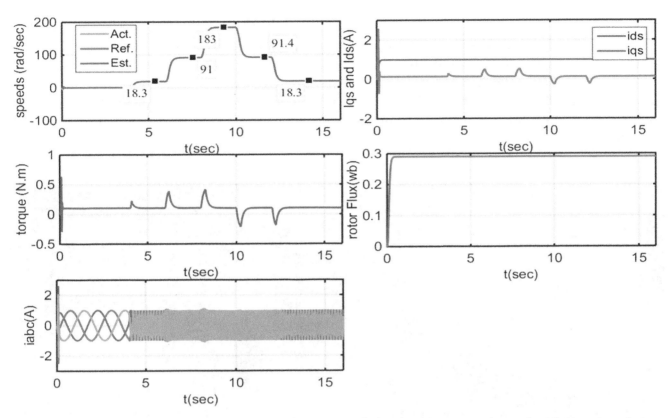

Figure 11. Simulation results of the acceleration and deceleration operation using the proposed sensorless drive system.

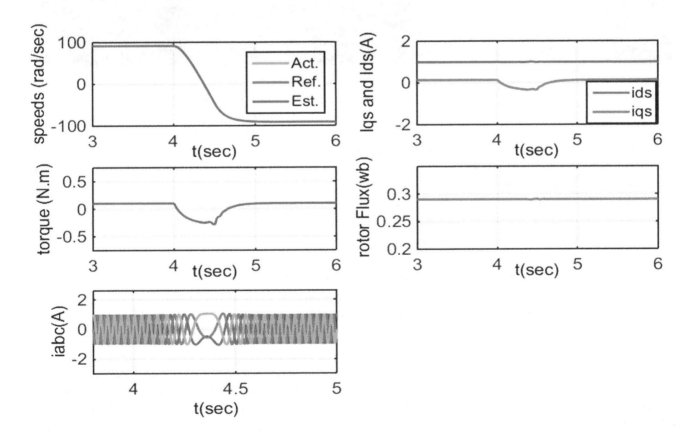

Figure 12. System performance when reversing the motor speed.

6.2. Practical Results

Figure 13 shows the experimental setup for the configured drive system. The drive system includes an induction motor; with the same parameters and data specification of the induction motor which used for simulating the proposed control scheme; linked with a digital control board (TMDSHVMTRPFCKIT from Texas Instruments with a TMS320F28035 control card) [32,35]. The complete control scheme has been programmed in the package of Code Composer Studio CCS from Texas Instruments.

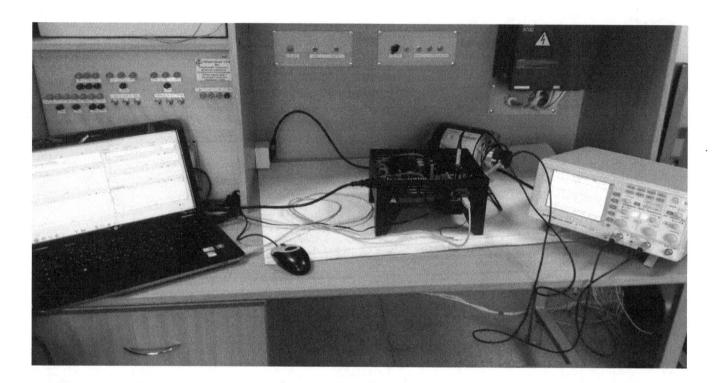

Figure 13. A photo of the experimental setup for induction motor drive.

To validate the effectiveness of the sensorless vector control of the induction motor drive, the experiments were accomplished at different values of the reference speed. Figures 14–16 show samples of the results when the proposed system is tested at reference speeds of 0.2 pu and 0.4 pu; the base speed is assumed to be 3600 rpm (So, when the reference speed is 0.5 pu, it is mean the speed equals 1800 rpm). The results proved that the drive system effectively works at an extensive range of speeds. In addition, the actual and estimated speeds have coincided. Moreover, from Figure 16, it is obviously seen that the current of the motor is sinusoidal.

Another case of study has been tested for more evaluating of the control scheme. In this case of study, the reference speed has been reversed from 0.4 pu to 0.2 pu in the reverse direction in a ramp variation. The results of this case of study are shown in Figures 17 and 18. Figure 17 shows the speed response of the control scheme. Moreover, the phase current is shown in Figure 18. The results show the control scheme has a good dynamic performance.

The last case of study has assumed many ramp changes in the references including reversing the speed. The rotor speed response is shown in Figure 19. The results have been plotted with the aid of CCS package and Digital Signal Processing (DSP). The results show the control scheme has a good dynamic performance.

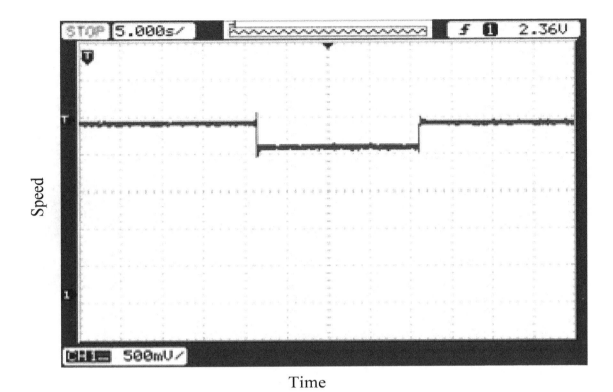

Figure 14. The transient performance of the entire drive system under variable speed from 0.4 pu to 0.2 pu to 0.4 pu.

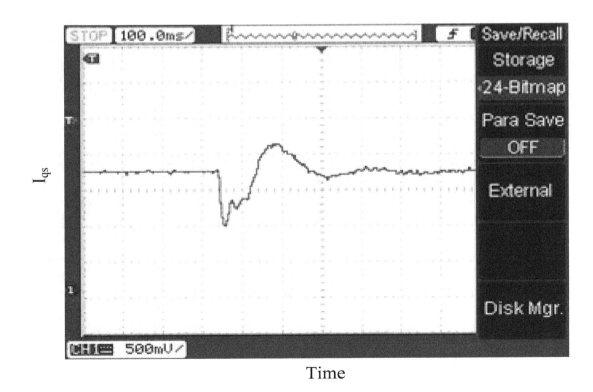

Figure 15. The transient performance of Iq$_S$ of the entire drive system under variable speed from 0.4 pu to 0.2 pu.

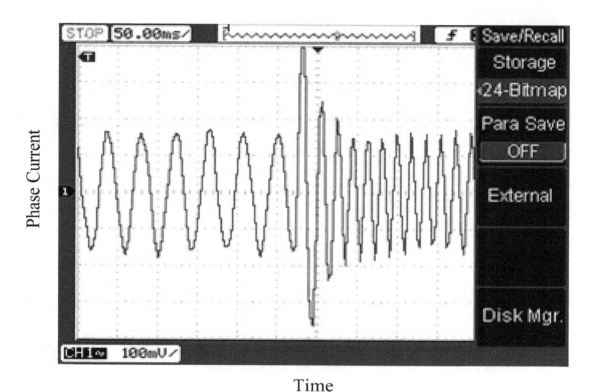

Figure 16. Current of phase (**a**) at reference speeds of 0.2 pu to 0.4 pu.

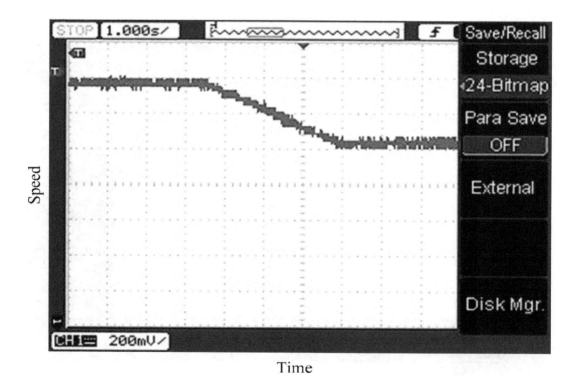

Figure 17. The transient performance of Iq$_S$ of the entire drive system under variable speed from 0.4 pu to 0.2 pu.

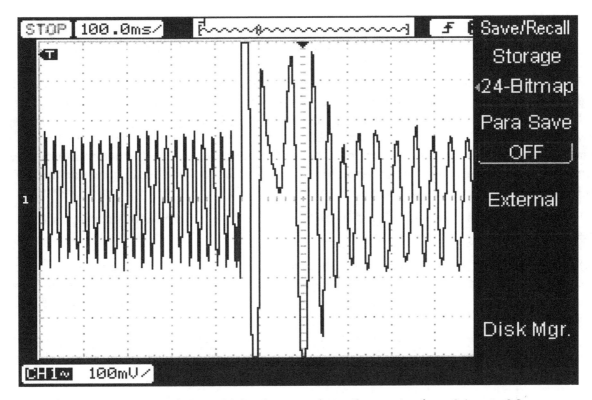

Figure 18. Current of phase (**a**) for the case of speed reversing from 0.4 pu to 0.2 pu.

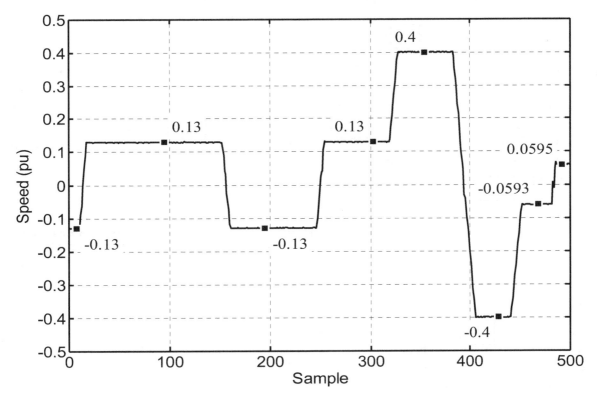

Figure 19. The transient performance of multi-variation in the rotor speed with reversing (The experimental data and measurements have been collected with the aid of the DSP and plotted using Matlab plot tool).

7. Conclusions

In this paper, H$_\infty$ theory has been proposed for designing an optimal robust speed controller for a field-oriented induction motor drive. The design problem of the H$_\infty$ controller was explained and derived in standard form with an assertion on the choice of weighting functions, which fulfills the optimal robustness and performance of the drive system. The proposed control strategy has many advantages: it is robust to plant uncertainties, and has a simple implementation and a fast response. Moreover, a robust motor speed estimator based on the MRAS is presented that estimates motor speed accurately for a sensorless IFO control system. The validation of the induction motor drive was performed using both simulated and experimental implementations. The main conclusions that can be drawn from the results in this study are as follows:

(1) The effectiveness of the considered induction motor drive system with the proposed controller has been demonstrated.

(2) Compared with a PI classical controller, the response of the proposed controller shows a reduced settling time in the case of a sudden change of the speed command in addition to smaller values of the maximum speed dip and overshoot as a result of the application and removal of stepped changes in load torque.

(3) The proposed controller achieved robust performance under stepped speed change commands or changes in load torque even when the parameters of the controlled system were varied.

(4) The forward-reverse operation of the drive is obtained by the robust MRAS speed estimator and guarantees the stability of the proposed sensorless control to the system at a speed of zero. Moreover, the presented speed estimator provides an accurate speed estimation regardless of the load conditions.

(5) Both simulated and real-world experimental results demonstrate that the proposed control drive system is capable of working at a wide range of motor speeds and that it exhibits good performance in both dynamic and steady-state conditions.

Further research work should consider the nonlinearity of the induction machine parameters tacking saturation and/or iron losses into consideration. Additionally, recent optimization techniques may be applied to determine the optimal weight functions for designing the controller. Moreover, the operation range should be expanded to study and analyse the operation of the control scheme in the field weakening region. Moreover, the estimation of the machine parameters may be an interesting research point for future work for improving the overall performance of the control scheme and speed estimator.

Author Contributions: A.A.Z.D. and H.H.A. developed the idea and the simulation models. A.A.Z.D. performed the experiments and analyzed the data. A.A.Z.D., A.-H.M.E.-S. and H.H.A. wrote the paper. A.A.Z.D., A.-H.M.E.-S. and M.A.E.S. contributed by drafting and making critical revisions. All the authors organized and refined the manuscript to its present form.

References

1. Vas, P. *Vector Control of AC Machines*; Oxford University Press: Oxford, UK, 1990; pp. 122–215.
2. Krishnan, R.; Doran, F.C. Study of parameter Sensitivity in High Performance Inverter Fed Induction Motor Drive Systems. *IEEE Trans. Ind. Appl.* **1987**, *IA-23*, 623–635. [CrossRef]
3. Vas, P. *Parameter Estimation, Condition Monitoring, and Diagnosis of Electrical Machines*; Oxford Science Publications: Oxford, UK, 1993.
4. Toliyat, H.A.; Levi, E.; Raina, M. A Review of RFO Induction Motor Parameter Estimation Techniques. *IEEE Trans. Energy Convers.* **2003**, *18*, 271–283. [CrossRef]

5. Minami, K.; Veles-Reyes, M.; Elten, D.; Verghese, G.C.; Filbert, D. Multi-Stage Speed and Parameter Estimation for Induction Machines. In Proceedings of the Record 22nd Annual IEEE Power Electronics Specialists Conference (IEEE PESC'91), Cambridge, MA, USA, 24–27 June 1991; pp. 596–604.

6. Roncero-Sánchez, P.L.; García-Cerrada, A.; Feliú, V. Rotor-Resistance Estimation for Induction Machines with Indirect-Field Orientation. *Control Eng. Pract.* **2007**, *15*, 1119–1133. [CrossRef]

7. Karayaka, H.B.; Marwali, M.N.; Keyhani, A. Induction Machines Parameter Tracking from Test Data Via PWM Inverters. In Proceedings of the Record of the 1997 IEEE Industry Applications Conference Thirty-Second IAS Annual Meeting, New Orleans, LA, USA, 5–9 October 1997; pp. 227–233.

8. Godoy, M.; Bose, B.K.; Spiegel, R.J. Design and performance evaluation of a fuzzy-logic-based variable-speed wind generation system. *IEEE Trans. Energy Convers.* **1997**, *33*, 956–964.

9. Munteau, I.; Cutululis, N.A.; Bratcu, A.I.; Ceanga, E. Optimization of variable speed wind power systems based on a LQG approach. In Proceedings of the IFAC Workshop on Control Applications of Optimisation—CAO'03 Visegrad, Visegrad, Hungary, 30 June–2 July 2003.

10. Hassan, A.A.; Mohamed, Y.S.; Yousef, A.M.; Kassem, A.M. Robust control of a wind driven induction generator connected to the utility grid. *Bull. Fac. Eng. Assiut Univ.* **2006**, *34*, 107–121.

11. Attaiaence, C.; Perfetto, A.; Tomasso, G. Robust postion control of DC drives by means of H_∞ controllers. *Proc. IEE Electr. Power Appl.* **1999**, *146*, 391–396. [CrossRef]

12. Naim, R.; Weiss, G.; Ben-Yakakov, S. H_∞ control applied to boost power converters. *IEEE Trans. Power Electron.* **1997**, *12*, 677–683. [CrossRef]

13. Lin, F.-J.; Lee, T.; Lin, C. Robust H_∞ controller design with recurrent neural network for linear synchronous motor drive. *IEEE Trans. Ind. Electron.* **2003**, *50*, 456–470.

14. Pohl, L.; Vesely, I. Speed Control of Induction Motor Using H_∞ Linear Parameter Varying Controller. *IFAC-PapersOnLine* **2016**, *49*, 74–79. [CrossRef]

15. Kao, Y.-T.; Liu, C.-H. Analysis and design of microprocessor-based vector-controlled induction motor drives. *IEEE Trans. Ind. Electron.* **1992**, *39*, 46–54. [CrossRef]

16. Prempain, E.; Postlethwaite, I.; Benchaib, A. A linear parameter variant H_∞ control design for an induction motor. *Control Eng. Pract.* **2002**, *10*, 633–644. [CrossRef]

17. Rigatos, G.; Siano, P.; Wira, P.; Profumo, F. Nonlinear H-infinity feedback control for asynchronous motors of electric trains. *Intell. Ind. Syst.* **2015**, *1*, 85–98. [CrossRef]

18. Riberiro, L.A.D.; Lima, A.M.N. Parameter Estimation of Induction Machines Under Sinusoidal PWM Excitation. *IEEE Trans. Energy Convers.* **1999**, *14*, 1218–1223. [CrossRef]

19. Moonl, S.; Keyhani, A. Estimation of Induction Machines Parameters from Standstill Time-Domain Data. *IEEE Trans. Ind. Appl.* **1994**, *30*, 1609–1615. [CrossRef]

20. Lima, A.M.N.; Jacobina, C.B.; Filho, E.B.D. Nonlinear Parameter Estimation of Steady-state Induction Machine Models. *IEEE Trans. Ind. Electron.* **1997**, *44*, 390–397. [CrossRef]

21. Yang, G.; Chin, T.H. Adaptive-Speed Identification Scheme for a Vector-Controlled Speed Sensorless Inverter-Induction Motor Drive. *IEEE Trans. Ind. Appl.* **1993**, *29*, 820–825. [CrossRef]

22. Kubota, H.; Matsuse, K. Speed Sensorless Field-Oriented Control of Induction Motor with Rotor Resistance Adaptation. *IEEE Trans. Ind. Appl.* **1994**, *30*, 1219–1224. [CrossRef]

23. Suwankawin, S.; Sangwongwanich, S. Design Strategy of an Adaptive Full Order Observer for Speed Sensorless Induction Motor Drives Tracking Performance and Stabilization. *IEEE Trans. Ind. Electron.* **2006**, *53*, 96–119. [CrossRef]

24. Al-Tayie, J.; Acarnley, P. Estimation of Speed, Stator Temperature and Rotor Temperature in Cage Induction Motor Drive Using the Extended Kalman Filter Algorithm. *Proc. IEE Electr. Power Appl.* **1997**, *144*, 301–309. [CrossRef]

25. Barut, M.; Bogosyan, O.S.; Gokasan, M. Switching EKF Technique for Rotor and Stator Resistance Estimation in Speed Sensorless Control of Induction Motors. *Energy Convers. Manag.* **2007**, *48*, 3120–3134. [CrossRef]

26. Jingchuan, L.; Longya, X.; Zhang, Z. An Adaptive Sliding-Mode Observer for Induction Motor sensorless Speed Control. *IEEE Trans. Ind. Appl.* **2005**, *41*, 1039–1046.

27. Rashed, M.; Stronach, A.F. A Stable Back-EMF MRAS-Based Sensorless Low Speed Induction Motor Drive Insensitive to Stator Resistance Variation. *IEE Proc. Electr. Power Appl.* **2004**, *151*, 685–693. [CrossRef]

28. Mohamed, Y.S.; El-Sawy, A.M.; Zaki, A.A. Rotor Resistance Identification for Speed Sensorless Vector

Controlled Induction Motor Drives Taking Saturation Into Account. *J. Eng. Sci. Assiut Univ.* **2009**, *37*, 393–412.

29. Middeton, R.H.; Goodwin, G.C. *Digital Control and Estimation*, 1st ed.; Prentice-Hall, Inc.: Englewood Cliffs, NJ, USA, 1990; Volume 1.

30. Blasco-Gimenez, R.; Asher, G.; Summer, M.; Bradley, K. Dynamic Performance Limitations for MRAS Based Sensorless Induction Motor Drives. Part 1: Stability Analysis for the Closed Loop Drive. *Proc. IEE-Electr. Power Appl.* **1996**, *143*, 113–122. [CrossRef]

31. Diab, A.A.; Khaled, A.; Elwany, M.A.; Hassaneen, B.M. Parallel estimation of rotor resistance and speed for sensorless vector controlled induction motor drive. In Proceedings of the 2016 17th International Conference of Young Specialists on Micro/Nanotechnologies and Electron Devices (EDM), Erlagol, Russia, 30 June–4 July 2016.

32. Diab, A.A. Real-Time Implementation of Full-Order Observer for Speed Sensorless Vector Control of Induction Motor Drive. *J. Control Autom. Electr. Syst.* **2014**, *25*, 639–648. [CrossRef]

33. Diab, A.A.Z. Implementation of a novel full-order observer for speed sensorless vector control of induction motor drives. *Electr. Eng.* **2016**, *99*, 907–921. [CrossRef]

34. Diab, A.A.Z. Novel robust simultaneous estimation of stator and rotor resistances and rotor speed to improve induction motor efficiency. *Int. J. Power Electron.* **2017**, *8*, 267–287. [CrossRef]

35. Texas Instruments C2000 Systems and Applications Team. High Voltage Motor Control and PFC (R1.1) Kit Hardware Reference Guide, v. 2. 2012. Available online: http://www.ti.com/tool/TMDSHVMTRPFCKIT (accessed on 31 January 2019).

6

Finite Control Set Model Predictive Control of Six-Phase Asymmetrical Machines

Pedro Gonçalves, Sérgio Cruz * and André Mendes

Department of Electrical and Computer Engineering, University of Coimbra, and Instituto de Telecomunicações, Pólo 2-Pinhal de Marrocos, P-3030-290 Coimbra, Portugal; pgoncalves@ieee.org (P.G.); amsmendes@ieee.org (A.M.)

* Correspondence: smacruz@ieee.org.

Abstract: Recently, the control of multiphase electric drives has been a hot research topic due to the advantages of multiphase machines, namely the reduced phase ratings, improved fault tolerance and lesser torque harmonics. Finite control set model predictive control (FCS-MPC) is one of the most promising high performance control strategies due to its good dynamic behaviour and flexibility in the definition of control objectives. Although several FCS-MPC strategies have already been proposed for multiphase drives, a comparative study that assembles all these strategies in a single reference is still missing. Hence, this paper aims to provide an overview and a critical comparison of all available FCS-MPC techniques for electric drives based on six-phase machines, focusing mainly on predictive current control (PCC) and predictive torque control (PTC) strategies. The performance of an asymmetrical six-phase permanent magnet synchronous machine is compared side-by-side for a total of thirteen PCC and five PTC strategies, with the aid of simulation and experimental results. Finally, in order to determine the best and the worst performing control strategies, each strategy is evaluated according to distinct features, such as ease of implementation, minimization of current harmonics, tuning requirements, computational burden, among others.

Keywords: multiphase electric drives; multiphase machines; six-phase machines; finite control set model predictive control; predictive current control; predictive torque control

1. Introduction

Rotating electrical machines with a number of phases higher than three ($n > 3$) are commonly referred to in the literature as multiphase machines [1]. Multiphase machines were first used in high power generation units during the 1920s due to the current limit of circuit breakers at that time and due to the size of the reactors needed to limit currents in the event of faults [2]. In the 1960s, it was demonstrated that an increase in the number of phases of electrical machines fed by voltage source inverters (VSIs) leads to an increase of the order of torque pulsations ($h = 2n$) and to a reduction of their magnitude [3]. Additionally, the increase in the number of phases also leads to a lower current or voltage per phase, decreasing the requirements of the power semiconductors ratings [4]. An electric drive with improved reliability based on a multiphase machine was first studied in the 1980s [5], where each phase was connected to an independent power converter and in the event of a fault in one or more phases, the drive could remain in operation with a reduced power rating. Since the decoupled control of flux and torque in multiphase machines only requires the regulation of two independent current components, regardless of the number of phases of the machine [6], multiphase machines provide additional degrees of freedom that can be used for several purposes without affecting the production of flux and torque [7]. The first works taking advantage of the additional degrees of freedom of multiphase machines were published in the 1990s, where the injection of current harmonics was used to enhance the torque developed by machines with concentrated windings [8,9]. Multimotor drives proposed in the 2000s is another application that takes advantage of the additional degrees of freedom, where a single n-phase VSI is able to drive independently up to $(n-1)/2$ machines if n is odd or up to $(n-2)/2$ machines if n is even, either connected in series or in parallel [10,11]. More recently, the additional degrees of freedom of multiphase machines are being used to provide: balancing of the dc-link capacitors of series-connected VSIs on the machine side [12]; unequal power sharing [13,14]; full-load test methods [15,16]; integrated battery charging for electric vehicles [17–19]; dynamic braking for non-regenerative electric drives [20,21]; and diagnosis of open-phase faults [22,23]. In addition to the reduced current or voltage ratings per phase, lower torque harmonics, improved fault-tolerant capabilities and additional degrees of freedom, multiphase machines also offer other advantages over their three-phase counterparts, namely: improved winding factors, reduced harmonic content in the magnetomotive force (MMF), lower rotor losses and lesser harmonics in the dc-link current [1,24–26]. Nowadays, electric drives based on multiphase machines are employed in a wide range of areas, such as aircraft [27,28], electric or hybrid vehicles [29], locomotive traction [30], high-speed elevators [31], ship propulsion [32], spacecraft [33] and wind energy applications [34–36].

In multiphase electric drives, n-phase machines are typically supplied by a n-phase VSI, whose power semiconductors are commanded by a control strategy in order to achieve variable speed operation [26]. Several control strategies have been reported in the literature over the years for multiphase electric drives, such as scalar or constant V/f control, field oriented control (FOC) and direct torque control (DTC) [25,26]. Scalar control regulates the speed of the machine by imposing a constant ratio between the amplitude and the frequency of the stator voltage [3]. Since the constant V/f control cannot control directly the currents, an unbalance in the machine can lead to the appearance of x-y current components with considerable magnitude [37]. Moreover, the reference voltages generated by scalar control are translated into command signals for the power semiconductors of the VSI using pulse width modulation (PWM) or space vector PWM (SV-PWM) techniques [7]. However, similarly to standard three-phase electric drives, scalar control cannot provide accurate control of the rotor speed of multiphase machines and leads to a poor dynamic performance [25]. On the other hand, both FOC and DTC schemes provide a decoupled control of the flux and torque, improving the control of the machine [38]. Typically, in FOC schemes the flux and torque of the machine are adjusted by regulating two independent current components with proportional-integral (PI) controllers, regardless of the number of phases of the machine, and the VSI control signals are synthesized with PWM or SV-PWM techniques [39,40]. In the case of DTC schemes, the flux and torque are controlled directly with hysteresis controllers and the control actuation is usually selected using a switching table [41,42].

In the last decade, finite control set model predictive control (FCS-MPC), along with control strategies such as FOC and DTC, has been proposed for the control of high-performance electric drives [43–45]. The main advantages of FCS-MPC over the classical control strategies are the improved dynamic performance, flexibility in the definition of control objectives and easy inclusion of constraints [46]. Since SV-PWM techniques can be hard to implement in multiphase drives [7], particularly for machines with a high number of phases or when multilevel converters are employed, FCS-MPC is also an attractive solution for multiphase drives since it does not require the use of a modulator [46]. FCS-MPC strategies use a discrete version of the system model to predict the future behavior of the controlled variables, considering a finite set of possible actuations of the power converters [47]. Typically, FCS-MPC strategies can be based on the application of a single switching state during a sampling period, referred to as optical switching vector MPC (OSV-MPC), or as an alternative, consider the application of an optimal switching sequence, known as optimal switching sequence (OSS-MPC) [48]. The control objectives of the FCS-MPC strategies are expressed in the form of a cost function, which evaluates the error between the controlled variables and their reference values. Hence, the optimal actuation is obtained by selecting, among the considered finite set of control actuations the one that leads to the minimum value of the cost function [46–48].

The FCS-MPC strategies available in the literature for multiphase drives are commonly classified according to their control objectives, such as predictive current control (PCC), predictive torque control (PTC) or predictive speed control (PSC) [49]. In the case of PCC schemes, the stator currents are the controlled variables, while the flux and torque are usually selected as the controlled variables in PTC strategies [50]. The PSC scheme eliminates the external PI speed loop present in PCC and PTC strategies although it requires the tuning of several weighting factors and depends on the mechanical parameters of the drive to estimate the load torque and predict the rotor speed [51]. Due to these limitations of PSC schemes, applications of both PCC and PTC strategies for multiphase drives are more popular among the research community and can be found in multiple publications [49,50,52–55].

Although several works have reported implementations of FCS-MPC strategies for electric drives based on multiphase machines in recent years, very few works attempted to review and compare these control strategies [50]. The publications [43,50,56,57] provide an overview of FCS-MPC strategies applied to five and six-phase machines, which are the simplest and the most addressed configurations in the literature [58]. However, these publications do not provide simulation or experimental results and lack a critical comparison among the considered FCS-MPC control strategies. On the other hand, a comparison between several FCS-MPC strategies applied to a six-phase PMSM drive was presented in Reference [49], although only simulation results were provided and the latest contributions in this field are missing.

This paper assembles in a single reference all published FCS-MPC strategies for electric drives based on six-phase machines. It presents a critical comparative study between the different FCS-MPC strategies, highlighting their advantages and drawbacks, being supported by a theoretical framework and by both simulation and experimental results obtained with a six-phase PMSM drive. Additionally, the paper includes a section providing an overview of the different topologies of multiphase electric drives and a section detailing the modeling of six-phase machine drives.

The paper is structured as follows—Section 2 provides an overview of the existing multiphase electric drives, Section 3 discusses the modeling of six-phase drives and Section 4 presents the theory behind the FCS-MPC strategies for the six-phase drives published so far. Moreover, Section 5 presents the simulation results of the reviewed FCS-MPC strategies, while Section 6 presents the experimental results for the same control strategies. Finally, Section 7 contains the main conclusions of this work.

2. Multiphase Electric Drives

Since n-phase machines are supplied by n-phase power converters in multiphase electric drives, the variable n is not restricted by the number of phases of the electric grid and can be selected according to the application [26]. Hence, this section provides an overview of the different types of n-phase machines and n-phase power converters reported in the literature. Since this paper provides an overview of the FCS-MPC strategies for six-phase drives ($n = 6$) in particular, this topology is analyzed in more detail in Section 2.3.

2.1. Types of Multiphase Electric Machines

The main difference between multiphase and standard three-phase machines is the configuration of the stator windings [58]. In multiphase machines, the stator windings are designed to have n phases and can be of distributed or concentrated type, depending on the number of stator slots per pole per phase [25]. Regarding the machine type in multiphase drives, the majority of the literature published in recent years has been focused on induction machines (IMs) and permanent magnet synchronous machines (PMSMs) [50,59]. In comparison to IMs, PMSMs provide a higher efficiency and power density, higher power factor and enhanced fault tolerance against open-phase faults [58,60].

Multiphase machines are typically classified according to the spatial displacement between phases and are denominated as symmetrical or asymmetrical machines [58]. In the symmetrical configuration, the stator windings of a n-phase machine are designed to have a displacement of $360/n$ electrical degrees between consecutive phases [61]. However, if n is an odd and non-prime number or if n

is an even number, that is, $n = \{6, 9, 12, 15, ...\}$, the stator windings of a n-phase machine can also be designed to have an asymmetrical configuration. In the asymmetrical configuration, the stator windings are associated in k sets of windings, each one with a phases ($n = a \cdot k$) spaced by $360/a$ electrical degrees and the k sets of windings are displaced by $\alpha = 180/n$ electrical degrees between them [62]. In the symmetrical configuration, the n-phases are usually wye-connected with a single neutral point, whereas in the asymmetrical configuration the a phases within each set of windings are wye-connected with k neutral points, which can be left isolated or connected to each other [59].

The asymmetrical configuration with k isolated neutral points is widely adopted since it restricts the circulation of zero-sequence currents (ZSC) [63] and provides isolation among the k sets of windings [26], although the number of independent currents is reduced to from $n - 1$ to $n - k$ in comparison with the single neutral point case [25]. Since the k sets of windings are isolated from each other, the use of coupling inductors is not necessary as in the case of high-power three-phase machines, where several power converters are associated in parallel to achieve high power ratings [60]. In the case of a fault in either the converter or the machine, a simple fault-tolerant control strategy can be adopted for machines with $n = a \cdot k$ phases and k isolated neutrals by simply deactivating the affected set of windings, while the drive is maintained in operation with a reduction in the power rating of $1/k \times 100\%$ [26]. On the other hand, the asymmetrical configuration with a single neutral configuration has additional $k - 1$ degrees of freedom, resulting in improved performance under fault-tolerant operation [62,64,65].

The majority of multiphase machines with an asymmetrical configuration are designed to have multiple sets of three-phase windings ($a = 3$) in order to maintain the compatibility with standard three-phase power converters [59,66]. Examples of application of these multiphase machines are the six-phase ($k = 2$) and twelve-phase generators ($k = 4$) used in wind energy applications [67], and the nine-phase machine used in high-speed elevators [31]. Although less common, the sets of windings can be arranged with a number of phases different from three, such as the fifteen-phase machine with three sets of five-phase windings ($a = 5$) reported in Reference [58] for a ship propulsion system.

2.2. Types of Power Converters

The n-phase power converters employed in multiphase drives can be of two types: n-phase VSIs or n-phase current source inverters (CSIs). Nowadays, n-phase VSIs are usually adopted in multiphase drives [58], while CSIs were used in some of the earlier multiphase drives [68,69]. Typically, two-level voltage source inverters (2L-VSIs) are used to drive multiphase machines in industrial applications, although multilevel topologies, such as the neutral-point-clamped (3L-NPC) converter can also be employed [7,70]. Other topologies of multiphase VSIs referred in the literature are the n-phase matrix converter and the n-phase H-bridge converter [26,66].

Multiphase power converters can be associated to supply the machine from one side or from both sides, being usually denominated as single or double-sided supply [7]. The single-sided configuration is the one typically employed in multiphase drives, since the stator windings in multiphase phase machines are commonly wye-connected into one or multiple stars [26]. In the double-sided configuration, the stator windings of the machine are supplied from both sides in an open winding configuration, increasing the number of levels of the phase voltage supplied by the VSIs and improving the fault-tolerant performance of the system [7,58]. However, the double-sided supply configuration requires twice the number of VSIs used in the single-sided configuration, increasing the complexity of the electric drive.

Typically, n-phase machines with multiple sets of a-phase windings ($n = a \cdot k$) are supplied from k VSIs, each one with a phases [59]. These k VSIs can be arranged into three configurations regarding the dc-link side: (i) a single dc-link; (ii) k isolated dc-links; (iii) k series-connected dc-links [58]. In healthy operation, both the single dc-link and the k isolated dc-links provide similar performance, the only difference is in fault-tolerant operation where the single dc-link provides better capabilities [71–73]. In spite of elevating the total dc-link voltage, the use of k series-connected dc-links requires a control strategy to guarantee the balance of the voltage of the dc-link capacitors and performs worse than the other configurations in fault-tolerant operation [12].

2.3. Particular Case: Six-Phase Drives

Among multiphase machines with multiple sets of three-phase windings, the asymmetrical six-phase machine with two isolated neutrals (2N) is the simplest and the most studied configuration [1,25,26,50,58,59,74]. In the literature, six-phase machines are also referred to as dual three-phase, dual-stator, double-star, quasi six-phase or split-phase machines [25,38]. The diagram of a typical six-phase drive is presented in Figure 1, where a six-phase machine (either an IM or a PMSM) with an asymmetrical winding configuration is supplied by two 2L-VSIs connected to a single dc-link.

Regarding the configuration of the stator windings of six-phase machines, the asymmetrical configuration is the most reported in the literature, where the two sets of three-phase windings are displaced by thirty electrical degrees ($\alpha = 30°$), as shown in Figure 1b [58]. The asymmetrical configuration provides a reduction of the MMF harmonic content and eliminates the torque harmonics of order $h = 6 \cdot m$, with m being an odd number [37,75,76]. Other values for the displacement between the two sets of windings, such as $\alpha = 0°$ and $\alpha = 60°$ (symmetrical configuration), do not provide a reduction of the harmonic content of the MMF and torque [25,76,77].

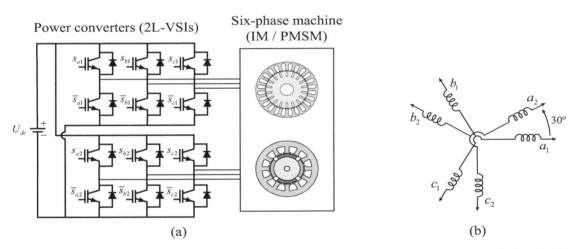

Figure 1. Six-phase asymmetrical drive: (**a**) power circuit; (**b**) winding arrangement of the six-phase asymmetrical machine with a 2N configuration.

Since in six-phase machines the two sets of windings are typically wye-connected, three neutral configurations are possible: (i) 2N; (ii) single isolated neutral (1N); and (iii) single neutral connected to the middle point of the dc-link bus or to an extra leg of the VSI, being termed as single non-isolated neutral (1NIN) in this paper [62]. The 2N configuration is often used since it provides a better dc-link voltage usage and avoids the circulation of ZSCs, leading to a better performance in steady-state operation, with lesser current harmonics, in comparison to the 1N and 1NIN configurations [58]. On the other hand, the 1N and 1NIN configurations proved to be advantageous in fault-tolerant operation [62,64,65,78], and in the enhancement of torque for the case of the 1NIN configuration [79,80].

3. System Model

This chapter presents the mathematical model of electric drives based on six-phase machines (both IMs and PMSMs) fed by two 2L-VSIs, which is required for the implementation of FCS-MPC strategies.

3.1. Introduction

In the literature, two distinct transformations are reported to model six-phase machines: the double d-q transformation and the variable space decomposition (VSD) transformation [38,50,81].

The double d-q transformation consists in the application of the Park transformation to both sets of windings [75,82] and was widely used in the first FOC and DTC strategies proposed for six-phase machines [25,38]. The VSD transformation is widely used nowadays not only in FCS-MPC but also in FOC and DTC strategies [50,58] since it is able to separate the current, flux linkage and voltage components responsible for the electromechanical energy conversion, mapped into the α-β subspace, from the remaining components, mapped into the x-y subspace, which can be used as additional degrees of freedom [7]. Moreover, the VSD transformation eliminates the coupling terms between the different subspaces in the model of six-phase machines, which are present when the double d-q transformation is used instead [81]. Additionally, the VSD transformation maps the current, flux and voltage harmonics of order $h = 12m \pm 1$ with $m = 1, 2, \ldots$ into the α-β subspace, while the harmonics of order $h = 6m \pm 1$ with $m = 1, 3, \ldots$, are mapped into the x-y subspace [83].

3.2. Two-Level Voltage Source Inverters

Considering a six-phase machine with a 2N configuration, the phase voltages depend on the switching state vector **s** of the 2L-VSIs defined in (1) and are calculated with (2):

$$\mathbf{s} = \begin{bmatrix} s_{a1} & s_{b1} & s_{c1} & s_{a2} & s_{b2} & s_{c2} \end{bmatrix}^{\mathrm{T}}, \tag{1}$$

$$\begin{bmatrix} u_{a1s} \\ u_{b1s} \\ u_{c1s} \\ u_{a2s} \\ u_{b2s} \\ u_{c2s} \end{bmatrix} = \frac{U_{dc}}{3} \begin{bmatrix} 2 & -1 & -1 & 0 & 0 & 0 \\ -1 & 2 & -1 & 0 & 0 & 0 \\ -1 & -1 & 2 & 0 & 0 & 0 \\ 0 & 0 & 0 & 2 & -1 & -1 \\ 0 & 0 & 0 & -1 & 2 & -1 \\ 0 & 0 & 0 & -1 & -1 & 2 \end{bmatrix} \cdot \mathbf{s}, \tag{2}$$

where $s_u = \{0, 1\}$ is the switching state of phase u, with $u \in \{a_1, b_1, c_1, a_2, b_2, c_2\}$. If $s_u = 1$, the top insulated gate bipolar transistor (IGBT) of phase u is ON and the bottom IGBT is OFF, while the opposite is true when $s_u = 0$. By applying the VSD transformation [83] defined in (3) to the phase voltages given by (2), the stator voltage components of the six-phase machine in the α-β, x-y and z_1-z_2 subspaces are calculated by (4):

$$\mathbf{T}_{vsd} = \frac{1}{3} \begin{bmatrix} 1 & \cos\left(\frac{2\pi}{3}\right) & \cos\left(\frac{4\pi}{3}\right) & \cos\left(\frac{\pi}{6}\right) & \cos\left(\frac{5\pi}{6}\right) & \cos\left(\frac{9\pi}{6}\right) \\ 0 & \sin\left(\frac{2\pi}{3}\right) & \sin\left(\frac{4\pi}{3}\right) & \sin\left(\frac{\pi}{6}\right) & \sin\left(\frac{5\pi}{6}\right) & \sin\left(\frac{9\pi}{6}\right) \\ 1 & \cos\left(\frac{4\pi}{3}\right) & \cos\left(\frac{2\pi}{3}\right) & -\cos\left(\frac{\pi}{6}\right) & -\cos\left(\frac{5\pi}{6}\right) & -\cos\left(\frac{9\pi}{6}\right) \\ 0 & \sin\left(\frac{4\pi}{3}\right) & \sin\left(\frac{2\pi}{3}\right) & \sin\left(\frac{\pi}{6}\right) & \sin\left(\frac{5\pi}{6}\right) & \sin\left(\frac{9\pi}{6}\right) \\ 1 & 1 & 1 & 0 & 0 & 0 \\ 0 & 0 & 0 & 1 & 1 & 1 \end{bmatrix}, \tag{3}$$

$$\begin{bmatrix} u_{\alpha s} & u_{\beta s} & u_{xs} & u_{ys} & u_{z1s} & u_{z2s} \end{bmatrix}^{\mathrm{T}} = \mathbf{T}_{vsd} \cdot \begin{bmatrix} u_{a1s} & u_{b1s} & u_{c1s} & u_{a2s} & u_{b2s} & u_{c2s} \end{bmatrix}^{\mathrm{T}}. \tag{4}$$

The sixty-four different possibilities for the switching state vector **s** result in forty-nine distinct voltage vectors mapped into the α-β and x-y subspaces simultaneously, as shown in Figure 2. The projections of the voltage vectors in z_1-z_2 are not considered in the model of six-phase machines with a 2N configuration since ZSCs cannot circulate [84]. The index of the voltage vectors represented in Figure 2 is obtained by the conversion of the binary number of vector **s** into a decimal number.

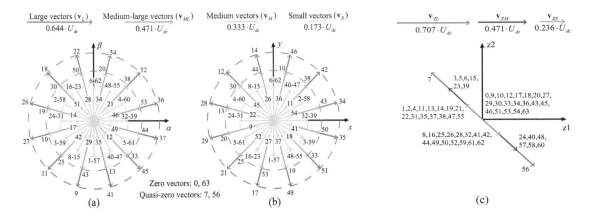

Figure 2. Voltage vectors of the six-phase machine in a stationary reference frame, mapped into the: (a) α-β; (b) x-y; (c) $z1$-$z2$ subspaces.

3.3. Six-Phase Induction Machine

In order to obtain the dynamic model of a six-phase IM in an arbitrary reference frame rotating at an angular speed ω_a, the following rotation matrix is used along with the VSD transformation:

$$\mathbf{R}(\theta_a) = \begin{bmatrix} \mathbf{T}_r(\theta_a) & \mathbf{0}_2 & \mathbf{0}_2 \\ \mathbf{0}_2 & \left(\mathbf{T}_r(\theta_a)\right)^{-1} & \mathbf{0}_2 \\ \mathbf{0}_2 & \mathbf{0}_2 & \mathbf{I}_2 \end{bmatrix}, \quad \mathbf{T}_r(\theta_a) = \begin{bmatrix} \cos(\theta_a) & \sin(\theta_a) \\ -\sin(\theta_a) & \cos(\theta_a) \end{bmatrix}, \tag{5}$$

where θ_a is the electrical angle of the arbitrary reference frame, with the d-axis aligned with the airgap, stator or rotor flux [38]. Matrix $R(\theta_a)$ rotates the α-β components in the counterclockwise direction in order to obtain the d-q components, while the x-y components are rotated in the clockwise direction in order to obtain the x'-y' components. This direction of rotation of the x-y components is adopted in recent works since it makes easier to control the unbalance of the machine [12,20,84,85].

Considering sinusoidally distributed windings, negligible saturation and symmetry between the different phases, the voltage equations of a six-phase IM in an arbitrary reference frame (rotating at an angular speed ω_a), obtained by the application of the VSD transformation along with (5) are given by [58,86]:

$$\begin{bmatrix} u_{ds} \\ u_{qs} \\ u_{xs'} \\ u_{ys'} \\ u_{z1s} \\ u_{z2s} \end{bmatrix} = R_s \begin{bmatrix} i_{ds} \\ i_{qs} \\ i_{xs'} \\ i_{ys'} \\ i_{z1s} \\ i_{z2s} \end{bmatrix} + \frac{\mathrm{d}}{\mathrm{dt}} \begin{bmatrix} \psi_{ds} \\ \psi_{qs} \\ \psi_{xs'} \\ \psi_{ys'} \\ \psi_{z1s} \\ \psi_{z2s} \end{bmatrix} + \omega_a \begin{bmatrix} 0 & -1 & 0 & 0 & 0 & 0 \\ 1 & 0 & 0 & 0 & 0 & 0 \\ 0 & 0 & 0 & 1 & 0 & 0 \\ 0 & 0 & -1 & 0 & 0 & 0 \\ 0 & 0 & 0 & 0 & 0 & 0 \\ 0 & 0 & 0 & 0 & 0 & 0 \end{bmatrix} \begin{bmatrix} \psi_{ds} \\ \psi_{qs} \\ \psi_{xs'} \\ \psi_{ys'} \\ \psi_{z1s} \\ \psi_{z2s} \end{bmatrix}, \tag{6}$$

$$\begin{bmatrix} u_{dr} \\ u_{qr} \end{bmatrix} = R_r \begin{bmatrix} i_{dr} \\ i_{qr} \end{bmatrix} + \frac{\mathrm{d}}{\mathrm{dt}} \begin{bmatrix} \psi_{dr} \\ \psi_{qr} \end{bmatrix} + (\omega_a - \omega_r) \begin{bmatrix} 0 & -1 \\ 1 & 0 \end{bmatrix} \begin{bmatrix} \psi_{dr} \\ \psi_{qr} \end{bmatrix}, \tag{7}$$

where $\{R_s, R_r\}$ are the equivalent resistance of the stator and rotor windings, respectively, ω_r is the rotor electric angular speed, the symbols u, i, ψ represent the voltage, current and flux linkage and indexes s and r stand for stator and rotor variables. The voltage equations of the six-phase IM can be written in the stationary or stator reference frame for $\omega_a = 0$, in the rotor reference frame for $\omega_a = \omega_r$, or in the synchronous reference frame for $\omega_a = \omega_s$, with ω_s being the synchronous angular speed.

Taking into account the effect of the mutual leakage inductance, which is non-negligible in six-phase machines with short-pitched windings [87], the relation between the flux linkage and the current components mapped into the d-q and x'-y' subspaces is given by [25,38]:

$$
\begin{bmatrix} \psi_{ds} \\ \psi_{qs} \\ \psi_{xs'} \\ \psi_{ys'} \\ \psi_{z1s} \\ \psi_{z2s} \end{bmatrix} = \begin{bmatrix} L_s & 0 & 0 & 0 & 0 & 0 \\ 0 & L_s & 0 & 0 & 0 & 0 \\ 0 & 0 & L_{ls} & 0 & 0 & 0 \\ 0 & 0 & 0 & L_{ls} & 0 & 0 \\ 0 & 0 & 0 & 0 & L_0 & 0 \\ 0 & 0 & 0 & 0 & 0 & L_0 \end{bmatrix} \begin{bmatrix} i_{ds} \\ i_{qs} \\ i_{xs'} \\ i_{ys'} \\ i_{z1s} \\ i_{z2s} \end{bmatrix} + \begin{bmatrix} M_m & 0 \\ 0 & M_m \\ 0 & 0 \\ 0 & 0 \\ 0 & 0 \\ 0 & 0 \end{bmatrix} \begin{bmatrix} i_{dr} \\ i_{qr} \end{bmatrix}, \tag{8}
$$

$$
\begin{bmatrix} \psi_{dr} \\ \psi_{qr} \end{bmatrix} = \begin{bmatrix} M_m & 0 & 0 & 0 & 0 & 0 \\ 0 & M_m & 0 & 0 & 0 & 0 \end{bmatrix} \begin{bmatrix} i_{ds} \\ i_{qs} \\ i_{xs'} \\ i_{ys'} \\ i_{z1s} \\ i_{z2s} \end{bmatrix} + \begin{bmatrix} L_r & 0 \\ 0 & L_r \end{bmatrix} \begin{bmatrix} i_{dr} \\ i_{qr} \end{bmatrix}. \tag{9}
$$

The inductance parameters in (8) and (9) are given by [88]:

$$
\begin{cases} L_s = L_{ls} + 2L_{lm} + 3L_m \\ L_0 = L_{ls} + L_{lm} \\ L_r = L_{lr} + 3L_m \\ M_m = 3L_m \end{cases}, \tag{10}
$$

where L_{ls} is the self leakage inductance of stator, L_{lm} is the mutual leakage inductance of the stator, L_m is the magnetizing inductance and L_{lr} is the self leakage inductance of the rotor. The torque developed by the six-phase IM is computed with [25]:

$$
t_e = 3p\left(\psi_{ds} \cdot i_{qs} - \psi_{qs} \cdot i_{ds}\right) = -3p\left(\psi_{dr} \cdot i_{qr} - \psi_{qr} \cdot i_{dr}\right) = 3pM_m\left(i_{dr} \cdot i_{qs} - i_{qr} \cdot i_{ds}\right). \tag{11}
$$

From (8) and (11), it becomes clear that the flux linkage components mapped into the x'-y' subspace do not contribute to the production of torque and only contribute to the stator leakage flux [87]. Since the equivalent impedance of the machine in the x'-y' subspace is very low, as it only depends on R_s and L_{ls}, it might lead to the circulation of large currents in this subspace, which contributes to the increase of the stator copper losses [89].

3.4. Six-Phase Permanent Magnet Synchronous Machine

Assuming sinusoidally distributed windings, negligible saturation and symmetry between the different phases, the dynamic model of a six-phase PMSM in the rotor reference frame (rotating at ω_r) obtained with the VSD transformation and rotation matrix (5) is defined by [58,76]:

$$
\begin{bmatrix} u_{ds} \\ u_{qs} \\ u_{xs'} \\ u_{ys'} \\ u_{z1s} \\ u_{z2s} \end{bmatrix} = R_s \begin{bmatrix} i_{ds} \\ i_{qs} \\ i_{xs'} \\ i_{ys'} \\ i_{z1s} \\ i_{z2s} \end{bmatrix} + \frac{d}{dt} \begin{bmatrix} \psi_{ds} \\ \psi_{qs} \\ \psi_{xs'} \\ \psi_{ys'} \\ \psi_{z1s} \\ \psi_{z2s} \end{bmatrix} + \omega_r \begin{bmatrix} 0 & -1 & 0 & 0 & 0 & 0 \\ 1 & 0 & 0 & 0 & 0 & 0 \\ 0 & 0 & 0 & 1 & 0 & 0 \\ 0 & 0 & -1 & 0 & 0 & 0 \\ 0 & 0 & 0 & 0 & 0 & 0 \\ 0 & 0 & 0 & 0 & 0 & 0 \end{bmatrix} \begin{bmatrix} \psi_{ds} \\ \psi_{qs} \\ \psi_{xs'} \\ \psi_{ys'} \\ \psi_{z1s} \\ \psi_{z2s} \end{bmatrix}, \tag{12}
$$

$$
\begin{bmatrix} \psi_{ds} \\ \psi_{qs} \\ \psi_{xs'} \\ \psi_{ys'} \\ \psi_{z1s} \\ \psi_{z2s} \end{bmatrix} = \begin{bmatrix} L_d & 0 & 0 & 0 & 0 & 0 \\ 0 & L_q & 0 & 0 & 0 & 0 \\ 0 & 0 & L_x & 0 & 0 & 0 \\ 0 & 0 & 0 & L_y & 0 & 0 \\ 0 & 0 & 0 & 0 & L_{01} & 0 \\ 0 & 0 & 0 & 0 & 0 & L_{02} \end{bmatrix} \begin{bmatrix} i_{ds} \\ i_{qs} \\ i_{xs'} \\ i_{ys'} \\ i_{z1s} \\ i_{z2s} \end{bmatrix} + \begin{bmatrix} \psi_{ds,PM} \\ \psi_{qs,PM} \\ \psi_{xs',PM} \\ \psi_{ys',PM} \\ \psi_{z1s,PM} \\ \psi_{z2s,PM} \end{bmatrix}, \tag{13}
$$

where $\{L_d, L_q, L_x, L_y, L_{01}, L_{02}\}$ are the equivalent inductances of the d, q, x', y', $z1$ and $z2$ axis, respectively and $\psi_{vs,PM}$ is the v-component of the stator flux linkage due to permanent magnets (PMs), with $v \in \{d, q, x', y', z1, z2\}$. Considering only the fundamental component of the stator flux linkage due to the PMs, the flux components in the d-q, x'-y' and $z1$-$z2$ subspaces are given by:

$$
\begin{bmatrix} \psi_{ds,PM} & \psi_{qs,PM} & \psi_{xs',PM} & \psi_{ys',PM} & \psi_{z1s,PM} & \psi_{z2s,PM} \end{bmatrix}^{\mathrm{T}} = \begin{bmatrix} \psi_{s,PM1} & 0 & 0 & 0 & 0 & 0 \end{bmatrix}^{\mathrm{T}}, \tag{14}
$$

where ψ_{PM1} is the peak value of the fundamental component of the stator flux linkage due to the PMs. The torque of a six-phase PMSM is calculated with (15) [81]:

$$
t_e = 3p \left[\psi_{s,PM1} i_{qs} + \left(L_d - L_q\right) i_{ds} i_{qs} \right]. \tag{15}
$$

Considering a six-phase PMSM with surface-mounted PMs (SPMSM), the inductance parameters are given by:

$$
\begin{cases} L_d = L_q = L_{dq} = L_{ls} + 2L_{lm} + 3L_m \\ L_x = L_y = L_{xy} = L_{ls} \\ L_{01} = L_{02} = L_0 = L_{ls} + L_{lm} \end{cases}, \tag{16}
$$

and the torque expression is reduced to [76]:

$$
t_e = 3p \left(\psi_{s,PM1} i_{qs} \right). \tag{17}
$$

Equations (12)–(17) show that only the d-q current components contribute to the production of torque in six-phase PMSMs with distributed windings and in the case of SPMSMs the torque depends only on the q-axis current component. On the other hand, the x'-y' current components are only limited by a small equivalent impedance, which can lead to the appearance of large x'-y' currents in six-phase PMSMs fed by VSIs [58].

4. Finite Control Set Model Predictive Control

Model predictive control (MPC) uses a model of the system to predict the future values of the output variables and selects a control actuation by minimizing a cost function, which defines the control objectives [47]. In the last decade, the increase in the computational power of real-time control platforms has made possible the application of MPC strategies to electric drives [46]. In the literature, MPC strategies are usually divided into two categories: CCS-MPC (continuous control set model predictive control) and FCS-MPC [43,44,57]. The FCS-MPC is usually preferred in the control of electric drives due to the easy inclusion of constraints and non-linearities in the cost function [48]. Due to the flexibility of FCS-MPC strategies, different control objectives can be set in the cost function, such as the reference tracking of current, torque, flux or speed. PCC and PTC are the most reported FCS-MPC variants for six-phase machine drives [50,58]. Although less common PSC was also proposed in Reference [51] to eliminate the speed PI controller present in PCC and PTC, although it requires the tuning of several weighting factors and depends on the mechanical parameters of the drive to estimate the load torque and predict the rotor speed. Hence, this paper is focused only on PCC and PTC variants.

4.1. Standard and Restrained Search Predictive Current Control

The standard predictive current control (S-PCC) strategy for electric drives based on six-phase IMs was introduced in References [90,91]. In order to predict the values of the stator currents for instant $k + h$, where h is the prediction horizon, the model of the six-phase IM (6)–(9) is discretized with the forward Euler method:

$$
\begin{bmatrix} i_{ds}^{k+h} \\ i_{qs}^{k+h} \\ i_{xs'}^{k+h} \\ i_{ys'}^{k+h} \end{bmatrix} = \begin{bmatrix} 1 - \frac{R_s T_s}{\sigma L_s} & \left(\frac{\omega_r}{\sigma} + \omega_k\right) T_s & 0 & 0 \\ -\left(\frac{\omega_r}{\sigma} + \omega_k\right) T_s & 1 - \frac{R_s T_s}{\sigma L_s} & 0 & 0 \\ 0 & 0 & 1 - \frac{R_s T_s}{L_{ls}} & -\omega_a T_s \\ 0 & 0 & \omega_a T_s & 1 - \frac{R_s T_s}{L_{ls}} \end{bmatrix} \begin{bmatrix} i_{ds}^{k+h-1} \\ i_{qs}^{k+h-1} \\ i_{xs'}^{k+h-1} \\ i_{ys'}^{k+h-1} \end{bmatrix} +
$$
$$
\begin{bmatrix} \frac{R_r M_m T_s}{\sigma L_r L_s} & \omega_r \frac{M_m T_s}{\sigma L_s} \\ -\omega_r \frac{M_m T_s}{\sigma L_s} & \frac{R_r M_m T_s}{\sigma L_r L_s} \\ 0 & 0 \\ 0 & 0 \end{bmatrix} \begin{bmatrix} i_{dr}^{k+h-1} \\ i_{qr}^{k+h-1} \end{bmatrix} + \begin{bmatrix} \frac{T_s}{\sigma L_s} & 0 & 0 & 0 \\ 0 & \frac{T_s}{\sigma L_s} & 0 & 0 \\ 0 & 0 & \frac{T_s}{L_{ls}} & 0 \\ 0 & 0 & 0 & \frac{T_s}{L_{ls}} \end{bmatrix} \begin{bmatrix} u_{ds}^{k+h-1} \\ u_{qs}^{k+h-1} \\ u_{xs'}^{k+h-1} \\ u_{ys'}^{k+h-1} \end{bmatrix}, \quad (18)
$$

where $\sigma = 1 - M_m^2 / (L_r L_s)$ and $\omega_k = \omega_a - \omega_r$. Since the rotor currents cannot be measured, they must be estimated either using an observer, such as the Luenberger observer or a Kalman filter [92,93] or using the past values of the measured variables [91,94]. In order to compensate the delay in the actuation, a prediction horizon of two samples ahead ($h = 2$) is usually selected in FCS-MPC strategies. Hence, the stator current components are predicted for instant $k + 2$ using (18) (with $h = 2$), which depend on the rotor current components at instant $k + 1$, given by:

$$
\begin{bmatrix} i_{dr}^{k+1} \\ i_{qr}^{k+1} \end{bmatrix} = \begin{bmatrix} 1 - \frac{R_r T_s}{\sigma L_r} & \left(\omega_a - \frac{\omega_r}{\sigma}\right) T_s \\ -\left(\omega_a - \frac{\omega_r}{\sigma}\right) T_s & 1 - \frac{R_r T_s}{\sigma L_r} \end{bmatrix} \begin{bmatrix} i_{dr}^k \\ i_{qr}^k \end{bmatrix} + \begin{bmatrix} \frac{M_m R_s T_s}{\sigma L_r L_s} & -\frac{\omega_r M_m T_s}{\sigma L_r} \\ \frac{\omega_r M_m T_s}{\sigma L_r} & \frac{M_m R_s T_s}{\sigma L_r L_s} \end{bmatrix} \begin{bmatrix} i_{ds}^k \\ i_{qs}^k \end{bmatrix}
$$
$$
\begin{bmatrix} -\frac{M_m T_s}{\sigma L_r} & 0 \\ 0 & -\frac{M_m T_s}{\sigma L_r} \end{bmatrix} \begin{bmatrix} u_{ds}^k \\ u_{qs}^k \end{bmatrix}. \quad (19)
$$

Alternatively, if a six-phase PMSM is used instead, the predictions of the stator current for instant $k + h$ are computed with:

$$
\begin{bmatrix} i_{ds}^{k+h} \\ i_{qs}^{k+h} \\ i_{xs'}^{k+h} \\ i_{ys'}^{k+h} \end{bmatrix} = \begin{bmatrix} 1 - \frac{R_s T_s}{L_d} & \frac{\omega_r L_q T_s}{L_d} & 0 & 0 \\ -\frac{\omega_r L_d T_s}{L_q} & 1 - \frac{R_s T_s}{L_q} & 0 & 0 \\ 0 & 0 & 1 - \frac{R_s T_s}{L_x} & -\frac{\omega_r L_y T_s}{L_x} \\ 0 & 0 & \frac{\omega_r L_x T_s}{L_y} & 1 - \frac{R_s T_s}{L_y} \end{bmatrix} \begin{bmatrix} i_{ds}^{k+h-1} \\ i_{qs}^{k+h-1} \\ i_{xs'}^{k+h-1} \\ i_{ys'}^{k+h-1} \end{bmatrix} +
$$
$$
\begin{bmatrix} \frac{T_s}{L_d} & 0 & 0 & 0 \\ 0 & \frac{T_s}{L_q} & 0 & 0 \\ 0 & 0 & \frac{T_s}{L_x} & 0 \\ 0 & 0 & 0 & \frac{T_s}{L_y} \end{bmatrix} \begin{bmatrix} u_{ds}^{k+h-1} \\ u_{qs}^{k+h-1} \\ u_{xs'}^{k+h-1} \\ u_{ys'}^{k+h-1} \end{bmatrix} - \begin{bmatrix} 0 \\ \frac{\omega_r T_s}{L_q} \psi_{s,PM1} \\ 0 \\ 0 \end{bmatrix}, \quad (20)
$$

The cost function of the S-PCC strategy is evaluated for forty-nine different voltage vectors (Figure 2) and is given by:

$$
g_c = \left(i_{ds}^* - i_{ds}^{k+2}\right)^2 + \left(i_{qs}^* - i_{qs}^{k+2}\right)^2 + \lambda_{xy}\left[\left(i_{xs'}^* - i_{xs'}^{k+2}\right)^2 + \left(i_{ys'}^* - i_{ys'}^{k+2}\right)^2\right], \quad (21)
$$

where λ_{xy} is the weighting factor that adjusts the relative importance of the reference tracking of the x'-y' current components over the d-q current components. The value of i_{ds}^* is regulated to impose rated flux in IMs, while in the case of SPMSMs, the value of i_{ds}^* is set to zero since the rated flux is produced by the PMs of the rotor [58]. For the operation above rated speed, the value of i_{ds}^* should be reduced in both cases in order to limit the level of the back-electromotive force (EMF), which increases proportionally with the rotor speed [78]. The value of i_{qs}^* can be set directly to regulate the torque of the machine or by a PI controller to regulate the speed of the machine. The voltage vector that minimizes the cost function (21) is selected for application during the next sampling period. Besides the last term of (21), which serves as a constraint to minimize the x-y current components, an additional constraint could be used to reduce the switching frequency, although it would require the tuning of a second weighting factor, which increases the complexity of the strategy. Although some PCC strategies consider the use of magnitude errors in the cost function, as in Reference [91], squared errors provide better reference tracking when the cost function has multiple terms, as stated in Reference [95].

In the case of IMs, the PCC strategies available in the literature use the model of the system in the stationary reference frame ($\omega_a = 0$) [90–92,94], while the PCC strategies for PMSMs use the model of the system in the synchronous reference frame ($\omega_a = \omega_r$), with the d-axis aligned with the flux due to the PMs [49,51]. It is important to note that although the control can be performed in both reference frames, using the synchronous reference frame can simplify the model of the system and avoid the extrapolation of current references to instant $k + 2$ used in (21), since both i_{ds} and i_{qs} are constant quantities during steady-state conditions in this frame [67]. On the other hand, the use of the stationary reference frame reduces the number of rotational transformations required in PCC strategies, decreasing their computational burden.

The general diagram of the S-PCC strategy for a six-phase IM drive, considering a stationary reference frame, is given in Figure 3, where the inverse of $\mathbf{T}_r$ defined in (5) is used to obtain the current references in the stationary reference frame [90,91]. Vectors $\mathbf{i}_s^{\alpha\beta}$ and $\mathbf{u}_s^{\alpha\beta}$ are the stator current and voltage vectors in the stationary reference frame, respectively and are defined as:

$$\mathbf{i}_s^{\alpha\beta} = \begin{bmatrix} i_{\alpha s} & i_{\beta s} & i_{xs} & i_{ys} & i_{z1s} & i_{z2s} \end{bmatrix}^{\mathrm{T}}, \qquad \mathbf{u}_s^{\alpha\beta} = \begin{bmatrix} u_{\alpha s} & u_{\beta s} & u_{xs} & u_{ys} & u_{z1s} & u_{z2s} \end{bmatrix}^{\mathrm{T}}. \quad (22)$$

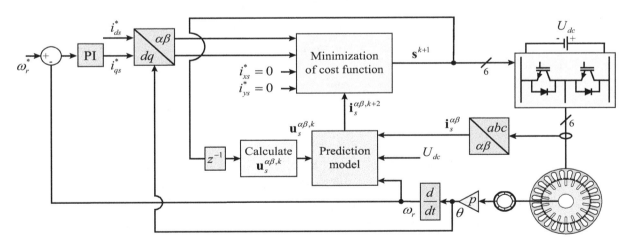

Figure 3. General diagram of the S-PCC strategy for six-phase IM drives.

The general diagram of the S-PCC strategy for six-phase PMSM drives, considering the model of the drive in the synchronous reference frame ($\omega_a = \omega_r$) is presented in Figure 4 [49]. Vectors $\mathbf{i}_s$ and $\mathbf{u}_s$ are the stator current and voltage vectors in the synchronous reference frame rotating at $\omega_a = \omega_r$, respectively and are defined as:

$$\mathbf{i}_s = \begin{bmatrix} i_{ds} & i_{qs} & i_{xs'} & i_{ys'} & i_{z1s} & i_{z2s} \end{bmatrix}^{\mathrm{T}}, \qquad \mathbf{u}_s = \begin{bmatrix} u_{ds} & u_{qs} & u_{xs'} & u_{ys'} & u_{z1s} & u_{z2s} \end{bmatrix}^{\mathrm{T}}. \quad (23)$$

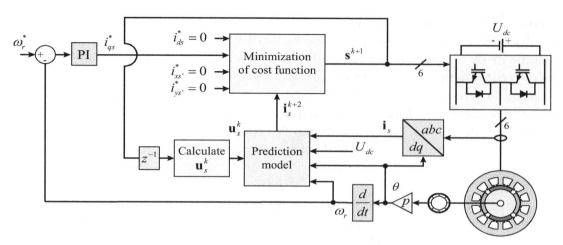

Figure 4. General diagram of the S-PCC strategy for six-phase PMSM drives.

The computational burden of the S-PCC strategy can be high for application in real-time platforms, especially if a multilevel VSI is employed [96]. The evaluation of only twelve large vectors in the S-PCC strategy was considered in References [90,91], however it led to the increase of the x'-y' current harmonics. As an alternative, a restrained search predictive current control (RS-PCC) strategy was proposed in Reference [97], which reduces the number of candidate switching states at each sampling instant. The RS-PCC algorithm imposes the following constraints: (i) the candidate voltage vectors can only generate commutations in two or less VSI legs; and (ii) A VSI leg cannot commute in two consecutive sampling periods. Thus, the RS-PCC algorithm reduces the number of voltage vectors for which the cost function (21) is evaluated from sixty-four to eleven voltage vectors if only two legs were commuted in the previous sampling period or to sixteen voltage vectors if only one leg was commuted [97].

4.2. Predictive Current Control Based on Pulse Width Modulation Schemes

A one-step modulation predictive current control (OSM-PCC) scheme was proposed in Reference [98], which optimizes the length of the voltage vectors in order to improve the performance of six-phase IMs at low speeds. This strategy considers only the twelve large voltage vectors in the α-β subspace and optimizes the length of the optimal vector by minimizing the function:

$$g_{cf}\left(\mathbf{v}_a, d_a\right) = \left(i_{ds}^* - d_a \cdot i_{ds,a}^{k+2} - (1 - d_a) \cdot i_{ds,0}^{k+2}\right)^2 + \left(i_{qs}^* - d_a \cdot i_{qs,a}^{k+2} - (1 - d_a) \cdot i_{qs,0}^{k+2}\right)^2 \qquad (24)$$

where d_a is the duty cycle of the optimal voltage vector with $d_a \in [0, 1]$, $\left\{i_{ds,0}^{k+2}, i_{qs,0}^{k+2}\right\}$ are the predicted d-q current components for instant $k + 2$ due to the application of a zero vector during T_s and $\left\{i_{ds,a}^{k+2}, i_{qs,a}^{k+2}\right\}$ are the predicted d-q current components for instant $k + 2$ due to the application of the optimal vector $\mathbf{v}_a$ during T_s. The minimization of (24) is performed by solving:

$$\frac{\partial g_{cf}\left(\mathbf{v}_a, d_a\right)}{\partial d_a} = 0, \qquad (25)$$

which yields:

$$d_a = \frac{\left(i_{ds}^* - i_{ds,0}^{k+2}\right)\left(i_{ds}^{k+2} - i_{ds,0}^{k+2}\right) + \left(i_{qs}^* - i_{qs,0}^{k+2}\right)\left(i_{qs}^{k+2} - i_{qs,0}^{k+2}\right)}{\left(i_{ds}^{k+2} - i_{ds,0}^{k+2}\right)^2 + \left(i_{qs}^{k+2} - i_{qs,0}^{k+2}\right)^2}. \qquad (26)$$

In order to minimize the x'-y' current harmonics and to provide a fixed switching frequency, a PWM-PCC strategy was proposed in Reference [99]. This strategy considers only thirteen voltage vectors (twelve large vectors in the α-β subspace and one zero vector) and since the PWM modulator is able to generate the optimal voltage vector with zero x-y components, the cost function is reduced to:

$$g_{cf} = \left(i_{ds}^* - i_{ds}^{k+2}\right)^2 + \left(i_{qs}^* - i_{qs}^{k+2}\right)^2. \tag{27}$$

However, it is important to mention that in order to generate zero x-y voltage components over a sampling period in the PWM-PCC strategy, the PWM modulator reduces the amplitude of the voltage vectors in the α-β subspace from $0.644 \cdot U_{dc}$ to $0.5 \cdot U_{dc}$, which is the limit of the linear modulation region [99]. An enhanced PWM-PCC (EPWM-PCC) was proposed in Reference [100], where the main difference in relation to Reference [99] is that the optimal voltage vector is firstly optimized in amplitude with (26), before being synthesized by the PWM modulator with zero voltage x-y components. Moreover, an extended range PWM-PCC (ERPWM-PCC) strategy that combines the EPWM-PCC approach for operation in the linear modulation region and the OSM-PCC method for operation in the overmodulation region was proposed in Reference [101] in order to improve the dc-link usage and the transient performance of six-phase machines.

The modulated PCC (M-PCC) proposed for six-phase IM drives in References [102–105] integrates a modulation technique in the control algorithm to reduce the x'-y' current components. This strategy considers that the α-β subspace is divided into forty-eight different sectors, which are defined by adjacent voltage vectors with the same amplitude. In order to calculate the duty cycles of the voltage vectors within each sector, the M-PCC strategy considers that the duty cycles of the zero and active vectors $\{d_z, d_i, d_j\}$ are inversely proportional to the value of the cost function (21) for the respective voltage vector, yielding [105]:

$$d_i = \frac{g_c(\mathbf{v}_z)\, g_c(\mathbf{v}_j)}{g_c(\mathbf{v}_z)\, g_c(\mathbf{v}_i) + g_c(\mathbf{v}_i)\, g_c(\mathbf{v}_j) + g_c(\mathbf{v}_z)\, g_c(\mathbf{v}_j)}, \tag{28}$$

$$d_j = \frac{g_c(\mathbf{v}_z)\, g_c(\mathbf{v}_i)}{g_c(\mathbf{v}_z)\, g_c(\mathbf{v}_i) + g_c(\mathbf{v}_i)\, g_c(\mathbf{v}_j) + g_c(\mathbf{v}_z)\, g_c(\mathbf{v}_j)}, \tag{29}$$

$$d_z = 1 - (d_i + d_j), \tag{30}$$

where $\{g_c(\mathbf{v}_z), g_c(\mathbf{v}_i), g_c(\mathbf{v}_j)\}$ are the values of the cost function (21) due to a zero voltage vector $\mathbf{v}_z$ and due to the active voltage vectors $\mathbf{v}_i$ and $\mathbf{v}_j$, respectively. The duty cycle d_z is equally divided among the two zero vectors $\mathbf{v}_0$ and $\mathbf{v}_{63}$, in order to achieve a fixed switching frequency. Finally, the M-PCC strategy determines the optimal sector by evaluating the cost function:

$$g_{cm} = g_c(\mathbf{v}_i)\, d_i + g_c(\mathbf{v}_j)\, d_j, \tag{31}$$

4.3. Predictive Current Control Based on Virtual Vectors

An innovative PCC strategy based on virtual vectors (VV-PCC) was proposed in References [106] to mitigate the current harmonics mapped into the x'-y' subspace of six-phase IM drives. The theory behind virtual vectors was initially introduced for the direct torque control (DTC) of five-phase [107,108] and six-phase [109] machines and consists in the creation of a new set of voltage vectors, denominated virtual vectors or synthetic vectors in the literature [110,111], with zero x-y voltage components. The twelve virtual vectors $\{\mathbf{v}_{v1}, ..., \mathbf{v}_{v12}\}$ with an amplitude of $0.598 \cdot U_{dc}$ shown in Figure 5 are created by the combination of one large and one medium-large vectors with the same phase in the α-β subspace (Figure 2), during a sampling period with the following duty cycles:

$$d_L = \sqrt{3} - 1 \approx 0.732, \qquad d_{MLI} = 1 - d_L \approx 0.268, \tag{32}$$

where d_L and d_{MLI} are the duty cycles of the large and medium-large vectors, respectively. Additionally, the zero virtual vector $\mathbf{v}_{v0}$ is obtained by the application of two zero vectors, $\mathbf{v}_0$ and

$\mathbf{v}_{63}$, with equal duty cycles. The VV-PCC strategy evaluates (27) for thirteen virtual vectors $\{\mathbf{v}_{v0}, ..., \mathbf{v}_{v12}\}$, and selects the virtual vector that minimizes the cost function for application during the next sampling period. The virtual vectors are synthesized with switching patterns centered to the middle of the sampling period as in References [110,111], in order to ease the implementation in digital controllers.

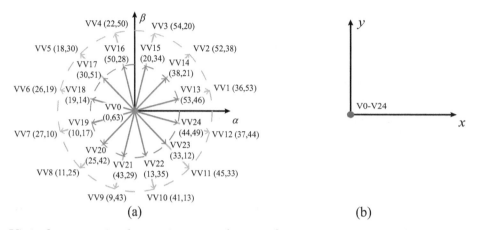

(a)   (b)

Figure 5. Virtual vectors in the stationary reference frame mapped into the: (**a**) α-β subspace; (**b**) x-y subspace.

To improve the performance of a six-phase PMSM drive at low speeds, a VV-PCC strategy based on an extended set of twenty-five virtual vectors (EVV-PCC) was proposed in Reference [112]. The extra twelve virtual vectors $\{\mathbf{v}_{v13}, ..., \mathbf{v}_{v24}\}$ shown in Figure 5 have an amplitude of $0.345 \cdot U_{dc}$ and are created by the combination of one medium-large and one small vector, with the same phase in the α-β subspace (Figure 2), during a sampling period with duty cycles given by:

$$d_{MLII} = \frac{\sqrt{3}}{3} \approx 0.577, \qquad d_S = 1 - d_{MLII} \approx 0.423, \qquad (33)$$

where d_{MLII} and d_S are the duty cycles of the medium-large and small vectors, respectively. To maintain a reduced computational burden, the authors of Reference [112] use a deadbeat approach with the aim to reduce the number of candidate virtual vectors from twenty-four to only two.

A PCC strategy based on the optimal amplitude of virtual vectors (OAVV-PCC) was introduced in Reference [113] with the aim to reduce current/torque ripples at low speeds. The OAVV-PCC strategy computes (27) for twelve virtual vectors $\{\mathbf{v}_{v1}, ..., \mathbf{v}_{v12}\}$, selects the vector $\mathbf{v}_{va}$ that provides the minimum value for the cost function and minimizes (34) in order to obtain the duty cycle d_a of vector $\mathbf{v}_{va}$ with (35):

$$g_{cf}(\mathbf{v}_{va}, d_a) = \left(i_{ds}^* - d_a \cdot i_{ds,a}^{k+2} - (1 - d_a) \cdot i_{ds,0}^{k+2}\right)^2 + \left(i_{qs}^* - d_a \cdot i_{qs,a}^{k+2} - (1 - d_a) \cdot i_{qs,0}^{k+2}\right)^2, \qquad (34)$$

$$d_a = \frac{\left(i_{ds}^* - i_{ds,0}^{k+2}\right)\left(i_{ds,a}^{k+2} - i_{ds,0}^{k+2}\right) + \left(i_{qs}^* - i_{qs,0}^{k+2}\right)\left(i_{qs,a}^{k+2} - i_{qs,0}^{k+2}\right)}{\left(i_{ds,a}^{k+2} - i_{ds,0}^{k+2}\right)^2 + \left(i_{qs,a}^{k+2} - i_{qs,0}^{k+2}\right)^2}, \qquad (35)$$

where d_a is bounded to the interval $[0, 1]$ and $\left\{i_{ds,a}^{k+2}, i_{qs,a}^{k+2}\right\}$ are the predicted d-q current components for instant $k + 2$ considering the application of $\mathbf{v}_{va}$ during T_s. The OAVV-PCC strategy uses a centered switching pattern to apply $\mathbf{v}_{va}$ during $d_a \cdot T_s$ and $\mathbf{v}_{v0}$ during $(1 - d_a) \cdot T_s$. Since vector $\mathbf{v}_{v0}$ is obtained by the application of two zero vectors $\mathbf{v}_0$ and $\mathbf{v}_{63}$ with equal application times $((1 - d_a) \cdot T_s/2)$, a fixed switching frequency is obtained [113].

In order to improve the reference tracking of the d-q current components of six-phase machines, a VV-PCC strategy based on the application of two virtual vectors over a sampling period (VV2-PCC) is suggested in Reference [114]. This strategy evaluates the cost function (27) for vectors $\{\mathbf{v}_{v0}, ..., \mathbf{v}_{v12}\}$ and selects the two adjacent active virtual vectors or one active and one zero virtual vector $\{\mathbf{v}_{vi}, \mathbf{v}_{vj}\}$ that lead to the smallest values in the cost function. The optimal values for the duty cycles of vectors $\{\mathbf{v}_{vi}, \mathbf{v}_{vj}\}$ are obtained by minimizing:

$$g_{cf}\left(\mathbf{v}_{vi}, d_i, \mathbf{v}_{vj}, d_j\right) = \left(i_{ds}^* - d_i \cdot i_{ds,i}^{k+2} - d_j \cdot i_{ds,j}^{k+2}\right)^2 + \left(i_{qs}^* - d_i \cdot i_{qs,i}^{k+2} - d_j \cdot i_{qs,j}^{k+2}\right)^2, \tag{36}$$

where $\{d_i, d_j\}$ are both limited to the interval $[0,1]$ and subjected to $d_i + d_j = 1$. The authors of Reference [114] evaluate (36) for a range of values of d_i from 0.5 to 1 with steps of 0.05 and with $d_j = 1 - d_i$, although an approach similar to that in References [113,115] can also be used to compute the optimal values for $\{d_i, d_j\}$. Finally, the VV2-PCC strategy has three switching possibilities: (i) application of only one active virtual vector (similarly to VV-PCC); (ii) application of a zero and an active virtual vector (similarly to OAVV-PCC); (iii) application of two active virtual vectors.

A PCC strategy based on virtual vectors with optimal amplitude and phase (OAPVV-PCC) that combines two active and one zero virtual vector during a sampling period is proposed in Reference [116]. This strategy applies an equivalent virtual vector optimized in both amplitude and phase to the machine, thus improving the reference tracking of the d-q current components in comparison to other PCC strategies based on virtual vectors. After selecting the two active virtual vectors $\{\mathbf{v}_{vi}, \mathbf{v}_{vj}\}$ from $\{\mathbf{v}_{v1}, ..., \mathbf{v}_{v12}\}$ that provide minimum values for (27), the OAPVV-PCC strategy optimizes first the phase and then the amplitude of the equivalent virtual vector to be applied. Considering that the equivalent virtual vector $\mathbf{v}_{vn}$ is defined as:

$$\mathbf{v}_{vn} = \mathbf{v}_{vi} \cdot d_i + \mathbf{v}_{vj} \cdot d_j, \tag{37}$$

with the duty cycles $\{d_i, d_j\}$ being subjected to the constraint $d_i + d_j = 1$, the minimization of (36) yields:

$$d_i = \frac{\left(i_{ds}^* - i_{ds,j}^{k+2}\right)\left(i_{ds,i}^{k+2} - i_{ds,j}^{k+2}\right) + \left(i_{qs}^* - i_{qs,j}^{k+2}\right)\left(i_{qs,i}^{k+2} - i_{qs,j}^{k+2}\right)}{\left(i_{ds,i}^{k+2} - i_{ds,j}^{k+2}\right)^2 + \left(i_{qs,i}^{k+2} - i_{qs,j}^{k+2}\right)^2}, \qquad d_j = 1 - d_i, \tag{38}$$

where $\{d_i, d_j\}$ are both limited to the interval $[0,1]$. The amplitude of $\mathbf{v}_{vn}$ is then optimized by minimizing:

$$g_{cf}\left(\mathbf{v}_{vn}, d_n\right) = \left(i_{ds}^* - d_n \cdot i_{ds,n}^{k+2} - (1-d_n) \cdot i_{ds,0}^{k+2}\right)^2 + \left(i_{qs}^* - d_n \cdot i_{qs,n}^{k+2} - (1-d_n) \cdot i_{qs,0}^{k+2}\right)^2, \tag{39}$$

which gives the duty cycle d_n:

$$d_n = \frac{\left(i_{ds}^* - i_{ds,0}^{k+2}\right)\left(i_{ds,n}^{k+2} - i_{ds,0}^{k+2}\right) + \left(i_{qs}^* - i_{qs,0}^{k+2}\right)\left(i_{qs,n}^{k+2} - i_{qs,0}^{k+2}\right)}{\left(i_{ds,n}^{k+2} - i_{ds,0}^{k+2}\right)^2 + \left(i_{qs,n}^{k+2} - i_{qs,0}^{k+2}\right)^2}, \tag{40}$$

where $\left\{i_{ds,n}^{k+2}, i_{qs,n}^{k+2}\right\}$ are the predicted d-q current components for instant $k+2$ due to the application of $\mathbf{v}_{vn}$ during T_s and $d_n \in [0,1]$. Finally, the equivalent virtual vector with both optimal amplitude and phase is defined as:

$$\mathbf{v}_{vn}' = \mathbf{v}_{vi} \cdot d_i' + \mathbf{v}_{vj} \cdot d_j', \tag{41}$$

where the duty cycles $\{d_i', d_j'\}$ are given by:

$$d_i' = d_i \cdot d_n, \qquad d_j' = d_j \cdot d_n, \tag{42}$$

with $0 < d'_i + d'_j < 1$. The virtual vectors $\{\mathbf{v}_{vi}, \mathbf{v}_{vj}, \mathbf{v}_{v0}\}$ are synthesized during the next sampling period with the duty cycles $\{d'_i, d'_i, d_0\}$, where $d_0 = 1 - d'_i - d'_j$, using a centered switching pattern as in Reference [116], leading to a fixed switching frequency.

4.4. Bi-Subspace Predictive Current Control Based on Virtual Vectors

Although virtual vectors impose zero x-y voltage components over a sampling period, x'-y' currents with considerable magnitude may continue to circulate in the stator windings due to machine asymmetries, deadtime effects in the power switches of the VSIs or, in the case of PMSMs, the back-EMF harmonics due to the non-sinusoidal shape of PMs [117–119]. Since the elimination of these current harmonics requires the application of non-zero x-y voltages, the concept of dual virtual vectors was introduced in Reference [120]. In opposition to the standard virtual vectors, the dual virtual vectors only contain x-y voltage components, hence the control of the x'-y' currents can be performed without disturbing the reference tracking of the d-q current components, which regulate the flux and torque of the machine. The dual virtual vectors are created by the combination of a large and a medium-large vector with the same phase in the x-y subspace and the duty cycles given by (32), resulting in twelve dual virtual vectors with an amplitude of $0.598 \cdot U_{dc}$ in the x-y subspace (stationary reference frame), as shown in Figure 6.

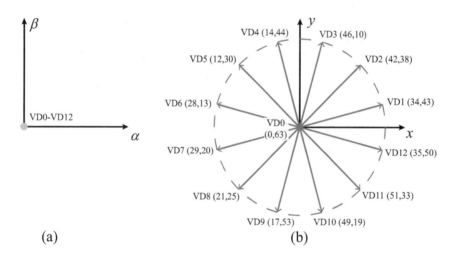

(a)   (b)

Figure 6. Dual virtual vectors in the stationary reference frame mapped into the: (**a**) α-β subspace; (**b**) x-y subspace.

The bi-subspace PCC strategy based on virtual vectors (BSVV-PCC) presented in Reference [120] aims to provide an accurate current control in both d-q and x'-y' subspaces. This strategy uses two FCS-MPC stages, where one regulates the d-q current components and the other regulates the x'-y' current components. The regulation of the d-q current components is performed as in the OAVV-PCC strategy, where the virtual vector $\mathbf{v}_{va}$ that minimizes (27) is optimized in amplitude by computing d_a with (35). Regarding the regulation of the x'-y' current components, the following cost function is evaluated for the twelve dual virtual vectors $\{\mathbf{v}_{dv1}, ..., \mathbf{v}_{dv12}\}$:

$$g_{cs} = \left(i^*_{xs'} - i^{k+2}_{xs'}\right)^2 + \left(i^*_{ys'} - i^{k+2}_{ys'}\right)^2, \tag{43}$$

where the values of $\{i^*_{xs'}, i^*_{ys'}\}$ are set to zero in order to minimize the x'-y' current components. Then, the duty cycle d_b of the optimal dual virtual vector $\mathbf{v}_{dvb}$ is obtained by minimizing:

$$g_{cs}\left(\mathbf{v}_{dvb}, d_b\right) = \left(i^*_{xs} - d_b \cdot i^{k+2}_{xs',b} - (1 - d_b) \cdot i^{k+2}_{xs',0}\right)^2 + \left(i^*_{ys} - d_b \cdot i^{k+2}_{ys',b} - (1 - d_b) \cdot i^{k+2}_{ys',0}\right)^2, \tag{44}$$

which results in:

$$d_b = \frac{\left(i^*_{xs'} - i^{k+2}_{xs',0}\right)\left(i^{k+2}_{xs',b} - i^{k+2}_{xs',0}\right) + \left(i^*_{ys'} - i^{k+2}_{ys',0}\right)\left(i^{k+2}_{ys',b} - i^{k+2}_{ys',0}\right)}{\left(i^{k+2}_{xs',b} - i^{k+2}_{xs',0}\right)^2 + \left(i^{k+2}_{ys',b} - i^{k+2}_{ys',0}\right)^2}, \tag{45}$$

where d_b belongs to the interval $[0,1]$ and $\left\{i^{k+2}_{xs',b}, i^{k+2}_{ys',b}\right\}$ are the predicted d-q current components for instant $k+2$ considering the application of $\mathbf{v}_{dvb}$ during T_s. Due to the voltage limitation of 2L-VSIs, the BSVV-PCC strategy imposes the following constraint to d_b:

$$\begin{cases} d'_b = 0, & d_b < 0 \\ d'_b = d_b, & 0 \le d_b \le 1 - d_a \ . \\ d'_b = 1 - d_a, & d_b > 1 - d_a \end{cases} \tag{46}$$

Finally, the vectors $\{\mathbf{v}_{va}, \mathbf{v}_{dvb}, \mathbf{v}_{v0}\}$ with duty cycles $\{d_a, d_b, d_0\}$ are applied to the machine in the next sampling period using centered switching patterns as described in Reference [120], thus leading to a fixed switching frequency.

4.5. Standard Predictive Torque Control

The standard predictive torque control (S-PTC) for six-phase IMs used in electric vehicles was presented in Reference [121]. Since in PTC schemes for IM drives the stator flux and torque are controlled directly in the stationary reference frame ($\omega_a = 0$), the stator current and rotor flux components are commonly selected as state variables [52,121]. Hence, from (6)–(9) and using the forward Euler discretization method, the following expressions are obtained:

$$\begin{bmatrix} i^{k+h}_{ds} \\ i^{k+h}_{qs} \\ i^{k+h}_{xs'} \\ i^{k+h}_{ys'} \end{bmatrix} = \begin{bmatrix} 1 - \frac{T_s}{\sigma\tau_s} - \frac{(1-\sigma)T_s}{\sigma\tau_r} & \omega_a T_s & 0 & 0 \\ -\omega_a T_s & 1 - \frac{T_s}{\sigma\tau_s} - \frac{(1-\sigma)T_s}{\sigma\tau_r} & 0 & 0 \\ 0 & 0 & 1 - \frac{R_s T_s}{L_{ls}} & -\omega_a T_s \\ 0 & 0 & \omega_a T_s & 1 - \frac{R_s T_s}{L_{ls}} \end{bmatrix} \begin{bmatrix} i^{k+h-1}_{ds} \\ i^{k+h-1}_{qs} \\ i^{k+h-1}_{xs'} \\ i^{k+h-1}_{ys'} \end{bmatrix} +$$

$$\begin{bmatrix} \frac{R_r M_m T_s}{\sigma L_s L_r^2} & \frac{\omega_r M_m T_s}{\sigma L_s L_r} \\ -\frac{\omega_r M_m T_s}{\sigma L_s L_r} & \frac{R_r M_m T_s}{\sigma L_s L_r^2} \\ 0 & 0 \\ 0 & 0 \end{bmatrix} \begin{bmatrix} \psi^{k+h-1}_{dr} \\ \psi^{k+h-1}_{qr} \end{bmatrix} + \begin{bmatrix} \frac{T_s}{\sigma L_s} & 0 & 0 & 0 \\ 0 & \frac{T_s}{\sigma L_s} & 0 & 0 \\ 0 & 0 & \frac{T_s}{L_{ls}} & 0 \\ 0 & 0 & 0 & \frac{T_s}{L_{ls}} \end{bmatrix} \begin{bmatrix} u^{k+h-1}_{ds} \\ u^{k+h-1}_{qs} \\ u^{k+h-1}_{xs'} \\ u^{k+h-1}_{ys'} \end{bmatrix}, \tag{47}$$

$$\begin{bmatrix} \psi^{k+h}_{dr} \\ \psi^{k+h}_{qr} \end{bmatrix} = \begin{bmatrix} 1 - \frac{T_s}{\tau_r} & \omega_k T_s \\ -\omega_k T_s & 1 - \frac{T_s}{\tau_r} \end{bmatrix} \begin{bmatrix} \psi^{k+h-1}_{dr} \\ \psi^{k+h-1}_{qr} \end{bmatrix} + \begin{bmatrix} \frac{M_m T_s}{\tau_r} & 0 \\ 0 & \frac{M_m T_s}{\tau_r} \end{bmatrix} \begin{bmatrix} i^{k+h-1}_{ds} \\ i^{k+h-1}_{qs} \end{bmatrix}, \tag{48}$$

where $\tau_s = L_s/R_s$ and $\tau_r = L_r/R_r$. The stator flux components are obtained from the stator current and rotor flux components with:

$$\begin{bmatrix} \psi^{k+h}_{ds} \\ \psi^{k+h}_{qs} \\ \psi^{k+h}_{xs'} \\ \psi^{k+h}_{ys'} \end{bmatrix} = \begin{bmatrix} \sigma L_s & 0 & 0 & 0 \\ 0 & \sigma L_s & 0 & 0 \\ 0 & 0 & L_{ls} & 0 \\ 0 & 0 & 0 & L_{ls} \end{bmatrix} \begin{bmatrix} i^{k+h}_{ds} \\ i^{k+h}_{qs} \\ i^{k+h}_{xs'} \\ i^{k+h}_{ys'} \end{bmatrix} + \begin{bmatrix} \frac{M_m}{L_r} & 0 \\ 0 & \frac{M_m}{L_r} \\ 0 & 0 \\ 0 & 0 \end{bmatrix} \begin{bmatrix} \psi^{k+h}_{dr} \\ \psi^{k+h}_{qr} \end{bmatrix}, \tag{49}$$

In order to select the optimal voltage vector, the S-PTC scheme evaluates the following cost function for forty-nine distinct voltage vectors:

$$g_t = \left(\frac{t^*_e - t^{k+2}_e}{t_n}\right)^2 + \left(\frac{\psi^*_s - \psi^{k+2}_s}{\psi_{sn}}\right)^2 + C_{dq} + C_{xy}, \tag{50}$$

where ψ_{sn} is the rated stator flux, t_n is the rated torque and t_e^{k+2} is calculated by (11) using the predictions of the current and stator flux components for instant $k+2$. The term ψ_s^{k+2} is defined as:

$$\psi_s^{k+2} = \sqrt{\left(\psi_{ds}^{k+2}\right)^2 + \left(\psi_{qs}^{k+2}\right)^2}.$$

(51)

The terms C_{dq} and C_{xy} in (50) are overcurrent constraints that penalize currents above a certain magnitude in both d-q and x'-y' subspaces:

$$\begin{cases} C_{dq} = 0, & i_{s,dq}^{k+2} \leq i_{s,dq}^{max} \\ C_{dq} = 10^5, & i_{s,dq}^{k+2} > i_{s,dq}^{max} \end{cases}, \qquad \begin{cases} C_{xy} = 0, & i_{s,xy}^{k+2} \leq i_{s,xy}^{max} \\ C_{xy} = 10^5, & i_{s,xy}^{k+2} > i_{s,xy}^{max} \end{cases},$$

(52)

where $\left\{i_{s,dq}^{max}, i_{s,xy}^{max}\right\}$ are the maximum values for the current amplitude in both d-q and x'-y' subspaces and $\left\{i_{s,dq}^{k+2}, i_{s,xy}^{k+2}\right\}$ are defined as:

$$i_{s,dq}^{k+2} = \sqrt{\left(i_{ds}^{k+2}\right)^2 + \left(i_{qs}^{k+2}\right)^2}, \qquad i_{s,xy}^{k+2} = \sqrt{\left(i_{xs'}^{k+2}\right)^2 + \left(i_{ys'}^{k+2}\right)^2}.$$

(53)

Finally, the voltage vector that minimizes (50) is applied to the six-phase IM during the next sampling period. As an example, the general diagram of the S-PTC strategy for six-phase IM drives, considering the model of the drive in the stationary reference frame ($\omega_a = 0$) is shown in Figure 7 [121].

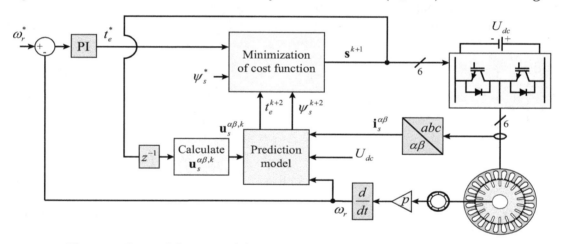

Figure 7. General diagram of the S-PTC strategy for six-phase IM drives.

An S-PTC strategy for six-phase PMSMs is presented in Reference [122], where the stator currents are predicted with (20) considering the synchronous reference frame ($\omega_a = \omega_r$) with the d-axis aligned with flux due to the PMs. The stator flux d-q and x'-y' components are calculated by:

$$\begin{bmatrix} \psi_{ds}^{k+h} \\ \psi_{qs}^{k+h} \\ \psi_{xs'}^{k+h} \\ \psi_{ys'}^{k+h} \end{bmatrix} = \begin{bmatrix} L_d & 0 & 0 & 0 \\ 0 & L_q & 0 & 0 \\ 0 & 0 & L_x & 0 \\ 0 & 0 & 0 & L_y \end{bmatrix} \begin{bmatrix} i_{ds}^{k+h} \\ i_{qs}^{k+h} \\ i_{xs'}^{k+h} \\ i_{ys'}^{k+h} \end{bmatrix} + \begin{bmatrix} \psi_{s,PM1} \\ 0 \\ 0 \\ 0 \end{bmatrix}.$$

(54)

The S-PTC strategy in Reference [122] uses a pre-selection process to reduce the number of candidate voltage vectors from forty-nine to only three, based on the angle of the stator flux in the α-β and x-y subspaces (stationary reference frame) and on the signal of the torque error. The following cost function is evaluated for the three candidate voltage vectors:

$$g_{tf} = \left(t_e^* - t_e^{k+2}\right)^2 + \lambda_\psi \left(\psi_s^* - \psi_s^{k+2}\right)^2,$$

(55)

where λ_ψ is a weighting factor. The voltage vector that minimizes (55) is applied to the six-phase PMSM during the next sampling period. The general diagram of the S-PTC strategy for six-phase PMSM drives in the synchronous reference frame ($\omega_a = \omega_r$) is shown in Figure 8 [122].

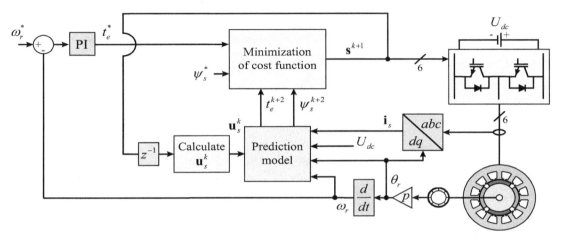

Figure 8. General diagram of the S-PTC strategy for six-phase PMSM drives.

4.6. Predictive Torque Control Based on the Duty Cycle Optimization of Voltage Vectors

An approach similar to the S-PTC, baptized as high robustness PTC (HR-PTC), which considers a discrete duty cycle optimization, is proposed in Reference [123]. Instead of considering the application of the optimal voltage vector $\mathbf{v}_a$ during the entire sampling period, this strategy finds the optimal duty cycle d_a of vector $\mathbf{v}_a$, with $d_a \in \{0, 0.2, 0.4, 0.6, 0.8, 1\}$, by minimizing:

$$g_{tf} = \left(t_e^* - d_a \cdot t_{e,a}^{k+2} - (1 - d_a) \cdot t_{e,0}^{k+2} \right)^2 + \lambda_\psi \left(\psi_s^* - d_a \cdot \psi_{s,a}^{k+2} - (1 - d_a) \cdot \psi_{s,0}^{k+2} \right)^2 \tag{56}$$

where $\left\{ t_{e,a}^{k+2}, \psi_{s,a}^{k+2} \right\}$ are the predicted torque and stator flux for instant $k + 2$ considering the application of vector $\mathbf{v}_a$ during T_s and $\left\{ t_{e,a}^{k+2}, \psi_{s,a}^{k+2} \right\}$ are the predicted torque and stator flux for instant $k + 2$ due to the application of a zero vector during T_s. The vector $\mathbf{v}_a$ is selected for application in the next sampling period during $d_a \cdot T_s$, where the value of d_a is selected from the minimization of (56).

A reduced cost function PTC (RCF-PTC) strategy was presented in Reference [124], where a deadbeat approach is used to determine the sector of the optimal voltage vector, reducing the number of candidates from forty-nine to only three. This strategy uses a model of the six-phase PMSM with stator current and stator flux components as state variables, hence the stator currents are predicted for instant $k + h$ with (20) and the d-q components of the stator flux are predicted for instant $k + h$ with:

$$\begin{bmatrix} \psi_{ds}^{k+h} \\ \psi_{qs}^{k+h} \end{bmatrix} = \begin{bmatrix} 1 & \omega_r T_s \\ -\omega_r T_s & 1 \end{bmatrix} \begin{bmatrix} \psi_{ds}^{k+h-1} \\ \psi_{qs}^{k+h-1} \end{bmatrix} + \begin{bmatrix} -R_s & 0 \\ 0 & -R_s \end{bmatrix} \begin{bmatrix} i_{ds}^{k+h-1} \\ i_{qs}^{k+h-1} \end{bmatrix} + \begin{bmatrix} 1 & 0 \\ 0 & 1 \end{bmatrix} \begin{bmatrix} u_{ds}^{k+h-1} \\ u_{qs}^{k+h-1} \end{bmatrix}. \tag{57}$$

In the RCF-PTC, the voltage components in the d-q subspace are obtained using a deadbeat approach, that is, considering $t_e^{k+2} = t_e^*$, and their angle in the stationary reference frame (α-β subspace) is used to select three candidate voltage vectors, one small, one medium-large and one large with the same phase. The amplitude of these three vectors is optimized by computing their duty cycle with:

$$d_a = \frac{\sqrt{\left(u_{\alpha s}^* \right)^2 + \left(u_{\beta s}^* \right)^2} \cdot \sqrt{\left(u_{\alpha s}^{k+1} \right)^2 + \left(u_{\beta s}^{k+1} \right)^2} \cdot \cos\left(\theta_v \right)}{\left(u_{\alpha s}^{k+1} \right)^2 + \left(u_{\beta s}^{k+1} \right)^2}, \tag{58}$$

where $\left\{u_{\alpha s}^*, u_{\beta s}^*\right\}$ are the α-β components of the stator voltage reference computed by the RCF-PTC strategy, $\left\{u_{\alpha s}^{k+1}, u_{\beta s}^{k+1}\right\}$ are the α-β components of the three candidate vectors and θ_v is the angle between the reference and candidate voltage vectors in the stationary reference frame (α-β subspace). Then, the voltage vector, among the three candidates, that provides minimal x'-y' current components is selected for application during the next sampling period. Thus, the cost function of the RCF-PTC is defined as [124]:

$$g_{fcs} = \left(\sqrt{i_{xs'}^{k+2} + i_{ys'}^{k+2}}\right)^2. \tag{59}$$

4.7. Predictive Torque Control Based on Virtual Vectors

In order to eliminate the stator flux weighting factor, a flux constrained PTC (FC-PTC) that calculates the stator flux references from the reference torque is presented in Reference [125]. Moreover, the FC-PTC strategy considers the use of virtual vectors $\{\mathbf{v}_{V0}, ..., \mathbf{v}_{V24}\}$ (Figure 5), hence the cost function for this strategy is defined as:

$$g_f = \left(\psi_{ds}^* - \psi_{ds}^{k+2}\right)^2 + \left(\psi_{qs}^* - \psi_{qs}^{k+2}\right)^2, \tag{60}$$

where the reference values of the d-q components of the stator flux $\left\{\psi_{ds}^*, \psi_{qs}^*\right\}$ are calculated with (61) considering $i_{ds}^* = 0$, which corresponds to maximum torque per ampere (MTPA) conditions in SPMSMs:

$$\begin{cases} \psi_{ds}^* = \psi_{s,PM1} \\ \psi_{qs}^* = \dfrac{L_q t_e^*}{3p\psi_{s,PM1}} \end{cases}. \tag{61}$$

As the computational burden of FC-PTC can be considerable for implementation in digital controllers, the authors of Reference [125] have used a look-up table in order to reduce the number of candidate virtual vectors from twenty-four to only six.

A multi-vector PTC (MV-PTC) scheme was proposed in Reference [126] with the aim to improve the steady-state operation of a six-phase PMSM drive. This strategy considers only twelve active virtual vectors $\{\mathbf{v}_{V1}, ..., \mathbf{v}_{V12}\}$ from Figure 5 and optimizes the amplitude of each one using:

$$T_a = \frac{t_e^* - t_e^{k+1} - \Delta t_{e,0} \cdot T_s}{\Delta t_{e,a} - \Delta t_{e,0}}, \tag{62}$$

where $\{\Delta t_{e,0}, \Delta t_{e,a}\}$ are the torque deviation due to the application of a zero and an active virtual vector, respectively and are defined as [126]:

$$\Delta t_{e,0} = t_{e,0}^{k+2} - t_e^{k+1}, \qquad \Delta t_{e,a} = t_{e,a}^{k+2} - t_e^{k+1}. \tag{63}$$

The MV-PTC strategy evaluates (55) for twelve virtual vectors with optimized amplitude and applies the optimal virtual vector in the next sampling period, combined with a zero virtual vector, leading to a fixed switching frequency.

5. Simulation Results

In order to assess and compare the performance of the different FCS-MPC strategies described in the previous section, several simulations results obtained with a six-phase PMSM drive are presented in this section. The 2L-VSIs were modelled in Matlab/Simulink using the ideal IGBT model from the Simscape Power Systems library and the six-phase PMSM was modeled using (6)–(11) with the parameters given in Table 1, where $\{P_s, U_s, I_s, n_n, t_n, \psi_{sn}\}$ are the rated values of the power, voltage, current, speed, torque and stator flux of the machine designed in Reference [76]. Since both the non-linearities of the power converters and the back-EMF harmonics contribute to the appearance of considerable x'-y' currents, the simulation model considers a deadtime of $t_d = 2.2$ μs in the power switches of the 2L-VSIs and also accounts for the 5th and 7th harmonics of the no-load flux linkage due to the PMs, whose amplitudes $\{\psi_{s,\text{PM5}}, \psi_{s,\text{PM7}}\}$ and phases $\{\phi_5, \phi_7\}$ are provided in Table 1.

Table 1. Parameters of the six-phase drive.

Parameter	Value	Parameter	Value	Parameter	Value	Parameter	Value
P_s (kW)	4	ψ_{sn} (mWb)	1013.8	$\psi_{s,\text{PM1}}$ (mWb)	980.4	U_{dc} (V)	650
U_s (V)	340	p	2	$\psi_{s,\text{PM5}}$ (mWb)	2.4	t_d (µs)	2.2
I_s (A)	3.4	R_s (Ω)	1.5	$\psi_{s,\text{PM7}}$ (mWb)	1.6	T_s (µs)	30, 40, 60, 100, 200
n_n (rpm)	1500	L_{dq} (mH)	53.8	ϕ_5 (deg)	1.3	λ_i	0.025
t_n (N.m)	28.4	L_{xy} (mH)	2.1	ϕ_7 (deg)	-12.7	λ_f	1000

To measure the performance of the six-phase PMSM drive under the considered FCS-MPC strategies, the following performance indicators are defined to quantify the reference tracking error of the current and stator flux components:

$$E_{i,v} = \frac{\frac{1}{N}\sum_{n=1}^{N}|i_{vs}^*(n) - i_{vs}(n)|}{\sqrt{2}\times I_s}\times 100\%, \tag{64}$$

$$E_{\psi,v} = \frac{\frac{1}{N}\sum_{n=1}^{N}|\psi_{vs}^*(n) - \psi_{vs}(n)|}{\psi_{sn}}\times 100\%, \tag{65}$$

where $v \in \{d, q, x', y'\}$ and N is the number of samples corresponding to a time window of 1 s. Moreover, the current harmonic distortion considering up to the fiftieth current harmonic is computed with:

$$\text{THD}_i = \frac{1}{6}\sum_{x=a1,...,c2}\frac{\sqrt{i_{xs,2}^2 + ... + i_{xs,50}^2}}{i_{xs,1}}\times 100\%, \tag{66}$$

where $i_{xs,h}$ is the h-order harmonic of the x-phase current. In order to account for all harmonic content of currents, the total waveform distortion of current is defined as:

$$\text{TWD}_i = \frac{1}{6}\sum_{x=a1,...,c2}\frac{\sqrt{I_{xs}^2 - i_{xs,1}^2}}{i_{xs,1}}\times 100\%, \tag{67}$$

where I_{xs} is the rms value of the current in phase x. The total waveform ripple of torque is calculated with:

$$\text{TWR}_t = \frac{\sqrt{T_e^2 - \bar{t}_e^2}}{|\bar{t}_e|}\times 100\%, \tag{68}$$

where T_e is the torque rms value and $\bar{t}_e$ is the mean value of torque.

To compare the PCC strategies considered in Section 4, the six-phase drive is simulated in Matlab/Simulink environment for operation at a constant speed of 750 rpm and rated load condition (motoring mode), which is obtained by setting $i_{qs} = 4.8$ A. Different values of T_s were considered in the PCC strategies in order to obtain a mean switching frequency of around 5 kHz. A speed of 750 rpm was selected to show the difference in the performance of the strategies capable of applying multiple voltage vectors or multiple virtual vectors during a sampling period from the remaining, which provide a much better performance at speed levels below the rated value. The simulation results obtained for the steady-state operation of the six-phase PMSM drive under the considered PCC strategies are presented in Figures 9–11, while the respective performance indicators are summarized in Table 2.

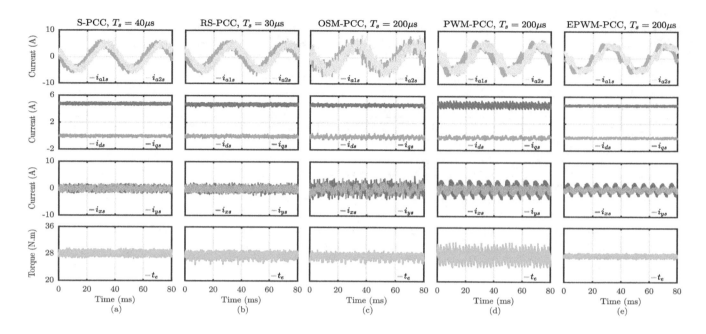

Figure 9. Simulation results for the PMSM drive operating at 750 rpm and rated load (motoring mode) for: (a) S-PCC; (b) RS-PCC; (c) OSM-PCC; (d) PWM-PCC and (e) EPWM-PCC.

The simulation results for the S-PCC strategy are presented in Figure 9a. Since the S-PCC only applies one out of sixty-four voltage vectors per sampling period and each voltage vector contains both α-β and x-y components, this strategy cannot completely suppress the x'-y' currents. Moreover, the value of λ_i could be increased to further minimize the x'-y' currents but this would degrade the reference tracking of the d-q currents, which regulate the flux and torque of the machine. The higher value obtained for the TWD$_i$ in comparison to the THD$_i$ in the case of the S-PCC (TWD$_i = 18.41\%$ and THD$_i = 4.23\%$) shows that the observed distortion in the currents is mainly of high frequency and is mostly mapped into the x'-y' subspace. The RS-PCC strategy provides a reduced mean switching frequency in comparison to the S-PCC by limiting the number of candidate voltage vectors, giving a slightly deteriorated performance even with a smaller value of T_s. On the other hand, the OSM-PCC strategy optimizes the length of the applied voltage vector by combining it with two zero vectors ($\mathbf{v}_0$ and $\mathbf{v}_{63}$) over a sampling period, resulting in a fixed switching frequency of $\bar{f}_{sw} = 1/T_s$. Hence, the value of T_s is increased to 200 µs to obtain a fixed value of $\bar{f}_{sw} = 5.0$ kHz, which worsens the performance of the system in comparison to the S-PCC, as shown in Figure 9c but greatly reduces the computational requirements of digital control platforms for the execution of this control strategy. The use of a PWM technique in the PWM-PCC strategy avoids the injection of x-y voltage components and guarantees a fixed switching frequency, as in the case of the OSM-PCC strategy. Since no x-y voltage components are applied to the machine, the x'-y' currents components cannot be regulated, that is, they are left in open-loop. This leads to the appearance of low-frequency current harmonics in the x'-y' subspace, as shown in Figure 9d, caused by the deadtime effect of the power switches and by the back-EMF harmonics. The EPWM-PCC optimizes the amplitude of the applied voltage vector in the α-β subspace, while guaranteeing the application of zero x-y voltage components over a sampling period. Hence, the EPWM-PCC strategy improves the reference tracking of the d-q currents and reduces the value of TWR$_t$ in comparison to the S-PCC, OSM-PCC and PWM-PCC strategies. However, as in the case of PWM-PCC, the EPWM-PCC strategy is not able to regulate the x'-y' currents, giving a high value for the THD$_i$.

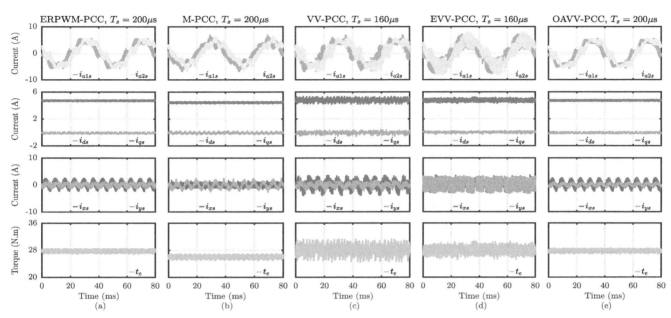

Figure 10. Simulation results for the PMSM drive operating at 750 rpm and rated load (motoring mode) for: (**a**) ERPWM-PCC; (**b**) M-PCC; (**c**) VV-PCC; (**d**) EVV-PCC and (**e**) OAVV-PCC.

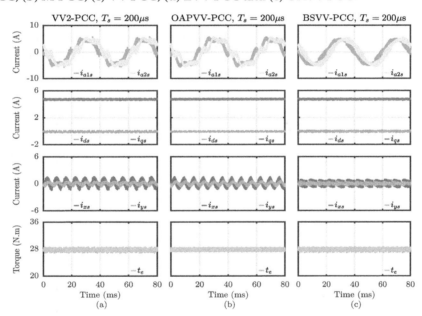

Figure 11. Simulation results for the PMSM drive operating at 750 rpm and rated load (motoring mode) for: (**a**) VV2-PCC; (**b**) OAPVV-PCC and (**c**) BSVV-PCC.

The simulation results for the drive operating with the ERPWM-PCC strategy are presented in Figure 10a. The performance of the drive in steady-state conditions for the considered point of operation is very similar to the one obtained with the EPWM-PCC strategy. However, the EPWM-PCC strategy can only apply a voltage vector with an amplitude of up to $0.5 \cdot U_{dc}$, which corresponds to the limit of the linear region of the PWM technique used. On the other hand, the ERPWM-PCC strategy is able to operate in both the linear and in the overmodulation regions (for an amplitude of the voltage vectors between $0.5 \cdot U_{dc}$ and $0.644 \cdot U_{dc}$), which not only improves the dc-link voltage usage but also the transient performance of the drive. The simulation results for a step in i_{qs}^* from 2.4 A to 4.8 A at $t = 10$ ms are shown in Figure 12 and validate the superior performance of the ERPWM-PCC strategy over the EPWM-PCC, obtaining a reduction of the rise time from 4 ms to 1.3 ms. However, when operating in the overmodulation region, the ERPWM-PCC strategy cannot guarantee the injection of zero x-y voltage components, as in the case of the operation in the linear region of modulation.

The simulation results for the steady-state operation of the drive under the M-PCC strategy are shown in Figure 10b. Differently from the PWM-PCC, EPWM-PCC and ERPWM-PCC strategies, the M-PCC strategy combines two active vectors and two zeros ($\mathbf{v}_0$ and $\mathbf{v}_{63}$) over a sampling period, which provides a fixed switching frequency but does not guarantee the application of zero x-y voltage components over a sampling period. The cost function of the M-PCC strategy evaluates the current errors in both subspaces and uses a weighting factor (λ_i) to determine the relative importance between the tracking of reference currents in both subspaces. Even when $\lambda_i = 0.025$ is selected, the current errors in the x-y subspace disturb the reference tracking of the d-q current components, as demonstrated by the increase in the values of $E_{i,d}$ and $E_{i,q}$ (Table 2), and a steady-state error is perceptible in both the q-axis current and torque, as shown in Figure 10b. An even smaller value for λ_i could be selected to reduce the steady-state errors in i_{qs} and in t_e but the amplitude of x-y current components would also increase. The VV-PCC strategy uses twelve active and one zero virtual vectors instead of standard fourty-nine voltage vectors to apply zero x-y voltage components to the machine. The results obtained for the VV-PCC strategy are presented in Figure 10c and are very similar to the ones obtained with the PWM-PCC strategy, however the virtual vectors have an amplitude of $0.598 \cdot U_{dc}$, which improves the dc-link voltage usage and the transient performance of the drive. The EVV-PCC strategy provides a decrease in the d-q currents errors and in the torque ripple in comparison to the VV-PCC strategy, as seen in Figure 10d, due to the addition of twelve small active virtual vectors with an amplitude of $0.345 \cdot U_{dc}$ to the control algorithm. The simulation results for the OAVV-PCC strategy are presented in Figure 10e and show a significant improvement in terms of torque ripple and d-q current errors during steady-state operation over the VV-PCC and EVV-PCC strategies. Since in the OAVV-PCC technique the selected virtual vector is combined with a zero virtual vector over a sampling period, the operation of the drive at low speeds is highly improved while maintaining a fixed switching frequency.

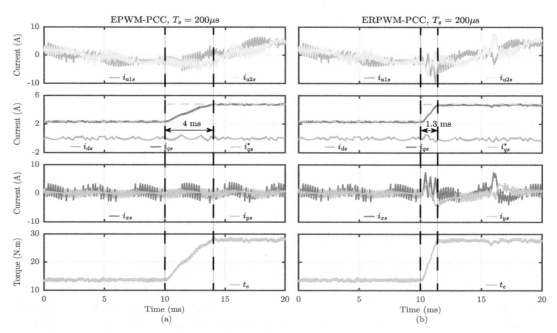

Figure 12. Simulation results for a step response in i_{qs}^* for the PMSM drive operating at 1300 rpm for: (**a**) EPWM-PCC and (**b**) ERPWM-PCC.

The simulation results for the drive operating under the VV2-PCC strategy are shown in Figure 11a. The obtained results show a performance similar to the one obtained with the OAVV-PCC strategy, however the torque ripple is slightly increased from to 1.18% to 1.33%. Although the VV2-PCC is able to apply one virtual vector and one zero virtual vector or two virtual vectors over a sampling period and in theory should provide lower current errors and lower torque ripple than the OAVV-PCC, this is not verified since the VV2-PCC strategy is only able to apply a finite set of values for the duty cycles of the two vectors, as discussed in Section 4.3. The simulation results for the drive operating under the OAPVV-PCC strategy are presented in Figure 11b and demonstrate a very good performance under steady-state operation in terms of tracking of the reference d-q current components and torque ripple. Since this strategy combines two active and one zero virtual vectors during a sampling period, the resultant voltage vector provides very low d-q current errors and the lowest value of TWR_t among the compared PCC strategies. As the PCC strategies based on PWM techniques, such as the EPWM-PCC and ERPWM-PCC, and the strategies based on virtual vectors, such as OAVV-PCC, VV2-PCC and OAPVV-PCC, do not apply x-y voltage components, those techniques cannot compensate the low frequency x'-y' current harmonics generated by the deadtime effects of the power switches and by the back-EMF harmonics. The simulation results for the BSVV-PCC strategy are presented in Figure 11c and show a significant reduction in the amplitude of the x'-y' current components. The BSVV-PCC strategy not only provides low current errors in both subspaces and low torque ripple but also provides the lowest values for the THD_i and TWD_i, among the compared control strategies.

Table 2. Performance indicators for the drive operating at 750 rpm and rated load (motoring mode) for the different PCC strategies.

Strategy	$E_{i,d}$ (%)	$E_{i,q}$ (%)	$E_{i,x}$ (%)	$E_{i,y}$ (%)	THD_i (%)	TWD_i (%)	TWR_t (%)	$\bar{f}_{sw}$ (kHz)
S-PCC	1.54	1.40	10.53	10.70	4.23	18.41	1.69	4.06
RS-PCC	1.93	2.77	11.16	10.22	4.71	19.64	2.53	3.32
OSM-PCC	3.50	2.89	25.58	22.91	27.47	43.60	2.22	5.00
PWM-PCC	2.95	4.30	18.35	10.89	18.64	27.42	5.15	5.00
EPWM-PCC	1.63	1.64	16.20	8.75	18.83	21.63	1.17	5.00
ERPWM-PCC	1.65	1.64	16.21	8.77	18.68	21.48	1.16	5.00
M-PCC	2.05	7.63	6.03	18.34	22.05	24.71	1.32	5.00
VV-PCC	3.00	3.88	18.77	9.87	20.65	26.38	4.74	5.27
EVV-PCC	1.72	2.98	14.97	15.99	14.29	29.32	3.57	5.00
OAVV-PCC	1.63	1.63	16.25	7.20	18.90	20.71	1.18	5.00
VV2-PCC	1.60	1.65	16.21	7.21	18.71	20.46	1.25	5.00
OAPVV-PCC	1.22	1.46	18.76	8.55	22.40	23.93	1.10	5.00
BSVV-PCC	1.34	1.55	5.47	2.71	3.66	9.37	1.18	5.00

The simulation results for the operation of the six-phase drive under PTC strategies are presented in Figure 13, while the corresponding performance indicators are given in Table 3. In comparison to the S-PCC, the S-PTC strategy provides lower torque ripple although with a higher current harmonic distortion, as seen in Figure 13a. The HR-PTC strategy is similar to the S-PTC but provides an optimization in amplitude of the selected voltage vector, by combining it with two zero vectors ($\mathbf{v}_0$ and $\mathbf{v}_{63}$). Since, each voltage vector contains both α-β and x-y current components, a large value for T_s leads to the appearance of large currents in the x-y subspace, thus a $T_s = 6\,0\,\mu s$ was chosen. From Figure 13b, the HR-PTC strategy provides higher current distortion and higher torque ripple than the S-PTC strategy, even with a higher mean switching frequency ($f_{sw} = 13.26$ kHz).

The RFC-PTC strategy, whose results are presented in Figure 13c, provides a lower torque ripple in comparison to the S-PTC and HR-PTC strategies, however it gives a higher value for the THD_i and leads to a high mean switching frequency ($f_{sw} = 10.0$ kHz).

The simulation results for the FC-PTC strategy are presented in Figure 13d and show a reduction in the x'-y' flux errors due to the use of virtual vectors, even with a higher sampling period ($T_s = 160\,\mu s$) in comparison to previous PTC strategies. The MV-PTC strategy improves the steady-state operation of the drive by optimizing the amplitude of the selected virtual vector, giving reduced flux errors and a low torque ripple for a fixed switching frequency of 5 kHz.

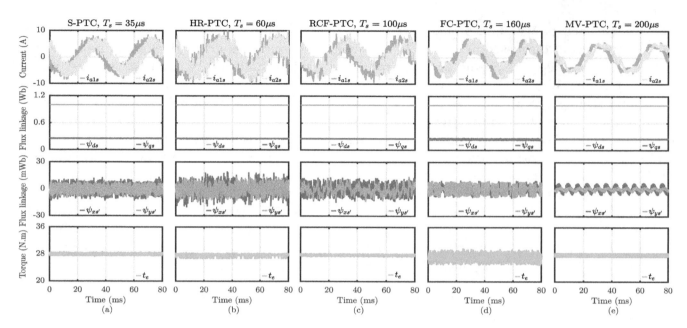

Figure 13. Simulation results for the PMSM drive operating at 750 rpm and rated load (motoring mode) for: (**a**) S-PTC; (**b**) HR-PTC; (**c**) RCF-PTC; (**d**) FC-PTC and (**e**) MV-PTC.

Table 3. Performance indicators for the drive operating at 750 rpm and rated load (motoring mode) for the different PTC strategies.

Strategy	$E_{\psi,d}$ (%)	$E_{\psi,q}$ (%)	$E_{\psi,x}$ (%)	$E_{\psi,y}$ (%)	THD_i (%)	THD_i (%)	TWR_t (%)	$\bar{f}_{sw}$ (kHz)
S-PTC	0.44	0.24	0.39	0.27	12.05	41.74	0.69	5.87
HR-PTC	0.42	0.43	0.48	0.35	16.59	55.04	1.06	13.26
RFC-PTC	0.23	0.40	0.50	0.36	26.63	53.56	0.59	10.00
FC-PTC	0.50	1.05	0.25	0.16	16.57	28.69	3.42	5.11
MV-PTC	0.47	0.35	0.22	0.07	18.58	20.47	1.06	5.00

6. Experimental Results

6.1. Experimental Setup

The experimental results presented in this section were obtained with a six-phase PMSM drive, with the same parameters as the ones given in Table 1 in Section 5. The 4 kW six-phase asymmetrical PMSM is supplied by two 2L-VSIs by Semikron (SKiiP 132 GD 120), which are fed by a dc-bus with a voltage level of 650 V. The speed of the six-phase PMSM is regulated by a mechanically coupled 7.5 kW IM fed by a commercial variable frequency converter. The rotor position of the PMSM is measured with an incremental encoder with 2048 ppr.

The PCC and PTC strategies are implemented in a digital control platform dS1103 by dSPACE and a cRIO-9066 by National Instruments is used to generate the switching patterns needed by the control strategies that: (i) optimize the amplitude of voltage vectors;(ii) require PWM techniques or (iii) consider the use of virtual vectors. In those control strategies, at the end of each sampling period, the dS1103 platform writes the six leg duty cycles in a digital port, which is read by the FPGA of the cRIO-9066. At the beginning of the next sampling period, the cRIO-9066 generates the switching signals for the 2L-VSIs with a symmetry to the middle of the sampling period. In order to maintain the processes of both platforms synchronized, an interrupt signal is generated at the beginning of each control cycle in the FPGA of the cRIO-9066, which determines the beginning of a new control cycle in the dS1103 platform. The experimental setup is shown in Figure 14.

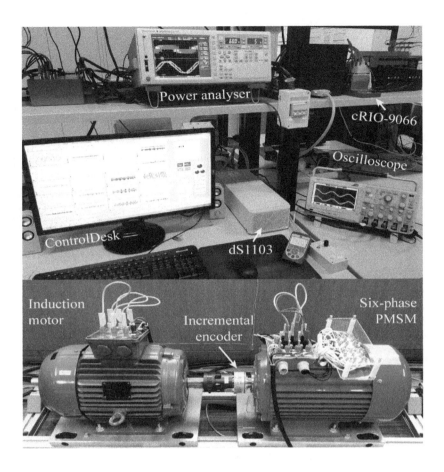

Figure 14. Experimental setup.

6.2. Obtained Results

The experimental results for the steady-state operation of the six-phase PMSM drive under the tested PCC strategies are shown in Figures 15–17, while the respective performance indicators are listed in Table 4, where $\bar{t}_{exe}$ is the mean execution time for each strategy. It is important to note that the execution times of the strategies that require the generation of custom switching patterns already contain the time required for the communication between the dS1103 and the cRIO-9066 platforms, which is around 15 μs. From Figure 15, it is shown that the RS-PCC strategy provides a worse performance than the S-PCC strategy in terms of current and torque ripples. However, the RS-PCC requires a lower execution time than the S-PCC strategy, which could be useful in digital control platforms with limited resources. As in the simulation results, both S-PCC and RS-PCC strategies give much higher values for the TWD_i, 27.87% and 39.30%, over the THD_i, 3.22% and 2.87%, meaning that the majority of the ripple observed in the phase currents is of high-frequency. The OSM-PCC strategy also gives a worse performance over previous strategies, increasing the TWD_i to 76.37%, but imposing a fixed switching frequency to the power switches of the inverters, which could ease the process of designing output filters for the six-phase machine. The use of a PWM technique in the PWM-PCC strategy leads to a reduction of the current ripple, mainly in the x'-y' currents, due to the imposition of mean zero x-y voltage components over a sampling period. However, since the PWM technique generates a fixed switching frequency of $\bar{f}_{sw} = 1/T_s$, the sampling period in the PWM-PCC strategy was set to $T_s = 200$ μs, which increases the d-q current errors and the torque ripple in comparison to previous strategies. The EPWM-PCC strategy provides a significant reduction in the torque ripple, that is, TWR_t decreased from 11.04% to 1.35%, due to the optimization in amplitude of the voltage vectors used by this strategy. Nonetheless, both PWM-PCC and EPWM-PCC strategies do not apply any x-y voltage components to the machine, meaning that the low order current harmonics mapped into the x'-y' subspace cannot be compensated.

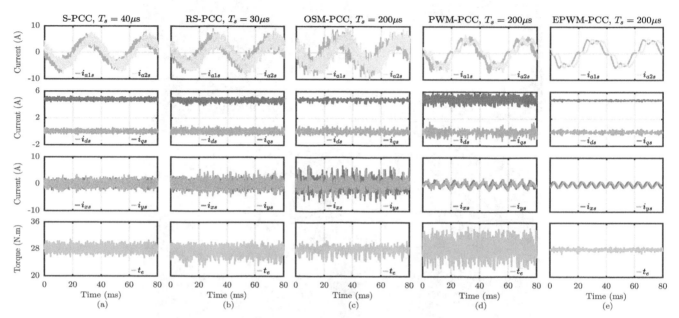

Figure 15. Experimental results for the PMSM drive operating at 750 rpm and rated load (motoring mode) for: (**a**) S-PCC; (**b**) RS-PCC; (**c**) OSM-PCC; (**d**) PWM-PCC and (**e**) EPWM-PCC.

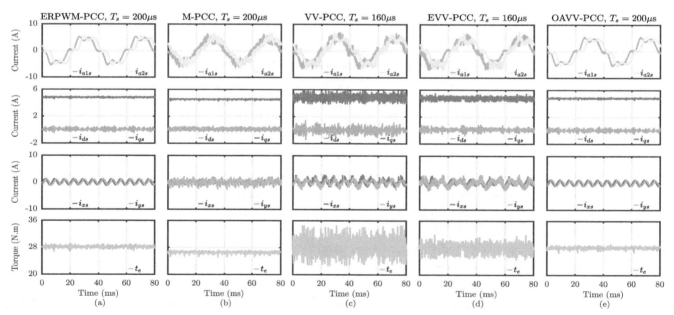

Figure 16. Experimental results for the PMSM drive operating at 750 rpm and rated load (motoring mode) for: (**a**) ERPWM-PCC; (**b**) M-PCC; (**c**) VV-PCC; (**d**) EVV-PCC and (**e**) OAVV-PCC.

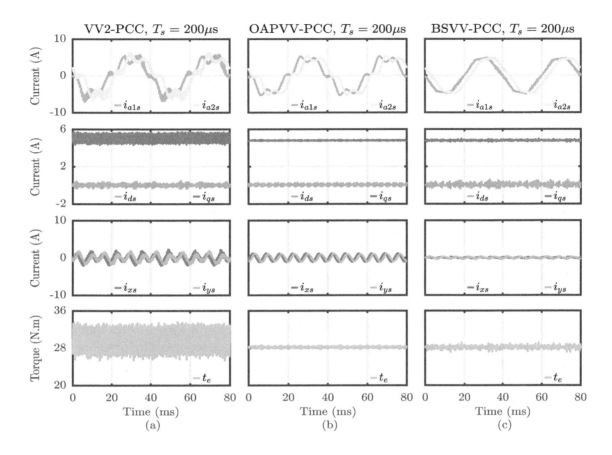

Figure 17. Experimental results for the PMSM drive operating at 750 rpm and rated load (motoring mode) for: **(a)** VV2-PCC; **(b)** OAPVV-PCC and **(c)** BSVV-PCC.

The experimental results for the operation of the six-phase drive under the strategies ERPWM-PCC, M-PCC, VV-PCC, EVV-PCC and OAVV-PCC are shown in Figure 16. In steady-state operation, the ERPWM-PCC strategy gives an equal performance to EPWM-PCC but since it is able to operate outside the linear modulation region (i.e., it can apply voltage vectors with a full amplitude of $0.644 \cdot U_{dc}$), the operation limits of the drive are increased and the performance of the drive during transients is enhanced. The M-PCC strategy integrates a different modulation strategy that combines two adjacent voltage vectors in the α-β subspace and two zero vectors ($\mathbf{v}_0$ and $\mathbf{v}_{63}$) in order to obtain a fixed switching frequency as with PWM-based PCC strategies. Although the M-PCC reduces the amplitude of the x'-y' current components, since the cost function of this strategy considers the reference current tracking errors in both subspaces, an optimal tracking of the current references in both subspaces is not possible and a steady-state error is observed in the q-axis current and in torque. In terms of computational requirements, the M-PCC strategy has a mean execution time of 78.0 μs, being the control strategy with higher computational requirements among the compared PCC techniques. The VV-PCC strategy considers the use of twelve large active and one zero virtual vectors, which avoids the application of x-y voltage components to the machine and presents a similar performance to the PWM-PCC strategy, although it provides higher dc-link voltage usage and leads to a mean switching frequency of 4.88 kHz. The EVV-PCC strategy manages to reduce the d-q current errors and torque ripple in comparison to the VV-PCC strategy, due to the inclusion of small virtual vectors. Although the EVV-PCC strategy is able to apply one out of twenty-five virtual vectors, this strategy uses a deadbeat approach to reduce the number of candidates to only two, thus providing a small execution time ($\bar{t}_{\mathrm{exe}} = 28.28$ μs). Since in the OAVV-PCC strategy the virtual vectors are optimized in amplitude, the d-q current errors and torque ripple are significantly reduced in comparison to VV-PCC and EVV-PCC strategies.

Table 4. Performance indicators for the drive operating at 750 rpm and rated load (motoring mode) for the PCC strategies.

Strategy	$E_{i,d}$ (%)	$E_{i,q}$ (%)	$E_{i,x}$ (%)	$E_{i,y}$ (%)	THD$_i$ (%)	TWD$_i$ (%)	TWR$_i$ (%)	$\bar{f}_{sw}$ (kHz)	$\bar{t}_{exe}$ (μs)
S-PCC	2.94	2.56	13.50	17.38	3.22	27.87	3.10	3.41	38.89
RS-PCC	4.22	4.61	18.83	22.64	2.87	39.30	4.75	3.77	25.25
OSM-PCC	5.96	3.35	50.37	36.17	23.77	76.37	4.51	5.00	35.22
PWM-PCC	7.52	9.29	15.34	15.00	22.76	29.37	11.04	5.00	29.99
EPWM-PCC	3.41	1.08	14.60	13.68	22.59	23.05	1.35	5.00	33.67
ERPWM-PCC	3.41	1.08	14.60	13.68	22.59	23.05	1.35	5.00	33.67
M-PCC	4.16	5.99	6.62	18.10	15.92	24.44	1.27	5.00	78.00
VV-PCC	7.14	8.47	18.01	15.39	24.09	31.87	10.00	4.88	29.54
EVV-PCC	4.71	4.41	15.28	20.62	25.35	32.04	5.41	4.85	28.28
OAVV-PCC	3.32	1.05	14.94	13.67	22.80	23.20	1.31	5.00	37.46
VV2-PCC	3.30	6.46	16.31	15.30	23.96	25.92	8.47	4.64	33.60
OAPVV-PCC	2.11	0.54	14.86	13.45	22.52	22.60	0.66	5.00	49.23
BSVV-PCC	3.28	1.03	2.16	3.40	4.61	6.38	1.30	5.00	35.45

The experimental results for the operation of the six-phase drive under strategies VV2-PCC, OAPVV-PCC and BSVV-PCC are shown in Figure 17. The VV2-PCC strategy combines two virtual vectors over a sampling period and offers a performance slightly worse than the one obtained with OAVV-PCC. The ripple in the d-q current components is due to a finite set of values that can be selected for the duty cycles of the two virtual vectors, as detailed in Section 4.3. The OAPVV-PCC strategy provides the lowest d-q current errors and torque ripple among the different PCC strategies. However, as in the case of PWM and virtual vector based PCC strategies, the low order harmonics in the x'-y' current components cannot be suppressed. On the other hand, the BSVV-PCC strategy is able to control both the d-q and x'-y' current components and provides the lowest x'-y' current errors and the lowest current harmonic distortion (TWD$_i$ = 6.38%) among all tested PCC strategies.

The experimental results for the operation of the six-phase drive under strategies S-PTC, HR-PTC, RFC-PTC, FC-PTC and MV-PTC are shown in Figure 18, while the corresponding performance indicators are given in Table 5. The obtained results show that although the S-PTC strategy provides a higher current harmonic distortion over the S-PCC strategy, 67.17% versus 27.87%, it gives a smaller torque ripple, 1.71% versus 3.10%. The HR-PTC strategy optimizes the amplitude of the optimal voltage vector, which would improve the steady-state performance of the drive. However, since it also increases the number of commutations of the power switches, the sampling period was set to $T_s = 60$ μs. Even with a higher switching frequency of 12.83 kHz, the HR-PTC gives the worst results in terms of current waveform distortion (TWD$_i$ = 89.36%) among the compared PTC strategies. The RFC-PTC strategy leads to a lower torque ripple in comparison to the previous PTC strategies, however it produces a high current distortion (TWD$_i$ = 77.02%), even with a high switching frequency of 10 kHz. On the other hand, the FC-PTC strategy reduces the ripple of the phase currents due to the use of virtual vectors. Moreover, the MV-PTC strategy optimizes the amplitude of the selected virtual vector, giving the lowest torque ripple (TWR$_t$ = 0.46%) for the compared PTC strategies, while maintaining a fixed switching frequency of 5 kHz.

Table 5. Performance indicators for the drive operating at 750 rpm and rated load (motoring mode) for all the PTC strategies.

Strategy	$E_{\psi,d}$ (%)	$E_{\psi,q}$ (%)	$E_{\psi,x}$ (%)	$E_{\psi,y}$ (%)	THD$_i$ (%)	TWD$_i$ (%)	TWR$_t$ (%)	$\bar{f}_{sw}$ (kHz)	$\bar{t}_{exe}$ (μs)
S-PTC	0.85	0.40	0.52	0.47	13.66	67.17	1.71	5.03	18.51
HR-PTC	0.97	0.66	0.72	0.59	17.00	89.36	2.87	12.83	43.35
RFC-PTC	0.52	0.23	0.64	0.51	23.00	77.02	1.02	10.00	34.34
FC-PTC	1.03	1.63	0.20	0.24	25.85	31.42	7.14	4.97	29.90
MV-PTC	0.77	0.10	0.19	0.18	22.96	23.20	0.46	5.00	34.01

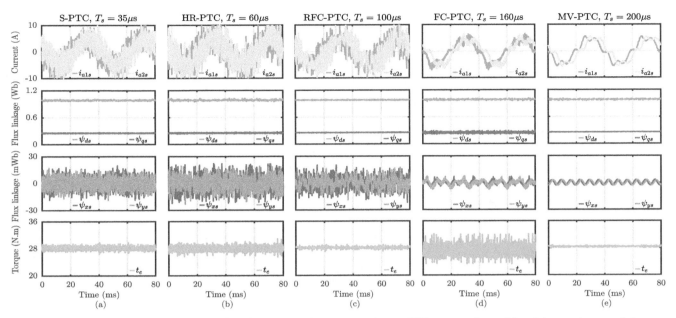

Figure 18. Experimental results for the PMSM drive operating at 750 rpm and rated load (motoring mode) for: (**a**) S-PTC; (**b**) HR-PTC; (**c**) RCF-PTC; (**d**) FC-PTC and (**e**) MV-PTC.

6.3. Comparison of Tested PCC and PCC Strategies

In order to summarize the merits and demerits of all tested PCC and PTC strategies applied to an electric drive based on a six-phase SPMSM, a comparison between these control strategies is given in Table 6. The following statements are defined to evaluate each control strategy:

- S1: The concept of the control strategy is simple and of easy implementation.
- S2: The control strategy produces a fixed switching frequency.
- S3: The high-frequency content of the phase currents (mapped into the x'-y' subspace) is minimized due to the use of a PWM technique or virtual vectors.
- S4: The low-frequency order harmonics of the phase currents (mapped into the x'-y' subspace) due to deadtime effects and back-EMF harmonics are suppressed by the control strategy.
- S5: No weighting factors need to be tuned.
- S6: The computational burden of the control strategy is low.
- S7: The control strategy gives a good performance at low speeds.
- S8: The control strategy provides full dc-bus usage, that is, it is able to apply a voltage vector with an amplitude up to $0.644 \cdot U_{dc}$ in the α-β subspace.
- S9: A separate and fast digital control platform (e.g. an FPGA) is not required to generate switching patterns for the power switches of the 2L-VSIs.

Each statement listed above is classified in Table 6 with a '+', when it is verified for the control strategy and with a '−' when the statement is not verified. Additionally, symbol '0' is employed in the case when the statement is not completely verified. For instance, control strategy M-PCC is not able to eliminate the high-frequency ripple of phase currents as other strategies based on PWM techniques or virtual vectors but still provides less current waveform distortion than S-PCC, RS-PCC, OSM-PCC, S-PTC, HR-PTC and RFC-PTC strategies. Moreover, in the case of statement S8, the control strategies based on virtual vectors are classified with a symbol '0' since the maximum amplitude of virtual vectors in the α-β subspace is $0.598 \cdot U_{dc}$, which is 7.14% smaller than the length of the large voltage vectors (Figure 2).

According to Table 6, the BSVV-PCC strategy is the best among the tested PCC and PTC strategies, since it verifies seven out of the nine statements given above. This strategy is simple and intuitive, gives a fixed switching frequency, minimizes both high-frequency and low-frequency harmonics of the x'-y' currents, does not require the tuning of weighting factors and provides good performance at low speeds. Moreover, the ERPWM-PCC and OAVV-PCC strategies are classified in second place fulfilling six out of nine statements, since they are not capable of suppressing the low-frequency harmonics of x'-y' currents and the ERPWM-PCC strategy requires a x-y weighting factor for the operation in the overmodulation region. Additionally, the MV-PTC strategy is the best strategy among the PTC strategies, verifying five out of nine statements. This strategy loses to BSVV-PCC in the complexity of the algorithm and in the inability to eliminate low order x'-y' current harmonics. On the other hand, the control strategies that provided the worst performance were the HR-PTC and RFC-PTC strategies, which did not comply with five out of nine statements given above. When testing these two strategies, lower values for the sampling time were used, giving high values for the mean switching frequency, in order to avoid excessive high-frequency ripple in the phase currents of the machine.

Table 6. Comparison between the PCC and PTC strategies applied to an electric drive based on a six-phase SPMSM.

Strategy	S1	S2	S3	S4	S5	S6	S7	S8	S9
S-PCC	+	−	−	0	−	+	−	+	+
RS-PCC	+	−	−	0	−	+	−	+	+
OSM-PCC	+	+	−	−	−	+	0	+	−
PWM-PCC	+	+	+	−	+	+	−	−	−
EPWM-PCC	+	+	+	−	+	+	+	−	−
ERPWM-PCC	+	+	+	−	0	+	+	+	−
M-PCC	−	+	0	0	−	−	+	+	−
VV-PCC	+	−	+	−	+	+	−	0	−
EVV-PCC	+	−	+	−	+	+	0	0	−
OAVV-PCC	+	+	+	−	+	+	+	0	−
VV2-PCC	0	−	+	−	+	0	+	0	−
OAPVV-PCC	0	+	+	−	+	0	+	0	−
BSVV-PCC	+	+	+	+	+	+	+	0	−
S-PTC	+	−	−	0	−	+	−	+	+
HR-PTC	+	−	−	−	−	0	0	+	−
RFC-PTC	−	−	−	−	+	+	0	+	−
FC-PTC	−	−	+	−	+	+	0	0	−
MV-PTC	−	+	+	−	+	+	+	0	−

Each statement is classified with: '+' (verified); '0' (not completely verified); and '−' (not verified).

6.4. Parameter Sensitivity

Since FCS-MPC strategies use a machine model to predict the future behavior of the controlled variables, the accuracy of these predictions depends on the equivalent parameters of the machine [127]. As these parameters can vary for different operating conditions, any parameter mismatch will cause an error in the predictions of the FCS-MPC algorithm and will lead to a deteriorated performance of the drive [128,129]. For the case of six-phase IM drives, only Reference [130] has studied the parameter sensitivity of the S-PCC strategy, while the parameter sensitivity of FCS-MPC strategies for six-phase PMSM drives remains uncovered. Hence, the parameter sensitivity of the control strategies BSVV-PCC, OAVV-PCC and MV-PTC to variations of ±30% in the values of parameters R_s and L_{dq} is analyzed in this section. Additionally, since the BSVV-PCC is able to control the x'-y' current components, the analysis of this strategy to variations in the parameter L_{xy} is also considered.

The parameter sensitivity of the OAVV-PCC strategy is tested experimentally for variations of ±30% in the parameters R_s and L_{dq}, for the operation of the six-phase PMSM drive at 750 rpm and rated load (motoring mode) and the obtained performance indicators are given in Table 7. The OAVV-PCC provides a slightly worse performance for errors of ±30% in the value of R_s, since the current errors, torque ripple and current harmonic distortion are marginally increased. In the case of an error of −30% in L_{dq}, the OAVV-PCC strategy still provides an acceptable performance, although in the case of a +30% error both the d-q current errors and torque ripple are heavily increased.

The performance indicators obtained for the MV-PTC strategy considering variations of ±30% in the parameters R_s and L_{dq} are given in Table 8. As in the previous case, only an error of +30% in the parameter L_{dq} provides a considerable degradation of the drive performance in terms of d-q current errors and torque ripple. Moreover, when an error of -30% is considered in L_{dq}, the d-axis flux error increases substantially in comparison to the normal case, while the indicators for the remaining cases of Table 8 only change marginally.

Table 9 contains the performance indicators for the BSVV-PCC strategy considering variations of $\pm30\%$ in the parameters R_s, L_{dq} and L_{xy}. Similarly to what was observed with the previous strategies, a mismatch in the value of R_s has a small impact on the performance of the six-phase drive. On the other hand, an error of $+30\%$ in L_{dq} negatively influences the performance of the drive, as shown by the high values of the d-q current errors and TWR_t. The errors in L_{xy} slightly increase the x-y current errors and consequently give higher values for the current harmonic distortion in comparison to the normal case. It is important to note that errors in L_{xy} do not affect significantly the value of TWR_t, since the x-y current components do not contribute to the production of torque.

Table 7. Parameter sensitivity of the OAVV-PCC strategy.

Strategy	$E_{i,d}$ (%)	$E_{i,q}$ (%)	$E_{i,x}$ (%)	$E_{i,y}$ (%)	THD_i (%)	TWD_i (%)	TWR_t (%)
Normal parameters	3.32	1.05	14.94	13.67	22.80	23.20	1.31
$0.7 \cdot R_s$	3.47	1.25	15.12	14.17	23.50	23.96	1.36
$1.3 \cdot R_s$	3.50	1.07	15.14	14.30	23.38	23.73	1.35
$0.7 \cdot L_{dq}$	4.13	0.96	14.99	14.05	23.21	23.28	1.11
$1.3 \cdot L_{dq}$	13.90	5.91	15.12	15.57	22.56	31.50	7.22

Table 8. Parameter sensitivity of the MV-PTC strategy.

Strategy	$E_{\psi,d}$ (%)	$E_{\psi,q}$ (%)	$E_{\psi,x}$ (%)	$E_{\psi,y}$ (%)	THD_i (%)	TWD_i (%)	TWR_t (%)
Normal parameters	0.77	0.10	0.19	0.18	22.96	23.20	0.46
$0.7 \cdot R_s$	3.57	0.12	0.19	0.17	21.85	22.10	0.31
$1.3 \cdot R_s$	0.86	0.13	0.20	0.18	23.24	23.52	0.47
$0.7 \cdot L_{dq}$	3.57	0.12	0.19	0.17	21.85	22.10	0.31
$1.3 \cdot L_{dq}$	2.38	0.70	0.19	0.19	23.09	25.04	3.24

Table 9. Parameter sensitivity of the BSVV-PCC strategy.

Strategy	$E_{i,d}$ (%)	$E_{i,q}$ (%)	$E_{i,x}$ (%)	$E_{i,y}$ (%)	THD_i (%)	TWD_i (%)	TWR_t (%)
Normal parameters	3.28	1.03	2.16	3.40	4.61	6.38	1.30
$0.7 \cdot R_s$	3.42	1.17	2.40	3.38	4.68	6.39	1.35
$1.3 \cdot R_s$	3.44	1.08	2.42	3.47	4.75	6.66	1.36
$0.7 \cdot L_{dq}$	3.55	0.89	2.22	3.41	4.66	5.89	1.09
$1.3 \cdot L_{dq}$	5.09	1.95	2.83	3.62	4.51	9.15	2.43
$0.7 \cdot L_{xy}$	3.62	1.15	2.62	3.96	5.41	6.61	1.41
$1.3 \cdot L_{xy}$	3.59	1.10	2.46	3.16	4.22	6.77	1.40

7. Conclusions

This paper has presented a critical comparative study of the FCS-MPC strategies available in the literature for electric drives based on six-phase asymmetrical machines, including a comprehensive overview of the theoretical background of these FCS-MPC strategies. It also assembles in a single reference the mathematical models of the six-phase drive topology, based on either IMs or PMSMs.

A total of thirteen PCC and five PTC strategies applied to a six-phase PMSM drive were compared side-by-side, with the aid of simulation and experimental results and their merits and shortcomings were discussed. In general, the PCC strategies favor a reduced harmonic content in the phase currents of the machine, while the PTC strategies produce a smaller torque ripple. The control strategies based on virtual vectors provide less high-frequency harmonic content in the x'-y' currents. Additionally, the control strategies based on virtual vectors optimized in amplitude or optimized in both amplitude and phase provide the lowest current or flux errors in the d-q subspace and improve the performance of the drive at low speeds. The low-frequency components of the x'-y' currents, due to deadtime effects in the power switches and due to the back-EMF harmonics in the case of six-phase PMSMs, were only suppressed by the BSVV-PCC strategy.

In the authors' opinion, this paper is useful to introduce FCS-MPC to control engineers or researchers working in the area of control of multiphase electric drives. Additionally, the paper can also help those who are already engaged in this field to select the best FCS-MPC strategy for their application, considering the merits and shortcomings of each strategy.

Author Contributions: Conceptualization, S.C. and P.G.; methodology, P.G. and S.C.; software, P.G.; validation, P.G.; formal analysis, P.G.; investigation, P.G.; resources, S.C. and A.M.; data curation, P.G.; writing—original draft preparation, P.G. and S.C.; writing—review and editing, P.G., S.C., and A.M.; visualization, S.C. and A.M.; supervision, S.C. and A.M.; project administration, P.G. and S.C.; funding acquisition, S.C. and A.M.

Abbreviations

1N	single isolated neutral
1NIN	single non-isolated neutral
2L-VSI	two-level voltage source inverter
2N	two isolated neutrals
3L-NPC	three-level neutral-point-clamped
BSVV-PCC	bi-subspace predictive current control based on virtual vectors
CSI	current source inverter
CCS-MPC	continuous control set model predictive control
DTC	direct torque control
EMF	electromotive force
EPWM-PCC	enhanced pulse width modulation predictive current control
ERPWM-PCC	extended range pulse width modulation predictive current control
EVV-PCC	predictive current control based on a extended set of virtual vectors
FC-PTC	flux constrained predictive torque control
FCS-MPC	finite control set model predictive control
FOC	field oriented control
FPGA	field-programmable gate array
HR-PTC	high robustness predictive torque control
IGBT	insulated gate bipolar transistor
IM	induction machine
M-PCC	modulated predictive current control
MMF	magnetomotive force
MPC	model predictive control
MTPA	maximum torque per ampere
MV-PTC	multi-vector predictive torque control
OAVV-PCC	predictive current control based on virtual vectors with optimal amplitude
OAPVV-PCC	predictive current control based on virtual vectors with optimal amplitude and phase
OSM-PCC	one step modulation model predictive control
OSS-MPC	optimal switching sequence model predictive control
OSV-MPC	optimal switching vector model predictive control
PCC	predictive current control
PI	proportional-integral
PM	permanent magnet
PMSM	permanent magnet synchronous machine
PSC	predictive speed control
PTC	predictive torque control
PWM	pulse width modulation
PWM-PCC	pulse width modulation predictive current control
RCF-PTC	reduced cost function predictive torque control
RS-PCC	restrained search predictive current control

S-PCC	standard predictive current control
S-PTC	standard predictive torque control
SPMSM	permanent magnet synchronous machine with surface-mounted permanent magnets
SV-PWM	space vector pulse width modulation
THD	total harmonic distortion
TWD	total waveform distortion
TWR	total waveform ripple
VV-PCC	predictive current control based on virtual vectors
VV2-PCC	predictive current control based on the application of two virtual vectors
VSD	variable space decomposition
VSI	voltage source inverter
ZSC	zero-sequence current

List of Symbols

General

C_{dq}, C_{xy}	overcurrent hard constraints for the d-q and x-y subspaces
d_i	duty cycles of vectors $\mathbf{v}_i$, $\mathbf{v}_{vi}$ or $\mathbf{v}_{dvi}$
$E_{i,d}$, $E_{i,q}$, $E_{i,x}$, $E_{i,y}$,	current error of d, q, x and y-axis components
$E_{\psi,d}$, $E_{\psi,q}$, $E_{\psi,x}$, $E_{\psi,y}$,	flux linkage error of d, q, x and y-axis components
$\bar{f}_{sw}$	mean switching frequency
g_c, g_{cf}, g_f	cost functions for PTC strategies
g_{tf}, g_{fcs}, g_{cm}, g_{cs}	cost functions for PCC strategies
i_{dr}, i_{qr}	d-q rotor current components
i_{ds}, i_{qs}, $i_{xs'}$, $i_{ys'}$, i_{z1s}, i_{z2s}	d-q, x'-y' and z1-z2 stator current components
$i_{ds,0}$, $i_{qs,0}$, $i_{xs',0}$, $i_{ys',0}$	d-q and x'-y' stator current components due to a zero vector
$i_{ds,i}$, $i_{qs,i}$, $i_{xs',i}$, $i_{ys',i}$	d-q and x'-y' stator current components due to vector $\mathbf{v}_i$, $\mathbf{v}_{vi}$ or $\mathbf{v}_{dvi}$
$\mathbf{i}_s$	stator current vector in a synchronous reference frame
$\mathbf{i}_s^{\alpha\beta}$	stator current vector in a stationary reference frame
$i_{s,dq}$, $i_{s,xy}$	current amplitude in the d-q and x-y subspaces
L_d, L_q, L_x, L_y, L_{z1}, L_{z2}	d, q, x, y, z1 and z2-axis inductances
L_{dq}, L_{xy}, L_0	d-q, x-y and z1-z2 subspace inductances
L_{lm}	stator mutual leakage inductance
L_{ls}, L_{lr}	stator and rotor self leakage inductances
L_s, L_r, L_m	stator, rotor and magnetizing inductances
p	number of pole-pairs
$\mathbf{s}$	switching state vector
s_{a1s}, s_{b1s}, s_{c1s}, s_{a2s}, s_{b2s}, s_{c2s}	phase switching states
t_e	electromagnetic torque
$t_{e,0}$	electromagnetic torque due to a vector $\mathbf{v}_{v0}$
$t_{e,i}$	electromagnetic torque due to a vector $\mathbf{v}_{vi}$
$\bar{t}_{exe}$	mean execution time
T_s	sampling period
THD_i, TWD_i	current total harmonic distortion and total waveform distortion
TWR_r	torque waveform ripple
$\mathbf{u}_s$	stator voltage vector in a synchronous reference frame
$\mathbf{u}_s^{\alpha\beta}$	stator voltage vector in a stationary reference frame
u_{a1s}, u_{b1s}, u_{c1s}, u_{a2s}, u_{b2s}, u_{c2s}	phase stator voltages
u_{dr}, u_{qr}	d-q rotor voltage components
u_{ds}, u_{qs}, $u_{xs'}$, $u_{ys'}$, u_{z1s}, u_{z2s}	d-q, x'-y' and z1-z2 stator voltage components
$u_{\alpha s}$, $u_{\beta s}$, u_{xs}, u_{ys}	α-β, x-y stator voltage components
$\mathbf{v}_i$	voltage vector with index i
$\mathbf{v}_{dvi}$	dual virtual vector with index i
$\mathbf{v}_{vi}$	virtual vector with index i
θ	rotor electrical position
λ_i, λ_f	weighting factors of current and flux
ψ_{dr}, ψ_{qr}	d-q rotor flux linkage components
ψ_{ds}, ψ_{qs}, $\psi_{xs'}$, $\psi_{ys'}$, ψ_{z1s}, ψ_{z2s}	d-q, x'-y' and z1-z2 stator flux linkage components
$\psi_{ds,PM}$, $\psi_{qs,PM}$, $\psi_{xs',PM}$, $\psi_{ys',PM}$, $\psi_{z1s,PM}$, $\psi_{z2s,PM}$	d-q, x'-y' and z1-z2 stator flux linkage components due to the PMs

ψ_s	flux linkage amplitude in the d-q subspace
$\psi_{s,0}$	flux linkage amplitude in the d-q subspace due to a zero vector
$\psi_{s,i}$	flux linkage amplitude in the d-q subspace due to a vector $\mathbf{v}_{vi}$
$\psi_{s,\text{PMi}}$	i-order harmonic component of the flux linkage due to the PMs
ω_a	electrical angular speed of an arbitrary reference frame
ω_s, ω_r	stator and rotor electrical angular speeds
$\mathbf{R}$	rotation matrix
$\mathbf{T}_{vsd}$	VSD transformation
U_{dc}	dc-link voltage

Subscripts

$d,q,x',y',z1,z2$	d, q, x, y, $z1$ and $z2$-axis quantities
s,r	stator and rotor quantities
n	rated value
α,β,x,y	d, q, x, y, $z1$ and $z2$-axis quantities

Superscripts

$*$	reference value
$k+2$	predicted quantity for instant $k+2$
$k+h$	predicted quantity for instant $k+h$

References

1. Singh, G. Multi-phase induction machine drive research—A survey. *Elect. Power Syst. Res.* **2002**, *61*, 139–147. [CrossRef]
2. Alger, P.L.; Freiburghouse, E.; Chase, D. Double windings for turbine alternators. *Trans. Am. Inst. Electr. Eng.* **1930**, *49*, 226–244. [CrossRef]
3. Ward, E.; Harer, H. Preliminary investigation of an invertor-fed 5-phase induction motor. *Proc. Inst. Electr. Eng.* **1969**, *116*, 980–984. [CrossRef]
4. Schiferl, R.; Ong, C. Six phase synchronous machine with AC and DC stator connections, Part I: Equivalent circuit representation and steady-state analysis. *IEEE Trans. Power App. Syst.* **1983**, *102*, 2685–2693. [CrossRef]
5. Jahns, T.M. Improved reliability in solid-state AC drives by means of multiple independent phase drive units. *IEEE Trans. Ind. Appl.* **1980**, *16*, 321–331. [CrossRef]
6. Parsa, L. On advantages of multi-phase machines. In Proceedings of the 31st Annual Conference of IEEE Industrial Electronics Society (IECON), Raleigh, NC, USA, 6–10 November 2005; pp. 1–6.
7. Levi, E. Advances in converter control and innovative exploitation of additional degrees of freedom for multiphase machines. *IEEE Trans. Ind. Electron.* **2016**, *63*, 433–448. [CrossRef]
8. Toliyat, H.A.; Lipo, T.A.; White, J.C. Analysis of a concentrated winding induction machine for adjustable speed drive applications. I. Motor analysis. *IEEE Trans. Energy Convers.* **1991**, *6*, 679–683. [CrossRef]
9. Toliyat, H.A.; Lipo, T.A.; White, J.C. Analysis of a concentrated winding induction machine for adjustable speed drive applications. II. Motor design and performance. *IEEE Trans. Energy Convers.* **1991**, *6*, 684–692. [CrossRef]
10. Gataric, S. A polyphase Cartesian vector approach to control of polyphase AC machines. In Proceedings of the IEEE Industry Application Society (IAS) Annual Meeting, Rome, Italy, 8–12 October 2000; Volume 3, pp. 1648–1654.
11. Jones, M.; Vukosavic, S.N.; Levi, E. Parallel-connected multiphase multidrive systems with single inverter supply. *IEEE Trans. Ind. Electron.* **2009**, *56*, 2047–2057. [CrossRef]
12. Che, H.S.; Levi, E.; Jones, M.; Duran, M.J.; Hew, W.P.; Rahim, N.A. Operation of a six-phase induction machine using series-connected machine-side converters. *IEEE Trans. Ind. Electron.* **2014**, *61*, 164–176. [CrossRef]
13. Zoric, I.; Jones, M.; Levi, E. Arbitrary power sharing among three-phase winding sets of multiphase machines. *IEEE Trans. Ind. Electron.* **2018**, *65*, 1128–1139. [CrossRef]
14. Subotic, I.; Dordevic, O.; Gomm, B.; Levi, E. Active and Reactive Power Sharing Between Three-Phase Winding Sets of a Multiphase Induction Machine. *IEEE Trans. Energy Convers.* **2019**, *34*, 1401–1410. [CrossRef]
15. Zabaleta, M.; Levi, E.; Jones, M. A novel synthetic loading method for multiple three-phase winding electric machines. *IEEE Trans. Energy Convers.* **2019**, *34*, 70–78. [CrossRef]
16. Abduallah, A.A.; Dordevic, O.; Jones, M.; Levi, E. Regenerative test for multiple three-phase machines with even number of neutral points. *IEEE Trans. Ind. Electron.* **2020**, *67*, 1684–1694. [CrossRef]

17. Subotic, I.; Bodo, N.; Levi, E.; Jones, M.; Levi, V. Isolated chargers for EVs incorporating six-phase machines. *IEEE Trans. Ind. Electron.* **2016**, *63*, 653–664. [CrossRef]

18. Subotic, I.; Bodo, N.; Levi, E.; Dumnic, B.; Milicevic, D.; Katic, V. Overview of fast on-board integrated battery chargers for electric vehicles based on multiphase machines and power electronics. *IET Elect. Power Appl.* **2016**, *10*, 217–229. [CrossRef]

19. Subotic, I.; Bodo, N.; Levi, E. Integration of six-phase EV drivetrains into battery charging process with direct grid connection. *IEEE Trans. Energy Convers.* **2017**, *32*, 1012–1022. [CrossRef]

20. Duran, M.J.; Gonzalez-Prieto, I.; Barrero, F.; Levi, E.; Zarri, L.; Mengoni, M. A simple braking method for six-phase induction motor drives with unidirectional power flow in the base-speed region. *IEEE Trans. Ind. Electron.* **2017**, *64*, 6032–6041. [CrossRef]

21. Gonzalez-Prieto, I.; Duran, M.J.; Barrero, F.J. Fault-tolerant control of six-phase induction motor drives with variable current injection. *IEEE Trans. Power Electron.* **2017**, *32*, 7894–7903. [CrossRef]

22. Duran, M.J.; Gonzalez-Prieto, I.; Rios-Garcia, N.; Barrero, F. A simple, fast, and robust open-phase fault detection technique for six-phase induction motor drives. *IEEE Trans. Power Electron.* **2018**, *33*, 547–557. [CrossRef]

23. Gonzalez-Prieto, I.; Duran, M.J.; Rios-Garcia, N.; Barrero, F.; Martin, C. Open-switch fault detection in five-phase induction motor drives using model predictive control. *IEEE Trans. Ind. Electron.* **2018**, *65*, 3045–3055. [CrossRef]

24. Williamson, S.; Smith, S. Pulsating torque and losses in multiphase induction machines. *IEEE Trans. Ind. Appl.* **2003**, *39*, 986–993. [CrossRef]

25. Levi, E.; Bojoi, R.; Profumo, F.; Toliyat, H.; Williamson, S. Multiphase induction motor drives—A technology status review. *IET Elect. Power Appl.* **2007**, *1*, 489–516. [CrossRef]

26. Levi, E. Multiphase electric machines for variable-speed applications. *IEEE Trans. Ind. Electron.* **2008**, *55*, 1893–1909. [CrossRef]

27. Cao, W.; Mecrow, B.C.; Atkinson, G.J.; Bennett, J.W.; Atkinson, D.J. Overview of electric motor technologies used for more electric aircraft (MEA). *IEEE Trans. Ind. Electron.* **2012**, *59*, 3523–3531.

28. Bojoi, R.; Cavagnino, A.; Tenconi, A.; Vaschetto, S. Control of shaft-line-embedded multiphase starter/generator for aero-engine. *IEEE Trans. Ind. Electron.* **2016**, *63*, 641–652. [CrossRef]

29. Bojoi, R.; Rubino, S.; Tenconi, A.; Vaschetto, S. Multiphase electrical machines and drives: A viable solution for energy generation and transportation electrification. In Proceedings of the IEEE International Conference and Exposition on Electrical and Power Engineering (EPE), Iasi, Romania, 20–22 October 2016; pp. 632–639.

30. Bojoi, R.; Cavagnino, A.; Cossale, M.; Tenconi, A. Multiphase starter generator for a 48-V mini-hybrid powertrain: Design and testing. *IEEE Trans. Ind. Appl.* **2016**, *52*, 1750–1758.

31. Jung, E.; Yoo, H.; Sul, S.K.; Choi, H.S.; Choi, Y.Y. A nine-phase permanent-magnet motor drive system for an ultrahigh-speed elevator. *IEEE Trans. Ind. Appl.* **2012**, *48*, 987–995. [CrossRef]

32. Liu, Z.; Wu, J.; Hao, L. Coordinated and fault-tolerant control of tandem 15-phase induction motors in ship propulsion system. *IET Electr. Power Appl.* **2017**, *12*, 91–97. [CrossRef]

33. Moraes, T.D.S.; Nguyen, N.K.; Semail, E.; Meinguet, F.; Guerin, M. Dual-multiphase motor drives for fault-tolerant applications: Power electronic structures and control strategies. *IEEE Trans. Power Electron.* **2017**, *33*, 572–580. [CrossRef]

34. Zhu, Z.; Hu, J. Electrical machines and power-electronic systems for high-power wind energy generation applications: Part II–power electronics and control systems. *COMPEL Int. J. Comput. Math. Elect. Electron. Eng.* **2012**, *32*, 34–71. [CrossRef]

35. Yaramasu, V.; Wu, B.; Sen, P.C.; Kouro, S.; Narimani, M. High-power wind energy conversion systems: State-of-the-art and emerging technologies. *Proc. IEEE* **2015**, *103*, 740–788. [CrossRef]

36. Prieto-Araujo, E.; Junyent-Ferré, A.; Lavernia-Ferrer, D.; Gomis-Bellmunt, O. Decentralized control of a nine-phase permanent magnet generator for offshore wind turbines. *IEEE Trans. Energy Convers.* **2015**, *30*, 1103–1112. [CrossRef]

37. Abbas, M.A.; Christen, R.; Jahns, T.M. Six-phase voltage source inverter driven induction motor. *IEEE Trans. Ind. Appl.* **1984**, *IA-20*, 1251–1259. [CrossRef]

38. Bojoi, R.; Farina, F.; Profumo, F.; Tenconi, A. Dual-three phase induction machine drives control—A survey. *IEEJ Trans. Ind. Appl.* **2006**, *126*, 420–429. [CrossRef]

39. Bojoi, R.; Lazzari, M.; Profumo, F.; Tenconi, A. Digital field-oriented control for dual three-phase induction motor drives. *IEEE Trans. Ind. Appl.* **2003**, *39*, 752–760. [CrossRef]

40. Singh, G.K.; Nam, K.; Lim, S. A simple indirect field-oriented control scheme for multiphase induction machine. *IEEE Trans. Ind. Electron.* **2005**, *52*, 1177–1184. [CrossRef]

41. Bojoi, R.; Farina, F.; Griva, G.; Profumo, F.; Tenconi, A. Direct torque control for dual three-phase induction motor drives. *IEEE Trans. Ind. Appl.* **2005**, *41*, 1627–1636. [CrossRef]

42. Hatua, K.; Ranganathan, V. Direct torque control schemes for split-phase induction machine. *IEEE Trans. Ind. Appl.* **2005**, *41*, 1243–1254. [CrossRef]

43. Wang, F.; Mei, X.; Rodriguez, J.; Kennel, R. Model predictive control for electrical drive systems-an overview. *CES Trans. Elect. Mach. Syst.* **2017**, *1*, 219–230.

44. Liu, C.; Luo, Y. Overview of advanced control strategies for electric machines. *Chin. J. Elect. Eng.* **2017**, *3*, 53–61.

45. Wang, F.; Zhang, Z.; Mei, X.; Rodríguez, J.; Kennel, R. Advanced control strategies of induction machine: Field oriented control, direct torque control and model predictive control. *Energies* **2018**, *11*, 120. [CrossRef]

46. Kouro, S.; Perez, M.A.; Rodriguez, J.; Llor, A.M.; Young, H.A. Model predictive control: MPC's role in the evolution of power electronics. *IEEE Ind. Electron. Mag.* **2015**, *9*, 8–21. [CrossRef]

47. Bordons, C.; Montero, C. Basic principles of MPC for power converters: Bridging the gap between theory and practice. *IEEE Ind. Electron. Mag.* **2015**, *9*, 31–43. [CrossRef]

48. Vazquez, S.; Rodriguez, J.; Rivera, M.; Franquelo, L.G.; Norambuena, M. Model predictive control for power converters and drives: Advances and trends. *IEEE Trans. Ind. Electron.* **2017**, *64*, 935–947. [CrossRef]

49. Gonçalves, P.F.; Cruz, S.M.; Mendes, A.M. Comparison of Model Predictive Control Strategies for Six-Phase Permanent Magnet Synchronous Machines. In Proceedings of the 44th Annual Conference of the IEEE Industrial Electronics Society (IECON), Washington, DC, USA, 21–23 October 2018; pp. 5801–5806.

50. Barrero, F.; Duran, M.J. Recent advances in the design, modeling, and control of multiphase machines—Part I. *IEEE Trans. Ind. Electron.* **2016**, *63*, 449–458. [CrossRef]

51. Ye, D.; Li, J.; Qu, R.; Lu, H.; Lu, Y. Finite set model predictive mtpa control with vsd method for asymmetric six-phase pmsm. In Proceedings of the IEEE International Electric Machines and Drives Conference (IEMDC), Miami, FL, USA, 21–24 May 2017; pp. 1–7.

52. Riveros, J.A.; Barrero, F.; Levi, E.; Durán, M.J.; Toral, S.; Jones, M. Variable-speed five-phase induction motor drive based on predictive torque control. *IEEE Trans. Ind. Electron.* **2013**, *60*, 2957–2968. [CrossRef]

53. Lim, C.S.; Levi, E.; Jones, M.; Rahim, N.A.; Hew, W.P. FCS-MPC-based current control of a five-phase induction motor and its comparison with PI-PWM control. *IEEE Trans. Ind. Electron.* **2014**, *61*, 149–163. [CrossRef]

54. Wu, X.; Song, W.; Xue, C. Low-Complexity Model Predictive Torque Control Method Without Weighting Factor for Five-Phase PMSM Based on Hysteresis Comparators. *Trans. Emerg. Sel. Top. Power Electron.* **2018**, *6*, 1650–1661. [CrossRef]

55. Prieto, I.G.; Zoric, I.; Duran, M.J.; Levi, E. Constrained Model Predictive Control in Nine-phase Induction Motor Drives. *IEEE Trans. Energy Convers.* **2019**, *34*, 1881–1889. [CrossRef]

56. Liu, Z.; Li, Y.; Zheng, Z. A review of drive techniques for multiphase machines. *CES Trans. Elect. Mach. Syst.* **2018**, *2*, 243–251. [CrossRef]

57. Tenconi, A.; Rubino, S.; Bojoi, R. Model Predictive Control for Multiphase Motor Drives—A Technology Status Review. In Proceedings of the IEEE International Power Electronic Conference (IPEC), Niigata, Japan, 20–24 May 2018; pp. 732–739.

58. Duran, M.J.; Levi, E.; Barrero, F. Multiphase electric drives: Introduction. In *Wiley Encyclopedia of Electrical and Electronics Engineering*; Webster, J., Ed.; John Wiley and Sons, Inc.: Hoboken, NJ, USA, 2017; pp. 1–26.

59. Duran, M.J.; Barrero, F. Recent advances in the design, modeling, and control of multiphase machines—Part II. *IEEE Trans. Ind. Electron.* **2016**, *63*, 459–468. [CrossRef]

60. Yaramasu, V.; Dekka, A.; Durán, M.J.; Kouro, S.; Wu, B. PMSG-based wind energy conversion systems: Survey on power converters and controls. *IET Elect. Power Appl.* **2017**, *11*, 956–968. [CrossRef]

61. Yepes, A.G.; Malvar, J.; Vidal, A.; López, O.; Doval-Gandoy, J. Current harmonics compensation based on multiresonant control in synchronous frames for symmetrical n-phase machines. *IEEE Trans. Ind. Electron.* **2015**, *62*, 2708–2720. [CrossRef]

62. Yepes, A.G.; Doval-Gandoy, J.; Baneira, F.; Perez-Estevez, D.; Lopez, O. Current harmonic compensation for n-phase machines with asymmetrical winding arrangement and different neutral configurations. *IEEE Trans. Ind. Appl.* **2017**, *53*, 5426–5439. [CrossRef]
63. Betin, F.; Capolino, G.A.; Casadei, D.; Kawkabani, B.; Bojoi, R.I.; Harnefors, L.; Levi, E.; Parsa, L.; Fahimi, B. Trends in electrical machines control: Samples for classical, sensorless, and fault-tolerant techniques. *IEEE Ind. Electron. Mag.* **2014**, *8*, 43–55. [CrossRef]
64. Che, H.S.; Duran, M.J.; Levi, E.; Jones, M.; Hew, W.P.; Rahim, N.A. Postfault operation of an asymmetrical six-phase induction machine with single and two isolated neutral points. *IEEE Trans. Power Electron.* **2014**, *29*, 5406–5416. [CrossRef]
65. Munim, W.N.W.A.; Duran, M.J.; Che, H.S.; Bermúdez, M.; González-Prieto, I.; Rahim, N.A. A unified analysis of the fault tolerance capability in six-phase induction motor drives. *IEEE Trans. Power Electron.* **2017**, *32*, 7824–7836. [CrossRef]
66. Reusser, C. Power Converter Topologies for Multiphase Drive Applications. In *Electric Power Conversion*; Gaiceanu, M., Ed.; IntechOpen: London, UK, 2018; pp. 1–22.
67. Yaramasu, V.; Wu, B. *Model Predictive Control of Wind Energy Conversion Systems*; IEEE: Hoboken, NJ, USA, 2017.
68. Andersen, E. 6-phase induction motors for current source inverter drives. In Proceedings of the 16th Annual Meeting IEEE Industry Applications Society, Philadelphia, PA, USA, 5–9 October 1981; pp. 607–618.
69. Gopakumar, K.; Sathiakumar, S.; Biswas, S.; Vithayathil, J. Modified current source inverter fed induction motor drive with reduced torque pulsations. *IEE Proc. B* **1984**, *131*, 159–164. [CrossRef]
70. Levi, E.; Bodo, N.; Dordevic, O.; Jones, M. Recent advances in power electronic converter control for multiphase drive systems. In Proceedings of the IEEE Workshop on Electrical Machines Design, Control and Diagnosis (WEMDCD), Paris, France, 11–12 March 2013; pp. 158–167.
71. Duran, M.J.; Prieto, I.G.; Bermudez, M.; Barrero, F.; Guzman, H.; Arahal, M.R. Optimal fault-tolerant control of six-phase induction motor drives with parallel converters. *IEEE Trans. Ind. Electron.* **2016**, *63*, 629–640. [CrossRef]
72. Gonzalez-Prieto, I.; Duran, M.J.; Che, H.; Levi, E.; Bermúdez, M.; Barrero, F. Fault-tolerant operation of six-phase energy conversion systems with parallel machine-side converters. *IEEE Trans. Power Electron.* **2016**, *31*, 3068–3079. [CrossRef]
73. Gonzalez-Prieto, I.; Duran, M.J.; Barrero, F.; Bermudez, M.; Guzman, H. Impact of postfault flux adaptation on six-phase induction motor drives with parallel converters. *IEEE Trans. Power Electron.* **2017**, *32*, 515–528. [CrossRef]
74. Fuchs, E.; Rosenberg, L. Analysis of an alternator with two displaced stator windings. *IEEE Trans. Power App. Syst.* **1974**, *93*, 1776–1786. [CrossRef]
75. Nelson, R.; Krause, P. Induction machine analysis for arbitrary displacement between multiple winding sets. *IEEE Trans. Power App. Syst.* **1974**, *PAS-93*, 841–848. [CrossRef]
76. Gonçalves, P.F.; Cruz, S.M.; Mendes, A.M. Design of a six-phase asymmetrical permanent magnet synchronous generator for wind energy applications. *J. Eng.* **2019**, *2019*, 4532–4536. [CrossRef]
77. Eldeeb, H.M.; Abdel-Khalik, A.S.; Kullick, J.; Hackl, C.M. Pre and Post-fault Current Control of Dual Three-Phase Reluctance Synchronous Drives. *IEEE Trans. on Ind. Electron.* **2019**. [CrossRef]
78. Eldeeb, H.M.; Abdel-Khalik, A.S.; Hackl, C.M. Post-Fault Full Torque-Speed Exploitation of Dual Three-Phase IPMSM Drives. *IEEE Trans. Ind. Electron.* **2019**, *66*, 6746–6756. [CrossRef]
79. Hu, Y.; Zhu, Z.; Odavic, M. Torque capability enhancement of dual three-phase PMSM drive with fifth and seventh current harmonics injection. *IEEE Trans. Ind. Appl.* **2017**, *53*, 4526–4535. [CrossRef]
80. Wang, K. Effects of harmonics into magnet shape and current of dual three-phase permanent magnet machine on output torque capability. *IEEE Trans. Ind. Electron.* **2018**, *65*, 8758–8767. [CrossRef]
81. Hu, Y.; Zhu, Z.; Odavic, M. Comparison of two-individual current control and vector space decomposition control for dual three-phase PMSM. *IEEE Trans. Ind. Appl.* **2017**, *53*, 4483–4492. [CrossRef]
82. Lipo, T. A d-q model for six phase induction machines. *Proc. Int. Conf. Elect. Mach.* **1980**, *2*, 860–867.
83. Zhao, Y.; Lipo, T.A. Space vector PWM control of dual three-phase induction machine using vector space decomposition. *IEEE Trans. Ind. Appl.* **1995**, *31*, 1100–1109. [CrossRef]
84. Eldeeb, H.M.; Abdel-Khalik, A.S.; Hackl, C.M. Dynamic modeling of dual three-phase IPMSM drives with different neutral configurations. *IEEE Trans. Ind. Electron.* **2019**, *66*, 141–151. [CrossRef]

85. Duran, M.J.; González-Prieto, I.; González-Prieto, A.; Barrero, F. Multiphase energy conversion systems connected to microgrids with unequal power-sharing capability. *IEEE Trans. Energy Convers.* **2017**, *32*, 1386–1395. [CrossRef]

86. Che, H.S.; Levi, E.; Jones, M.; Hew, W.P.; Rahim, N.A. Current control methods for an asymmetrical six-phase induction motor drive. *IEEE Trans. Power Electron.* **2014**, *29*, 407–417. [CrossRef]

87. Hadiouche, D.; Razik, H.; Rezzoug, A. On the modeling and design of dual-stator windings to minimize circulating harmonic currents for VSI fed AC machines. *IEEE Trans. Ind. Appl.* **2004**, *40*, 506–515. [CrossRef]

88. Che, H.S.; Abdel-Khalik, A.S.; Dordevic, O.; Levi, E. Parameter estimation of asymmetrical six-phase induction machines using modified standard tests. *IEEE Trans. Ind. Electron.* **2017**, *64*, 6075–6085. [CrossRef]

89. Tessarolo, A.; Bassi, C. Stator harmonic currents in VSI-fed synchronous motors with multiple three-phase armature windings. *IEEE Trans. Energy Convers.* **2010**, *25*, 974–982. [CrossRef]

90. Arahal, M.; Barrero, F.; Toral, S.; Duran, M.; Gregor, R. Multi-phase current control using finite-state model-predictive control. *Control Eng. Pract.* **2009**, *17*, 579–587. [CrossRef]

91. Barrero, F.; Arahal, M.R.; Gregor, R.; Toral, S.; Durán, M.J. A proof of concept study of predictive current control for VSI-driven asymmetrical dual three-phase AC machines. *IEEE Trans. Ind. Electron.* **2009**, *56*, 1937–1954. [CrossRef]

92. Recalde, R.I.G. The asymmetrical dual three-phase induction machine and the mbpc in the speed control. In *Induction Motors: Modelling and Control*; Araujo, R., Ed.; IntechOpen: London, UK, 2012; pp. 385–400.

93. Rodas, J.; Gregor, R.; Takase, Y.; Moreira, H.; Riveray, M. A comparative study of reduced order estimators applied to the speed control of six-phase generator for a WT applications. In Proceedings of the 39th Annual Conference of the IEEE Industrial Electronics Society, Vienna, Austria, 10–13 November 2013; pp. 5124–5129.

94. Dasika, J.D.; Qin, J.; Saeedifard, M.; Pekarek, S.D. Predictive current control of a six-phase asymmetrical drive system based on parallel-connected back-to-back converters. In Proceedings of the IEEE Energy Conversion Congress and Exposition (ECCE), Raleigh, NC, USA, 15–20 September 2012; pp. 137–141.

95. Rodriguez, J.; Cortes, P. *Predictive Control of Power Converters and Electrical Drives*; John Wiley & Sons: Hoboken, NJ, USA, 2017.

96. Durán, M.J.; Barrero, F.; Prieto, J.; Toral, S. Predictive current control of dual three-phase drives using restrained search techniques and multi level voltage source inverters. In Proceedings of the IEEE International Symposium on Industrial Electronics (ISIE), Bari, Italy, 4–7 July 2010; pp. 3171–3176.

97. Duran, M.J.; Prieto, J.; Barrero, F.; Toral, S. Predictive current control of dual three-phase drives using restrained search techniques. *IEEE Trans. Ind. Electron.* **2011**, *58*, 3253–3263. [CrossRef]

98. Barrero, F.; Arahal, M.R.; Gregor, R.; Toral, S.; Durán, M.J. One-step modulation predictive current control method for the asymmetrical dual three-phase induction machine. *IEEE Trans. Ind. Electron.* **2009**, *56*, 1974–1983. [CrossRef]

99. Gregor, R.; Barrero, F.; Toral, S.; Duran, M.; Arahal, M.; Prieto, J.; Mora, J. Predictive-space vector PWM current control method for asymmetrical dual three-phase induction motor drives. *IET Elect. Power Appl.* **2010**, *4*, 26–34. [CrossRef]

100. Barrero, F.; Prieto, J.; Levi, E.; Gregor, R.; Toral, S.; Durán, M.J.; Jones, M. An enhanced predictive current control method for asymmetrical six-phase motor drives. *IEEE Trans. Ind. Electron.* **2011**, *58*, 3242–3252. [CrossRef]

101. Prieto, J.; Barrero, F.; Lim, C.S.; Levi, E. Predictive current control with modulation in asymmetrical six-phase motor drives. In Proceedings of the International Power Electronics and Motion Control Conference (EPE/PEMC), Novi Sad, Serbia, 4–6 September 2012; pp. 1–8.

102. Ayala, M.; Gonzalez, O.; Rodas, J.; Gregor, R.; Rivera, M. Predictive control at fixed switching frequency for a dual three-phase induction machine with Kalman filter-based rotor estimator. In Proceedings of the IEEE International Conference on Automatica (ICA-ACCA), Curico, Chile, 19–21 October 2016; pp. 1–6.

103. Ayala, M.; Rodas, J.; Gregor, R.; Doval-Gandoy, J.; Gonzalez, O.; Saad, M.; Rivera, M. Comparative study of predictive control strategies at fixed switching frequency for an asymmetrical six-phase induction motor drive. In Proceedings of the IEEE International Electrical Machines and Drives Conference (IEMDC), Miami, FL, USA, 21–24 May 2017.

104. Gonzalez, O.; Ayala, M.; Rodas, J.; Gregor, R.; Rivas, G.; Doval-Gandoy, J. Variable-speed control of a six-phase induction machine using predictive-fixed switching frequency current control techniques. In Proceedings of the IEEE International Symposium on Power Electron. for Distributed Generation Syst. (PEDG), Charlotte, NC, USA, 25–28 June 2018; pp. 1–6.

105. Gonzalez, O.; Ayala, M.; Doval-Gandoy, J.; Rodas, J.; Gregor, R.; Rivera, M. Predictive-Fixed Switching Current Control Strategy Applied to Six-Phase Induction Machine. *Energies* **2019**, *12*, 2294. [CrossRef]

106. Gonzalez-Prieto, I.; Duran, M.J.; Aciego, J.J.; Martin, C.; Barrero, F. Model predictive control of six-phase induction motor drives using virtual voltage vectors. *IEEE Trans. Ind. Electron.* **2018**, *65*, 27–37. [CrossRef]

107. Gao, L.; Fletcher, J.E.; Zheng, L. Low-speed control improvements for a two-level five-phase inverter-fed induction machine using classic direct torque control. *IEEE Trans. Ind. Electron.* **2011**, *58*, 2744–2754. [CrossRef]

108. Zheng, L.; Fletcher, J.E.; Williams, B.W.; He, X. A novel direct torque control scheme for a sensorless five-phase induction motor drive. *IEEE Trans. Ind. Electron.* **2011**, *58*, 503–513. [CrossRef]

109. Hoang, K.D.; Ren, Y.; Zhu, Z.Q.; Foster, M. Modified switching-table strategy for reduction of current harmonics in direct torque controlled dual-three-phase permanent magnet synchronous machine drives. *IET Elect. Power Appl.* **2015**, *9*, 10–19. [CrossRef]

110. Ren, Y.; Zhu, Z.Q. Reduction of both harmonic current and torque ripple for dual three-phase permanent-magnet synchronous machine using modified switching-table-based direct torque control. *IEEE Trans. Ind. Electron.* **2015**, *62*, 6671–6683. [CrossRef]

111. Pandit, J.K.; Aware, M.V.; Nemade, R.V.; Levi, E. Direct torque control scheme for a six-phase induction motor with reduced torque ripple. *IEEE Trans. Power Electron.* **2017**, *32*, 7118–7129. [CrossRef]

112. Luo, Y.; Liu, C. Elimination of harmonic currents using a reference voltage vector based-model predictive control for a six-phase PMSM motor. *IEEE Trans. Power Electron.* **2019**, *34*, 6960–6972. [CrossRef]

113. Gonçalves, P.; Cruz, S.; Mendes, A. Predictive current control based on variable amplitude virtual vectors for six-phase permanent magnet synchronous machines. In Proceedings of the 20th International Conference on Industrial Technology (ICIT), Melbourne, Australia, 13–15 February 2019; pp. 310–316.

114. Aciego, J.J.; Prieto, I.G.; Duran, M.J. Model predictive control of six-phase induction motor drives using two virtual voltage vectors. *Trans. Emerg. Sel. Top. Power Electron.* **2019**, *7*, 321–330. [CrossRef]

115. Gonçalves, P.; Cruz, S.; Mendes, A. Fixed and Variable Amplitude Virtual Vectors for Model Predictive Control of Six-Phase PMSMs with Single Neutral Configuration. In Proceedings of the 20th International Conference on Industrial Technology (ICIT), Melbourne, Australia, 13–15 February 2019; pp. 310–316.

116. Gonçalves, P.; Cruz, S.; Mendes, A. Predictive Current Control of Six-Phase Permanent Magnet Synchronous Machines Based on Virtual Vectors with Optimal Amplitude and Phase. In Proceedings of the 2nd International Conference on Smart Energy Systems and Technologies (SEST), Porto, Portugal, 9–11 September 2019; pp. 1–6.

117. Hu, Y.; Zhu, Z.Q.; Liu, K. Current control for dual three-phase permanent magnet synchronous motors accounting for current unbalance and harmonics. *IEEE Trans. Emerg. Sel. Top. Power Electron.* **2014**, *2*, 272–284.

118. Karttunen, J.; Kallio, S.; Peltoniemi, P.; Silventoinen, P. Current harmonic compensation in dual three-phase PMSMs using a disturbance observer. *IEEE Trans. Ind. Electron.* **2016**, *63*, 583–594. [CrossRef]

119. Karttunen, J.; Kallio, S.; Honkanen, J.; Peltoniemi, P.; Silventoinen, P. Partial current harmonic compensation in dual three-phase PMSMs considering the limited available voltage. *IEEE Trans. Ind. Electron.* **2017**, *64*, 1038–1048. [CrossRef]

120. Gonçalves, P.F.; Cruz, S.M.; Mendes, A.M. Bi-subspace predictive current control of six-phase PMSM drives based on virtual vectors with optimal amplitude. *IET Elect. Power Appl.* **2019**, *13*, 1672–1683. [CrossRef]

121. Prieto, J.; Riveros, J.A.; Bogado, B.; Barrero, F.; Toral, S.; Cortés, P. Electric propulsion technology based in predictive direct torque control and asymmetrical dual three-phase drives. In Proceedings of the 13th International Conference on Intelligent Transportation Systems (ITSC), Madeira Island, Portugal, 19–22 September 2010; pp. 397–402.

122. Luo, Y.; Liu, C. A simplified model predictive control for a dual three-phase PMSM with reduced harmonic currents. *IEEE Trans. Ind. Electron.* **2018**, *65*, 9079–9089. [CrossRef]

123. Luo, Y.; Liu, C. Model Predictive Control for a Six-Phase PMSM with High Robustness Against Weighting Factor Variation. *IEEE Trans Ind. Appl.* **2019**, *55*, 2781–2791. [CrossRef]

124. Luo, Y.; Liu, C. Model predictive control for a six-phase PMSM motor with a reduced-dimension cost function. *IEEE Trans. Ind. Electron.* **2020**, *67*, 969–979. [CrossRef]

125. Luo, Y.; Liu, C. A flux constrained predictive control for a six-phase PMSM motor with lower complexity. *IEEE Trans. Ind. Electron.* **2019**, *66*, 5081–5093. [CrossRef]

126. Luo, Y.; Liu, C. Multi-Vectors Based Model Predictive Torque Control for a Six-Phase PMSM Motor with Fixed Switching Frequency. *IEEE Trans. Energy Convers.* **2019**, *34*, 1369–1379. [CrossRef]

127. Martín, C.; Bermúdez, M.; Barrero, F.; Arahal, M.R.; Kestelyn, X.; Durán, M.J. Sensitivity of predictive controllers to parameter variation in five-phase induction motor drives. *Control Eng. Pract.* **2017**, *68*, 23–31. [CrossRef]

128. Siami, M.; Khaburi, D.A.; Rodriguez, J. Torque ripple reduction of predictive torque control for PMSM drives with parameter mismatch. *IEEE Trans. Ind. Electron.* **2017**, *32*, 7160–7168. [CrossRef]

129. Abdelrahem, M.; Hackl, C.M.; Kennel, R.; Rodriguez, J. Efficient Direct-Model Predictive Control with Discrete-Time Integral Action for PMSGs. *IEEE Trans. Energy Convers.* **2019**, *34*, 1063–1072. [CrossRef]

130. Bogado, B.; Barrero, F.; Arahal, M.; Toral, S.; Levi, E. Sensitivity to electrical parameter variations of predictive current control in multiphase drives. In Proceedings of the 39th Annual Conference of the IEEE Industrial Electronics Society (IECON), Vienna, Austria, 10–13 November 2013; pp. 5215–5220.

7

A Novel Magnet-Axis-Shifted Hybrid Permanent Magnet Machine for Electric Vehicle Applications

Ya Li, Hui Yang *, Heyun Lin, Shuhua Fang and Weijia Wang

School of Electrical Engineering, Southeast University, Nanjing 210096, China; seueelab_ly@163.com (Y.L.); hyling@seu.edu.cn (H.L.); shfang@seu.edu.cn (S.F.); seueelab_wwj@163.com (W.W.)
* Correspondence: huiyang@seu.edu.cn.

Abstract: This paper proposes a novel magnet-axis-shifted hybrid permanent magnet (MAS-HPM) machine, which features an asymmetrical magnet arrangement, i.e., low-cost ferrite and high-performance NdFeB magnets, are placed in the two sides of a "▽"-shaped rotor pole. The proposed magnet-axis-shift (MAS) effect can effectively reduce the difference between the optimum current angles for maximizing permanent magnet (PM) and reluctance torques, and hence the torque capability of the machine can be further improved. The topology and operating principle of the proposed MAS-HPM machine are introduced and are compared with the BMW i3 interior permanent magnet (IPM) machine as a benchmark. The electromagnetic characteristics of the two machines are investigated and compared by finite element analysis (FEA), which confirms the effectiveness of the proposed MAS design concept for torque improvement.

Keywords: hybrid permanent magnet; interior permanent magnet (IPM) machine; magnet-axis-shifted; reluctance torque

---

1. Introduction

Due to their high torque/power density, high efficiency and excellent flux weakening capability, interior permanent magnet (IPM) machines are considered as competitive candidates for electric vehicles (EVs) [1]. In order to improve the reluctance torque and reduce the magnet usage, multi-layer IPM machines are widely employed in EV applications, such as the BMW i3 traction machine [2]. However, for the conventional IPM machines, the optimum current angle for maximizing reluctance and permanent magnet (PM) torques basically differs by a 45° electrical angle, which results in a relatively low utilization ratio of the two torque components. Consequently, in order to deal with this issue, hybrid rotor [3–7], dual rotor [8] and asymmetrical permanent magnet (PM)-assisted synchronous reluctance machines [9] have been recently developed. The constant power-maintaining capabilities of the hybrid rotor configurations are investigated by adopting the parameter equivalent circuit method, which shows that the hybrid rotor topologies have more degrees of freedom for a given constant power operating range [10]. Moreover, the theoretical analysis demonstrates that the PM usages of synchronous machines can be reduced by about 10% with the reluctance axis shifted by a displacement angle of about 60° [11]. The hybrid synchronous machines with a displaced reluctance axis are comparatively studied with conventional pure PM and electrically excited synchronous machines [12], which demonstrates that the hybrid topologies exhibit higher torque and high-efficiency operating range. In addition, the effects of shifting the PM axis with respect to the reluctance axis in PM machines are investigated [13], showing that the asymmetric salient PM machine exhibits higher torque and constant power speed range [14]. Nevertheless, the hybrid and dual rotor machines suffer from complicated structures, while the latter asymmetrical one is characterized by shifts of both magnet and reluctance axes that require relatively sophisticated computational design efforts.

Recently, in order to reduce the use of the rare-earth NdFeB magnets, the hybrid PM concept has been proposed and developed in rotor PM [15,16] and stator PM [17–22] configurations. Compared with the structure of conventional spoke-type magnets, the proposed hybrid PM topology exhibits better field weakening capability and lower total cost [15]. Besides, compared with a double-layer PM structure, the U-shaped configuration has good irreversible demagnetization withstanding capability [16]. Due to the variable magnetization state of the low-coercive-force AlNiCo magnets, the flexible air-gap flux adjustment and wide operating range with high efficiency can be readily achieved in stator hybrid PM machines [16–22]. A novel magnet-axis-shifted hybrid PM (MAS-HPM) machine combined with the asymmetric and hybrid PM concepts is proposed in this paper.

The purpose of this paper is to propose an MAS-HPM machine for torque performance improvement. The proposed configuration features an asymmetrical PM arrangement, i.e., low-cost ferrite and high-performance NdFeB magnets, which significantly reduces the difference of the optimum current angle for maximizing PM and reluctance torques. Hence, the torque capability can be further improved. In order to validate the merits of the magnet-axis-shift (MAS) effect, the IPM machine of an BMW i3 vehicle is used as a benchmark. The basic electromagnetic characteristics of the two machines are comparatively investigated, which confirms the validity of the proposed MAS design concept.

2. Machine Topologies and Magnet-Axis-Shift Principle

2.1. Machine Topologies

The topologies of the benchmark 2016 BMW i3 IPM machine and the proposed MAS-HPM machine are shown in Figure 1a,b, respectively. The main design parameters are tabulated in Table 1. It should be noted that the proposed machine shares the same inverter power ratings, stator structure, active stack length and air-gap length with the BMW i3 IPM machine. Meanwhile, in order to make a fair comparison, the rare-earth PM usages are identical in the two structures. The main difference between the two machines lies in the fact that two kinds of PM, i.e., low-cost ferrite and high-performance NdFeB magnets, are simultaneously employed in the developed MAS-HPM machine to achieve the MAS effect. The total costs of the magnets are given in Table 1. Due to the additional ferrite magnets, the proposed machine has a slightly higher total cost of magnets than the BMW i3 IPM counterpart. However, compared with the BMW i3 IPM machine, the ratio of the peak torque to the total cost of magnets in the MAS-HPM configuration is increased by about 7.81%, which indicates that the torque capability can be improved by 7.81% at the same cost of magnets.

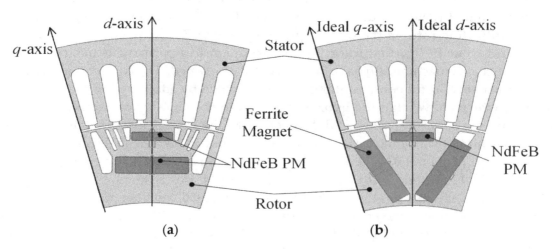

Figure 1. Machine topologies. (**a**) BMW i3 interior permanent magnet (IPM) machine. (**b**) The proposed magnet-axis-shifted hybrid permanent magnet (MAS-HPM) machine.

Table 1. Main design parameters of the machines.

Items	BMW i3 IPM	MAS-HPM
Stator slot number	72	
Rotor pole pair number	6	
Stator outer radius (mm)	121	
Air-gap length (mm)	0.7	
Rotor outer radius (mm)	89.3	
Active stack length (mm)	132	
Peak current (A)	530	
Steel grade	TKM330-35	
NdFeB grade	N35EH	
Ferrite magnet grade	-	AC-12
NdFeB PM volume (mm³)	24,816	
Ferrite magnet volume (mm³)	-	19,565
Total cost of magnets (¥)	51.3	53.1
Peak torque (Nm)	269.34	300.51
Peak torque/total cost of magnets (Nm/¥)	5.25	5.66
Working temperature (°C)	100	

The d and q-axes' equivalent electrical circuits are illustrated in Figure 2. In the synchronous reference frame, the voltage equations for the PM synchronous machine are expressed as:

$$\begin{cases} u_d = Ri_d - \omega\psi_q \\ u_q = Ri_q + \omega\psi_d \end{cases},\tag{1}$$

where R is the stator resistance, ω is the electric frequency, ψ_d and ψ_q are the d and q-axes' flux linkages, respectively. i_d and i_q are the d and q-axes' current, respectively.

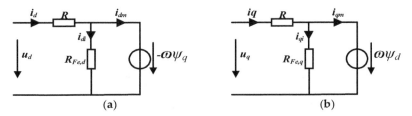

Figure 2. d and q-axes' equivalent electrical circuits. (a) d-axis. (b) q-axis.

By the application of Kirchhoff's voltage and current laws to both d and q-axes, the four equations can be obtained as:

$$\begin{cases} R_{Fe,d}i_{di} + \omega\psi_q = 0 \\ -R_{Fe,q}i_{qi} + \omega\psi_d = 0 \\ i_d - i_{di} - i_{dm} = 0 \\ i_q - i_{qi} - i_{qm} = 0 \end{cases},\tag{2}$$

where $R_{Fe,d}$ and $R_{Fe,q}$ are the iron losses resistances in d and q-axes, respectively. i_{di} and i_{qi} are the iron losses currents in d and q-axes, respectively. i_{dm} and i_{qm} are the d and q-axes' magnetization currents, respectively.

2.2. MAS Principle

The total torque T_{total} of an IPM machine, including the PM torque T_{PM} and the reluctance torque T_r, can be expressed as [23]:

$$T_{total} = \frac{3}{2}p_r\psi_f i_s \cos\beta + \frac{3}{4}p_r(L_d - L_q)i_s^2 \sin 2\beta = T_{PM} + T_r,\tag{3}$$

$$T_{PM} = \frac{3}{2}p_r\psi_f i_s \cos\beta,\tag{4}$$

$$T_r = \frac{3}{4} p_r (L_d - L_q) i_s^2 \sin 2\beta, \tag{5}$$

where p_r, ψ_f, i_s, L_d and L_q are the rotor pole pair number, the PM flux linkage, the phase current and the d- and q-axes' inductances, respectively. β is the current angle, which is defined as the angle between the phase current and open-circuit back electro-motive force (EMF) [24].

From Equations (3)–(5), it can be found that the optimum current angle for T_r is theoretically twice that for T_{PM}. If the difference between the optimum current angles for maximizing the two kinds of torques can be reduced, the torque capability of the machine will be improved. To achieve this goal, this paper proposes an asymmetrical PM arrangement by employing the HPM configuration, i.e., low-cost ferrite and high-performance NdFeB magnets, which is termed as the MAS effect. In this case, the magnet axis is shifted while the reluctance axis is unchanged due to the symmetrical rotor configuration. Thus, the difference of the current angles γ_s when both T_{PM} and T_r reach the maximum can be reduced, which can be defined as:

$$\gamma_s = \beta_{PM} - \beta_R, \tag{6}$$

where β_R and β_{PM} are the optimum current angles for the reluctance and PM torques, respectively.

The flux density distributions of the two machines are calculated by finite element analysis (FEA) and illustrated in Figure 3. It can be seen that the d-axis shifted by an angle α_s in the proposed machine under the open-circuit condition, as shown in Figure 3a, which confirms the MAS effect. The reluctance d and q-axes are not changed in the two machines, as shown in Figure 3b, which is mainly attributed to the design of the symmetrical flux barriers in the two rotor configurations. The flux density distributions of the two machines at the peak current load condition are given in Figure 3c. Due to dual excited armature windings and PMs, the two machines under the load condition have higher flux densities than at other operating conditions.

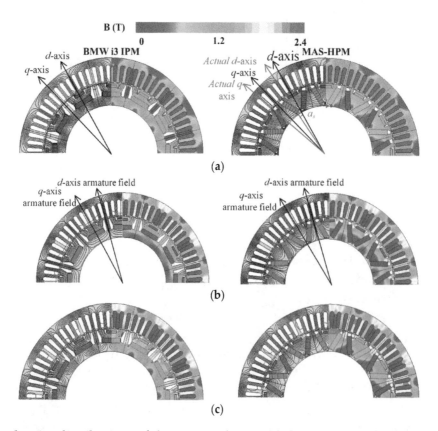

Figure 3. Flux density distributions of the two machines. (**a**) Open-circuit. (**b**) Reluctance axis with only armature windings excited. (**c**) Peak current load condition.

To clearly understand the MAS effect, the open-circuit air-gap flux density waveforms are given in Figure 4. Compared with the d-axis in the BMW i3 IPM machine, the displacement of the actual d-axis occurred in the proposed topology, which means that the magnet and reluctance axes grow closer by using the HPM configuration. Consequently, the resultant current angles for optimizing the reluctance and PM torques are closer, which enables the torque improvement. Moreover, the fundamental amplitude of the air-gap flux density in the MAS-HPM machine are found to be 53.70% higher than that of the BMW i3 IPM machine, as reflected in Figure 4b. Due to the asymmetrical PM configuration, larger high-order harmonics of the air-gap flux density are observed in the MAS-HPM machine.

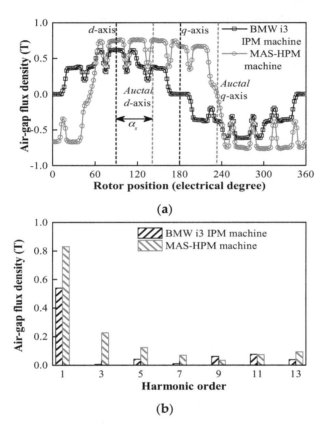

(a)

(b)

Figure 4. Open-circuit air-gap flux density. (**a**) Waveforms. (**b**) Spectra.

3. Electromagnetic Performance Comparison

In order to validate the MAS effect, the basic electromagnetic characteristics of the proposed MAS-HPM machine are comparatively studied with those of the BMW i3 IPM machine in this section. In order to reduce the computational time, 1:12 scale models are adopted for the two machines. The simulation time is 2.5 h.

3.1. Open-Circuit Performance

The back EMF waveforms of the two investigated machines are shown in Figure 5. Compared with the BMW i3 IPM machine, the proposed configuration exhibits a 53.54% higher back-EMF fundamental amplitude, which indicates that the magnet torque can be effectively improved by using the HPM configuration. In addition, the cogging torque waveforms of the two machines are shown in Figure 6, which experience the same periods due to the same numbers of stator slots and rotor poles. Because the air-gap flux density contains larger high-order harmonics, as shown in Figure 3b, the MAS-HPM structure has a higher cogging torque amplitude. The ratios of the cogging torque amplitudes to the corresponding peak torque values in BMW i3 IPM and MAS-HPM machines are 0.73% and 2.04%, respectively, which are lower than the acceptable value of 2.5%.

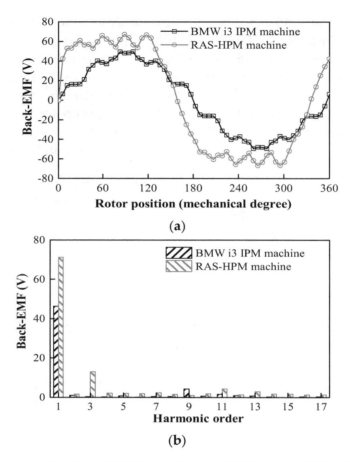

Figure 5. Back electro-motive force (EMF) waveforms at 3000 rpm. (**a**) Waveforms. (**b**) Spectra.

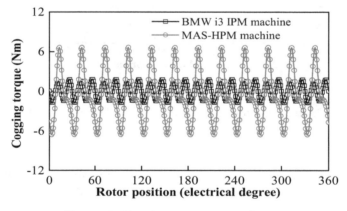

Figure 6. Cogging torque waveforms.

3.2. Torque Characteristics

The torque versus current angle characteristics of the two machines are illustrated in Figure 7. The PM and reluctance torques are separated by using the frozen permeability method [25]. It can be seen that the γ_s of the proposed MAS-HPM machine is smaller than that of the BMW i3 machine. As a result, a higher torque capability can be obtained in the HPM case, as evidenced in Figure 8. Moreover, due to the MAS effect, the ripple patterns of the PM and reluctance torques of the proposed machine are different, which results in a torque ripple offset effect. Hence the HPM configuration exhibits 55.99% lower torque ripple than the BMW i3 IPM machine, as shown in Figure 8b. The average torques versus phase current curves of the two machines are shown in Figure 9. It can be observed that the MAS-HPM machine has a higher torque capability regardless of the applied loads. As a whole, the feasibility of the proposed MAS-HPM design for torque performance improvement is clearly confirmed.

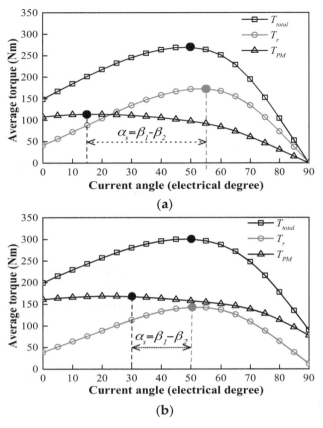

Figure 7. Torque versus current angle characteristics. (**a**) BMW i3 IPM machine. (**b**) MAS-HPM machine.

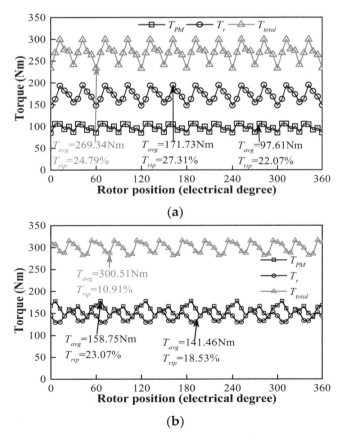

Figure 8. Steady torque. (**a**) BMW i3 IPM machine. (**b**) MAS-HPM machine.

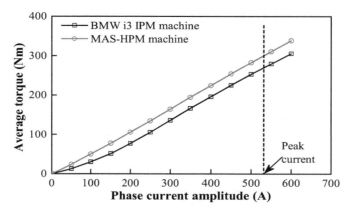

Figure 9. Average torque versus phase current curves.

3.3. Torque/Power versus Speed Curves

The torque and power versus speed curves of the two machines are illustrated in Figure 10. It can be seen that the MAS-HPM machine exhibits higher torque and power than the BMW i3 IPM machine over the whole operating range, consequently achieving a better high-speed constant power-maintaining capability.

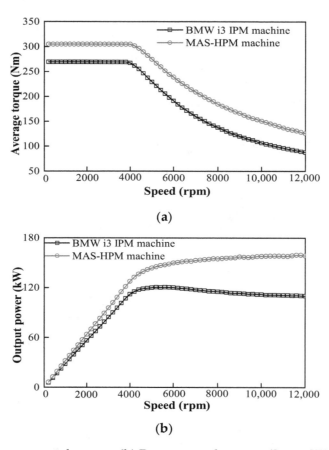

Figure 10. (**a**) Torque–speed curves. (**b**) Power–speed curves. (I_{rms} = 375 A, U_{dc} = 360 V).

3.4. Irreversible Demagnetization

The flux density distributions of the magnets are illustrated in Figure 11. When the working temperature is set as 100 °C, the knee points of ferrite and NdFeB magnets are −0.15 and −0.6 T, respectively. It can be observed that the irreversible demagnetization of ferrite and NdFeB magnets does not occur. In order to quantitatively illustrate the flux density variations of magnets, five typical points are selected in three magnets, as shown in Figure 11. The corresponding flux density variations of the typical five points on magnets are given in Figure 12. It can be seen that the working points of ferrite and NdFeB magnets are greater than the respective knee points, which indicates that good demagnetization withstanding capability can be achieved.

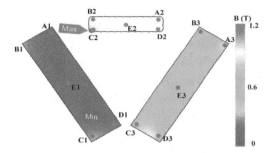

Figure 11. Flux density distributions of ferrite and NdFeB magnets.

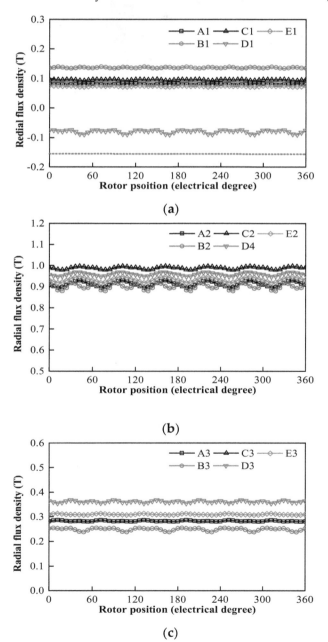

Figure 12. Variations of the working points of the five points on the ferrite and NdFeB magnets. (**a**) Ferrite magnet. (**b**) First-layer NdFeB magnet. (**c**) Second-layer NdFeB magnet.

3.5. Rotor Mechanical Analyses

The rotor mechanical strengths of the two machines are investigated at the maximum speed of 12,000 rpm in this section. The von Mises stress maps are shown in Figure 13. It can be observed that the peak stress of the MAS-HPM machine (268.8 MPa) is slightly lower than that of the BMW i3 IPM machine (282.4 MPa), which are both lower than the threshold yield value (396 MPa). Due to the differences in mesh subdivision, it can be observed that the mismatch between the maximal values occurs at the two sides of the symmetrical configurations. However, the difference in stress values between the points of the symmetrical structure is very small and thus negligible, as shown in Figure 13. As a result, it was confirmed that the proposed rotor configuration can withstand a larger centrifugal force at the maximum speed of 12,000 rpm.

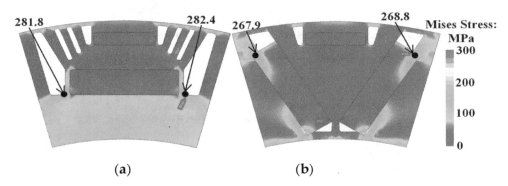

Figure 13. Rotor von Mises stress distributions at the maximum speed (12,000 rpm). (**a**) BMW i3 IPM machine. (**b**) MAS-HPM machine.

3.6. Loss and Efficiency

The two machines have the same stators and windings, which indicates that the same copper losses can be achieved. The iron loss and efficiency are calculated and illustrated in Figures 14 and 15, respectively. The iron loss p_i, eddy-current loss p_e and hysteresis loss p_h in the laminated core are calculated as follows [26]:

$$p_i = p_h + p_e, \tag{7}$$

$$p_e = \sum_n \left\{ \int_{iron} K_e D(nf)^2 (B_{r,n}^2 + B_{\theta,n}^2) dv \right\}, \tag{8}$$

$$p_h = \sum_n \left\{ \int_{iron} K_h D(nf)(B_{r,n}^2 + B_{\theta,n}^2) dv \right\}, \tag{9}$$

where K_e and K_h are the experimental constants obtained by the Epstein frame test of the core material, D is the density of the steel sheets, f is the fundamental frequency, $B_{r,n}$ and $B_{\theta,n}$ are the radial and tangential components of the flux density at each finite element.

The copper loss p_{cu} and efficiency η can be calculated by:

$$p_{cu} = 3R_a I^2 \tag{10}$$

$$\eta = \frac{\omega T}{\omega T + p_i + p_{cu}} \times 100\%, \tag{11}$$

where R_a, I, and ω are the armature winding resistance, the phase current and the mechanical angular velocity, respectively.

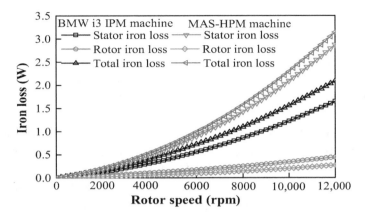

Figure 14. Iron losses versus speed curves.

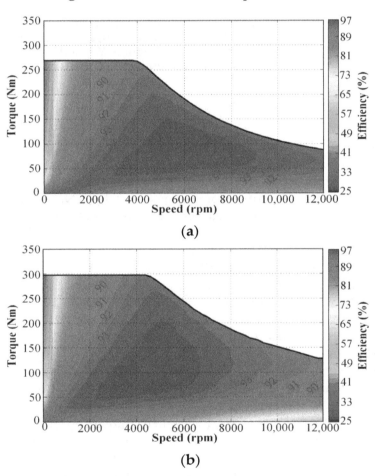

Figure 15. Efficiency maps. (**a**) BMW i3 IPM machine. (**b**) MAS-HPM machine.

The iron losses of the two machines under different speeds are given in Figure 14. It can be observed that the stator iron losses dominate the total iron losses in both machines at the rated load. The iron losses of the two machines are very close when the speed is lower than 4000 rpm. However, due to higher harmonics, the HPM structure produces a larger iron loss than the BMW i3 IPM machine when the speed exceeds 4000 rpm. Furthermore, the efficiency maps of the two cases are illustrated in Figure 15. The maximum efficiency of the proposed MAS-HPM machine (95.79%) is slightly higher than that of the BMW i3 IPM (95.57%). Due to the higher iron loss in high speed range, the proposed structure shows a relatively lower efficiency when the speed exceeds 10,000 rpm. Nevertheless, the MAS-HPM machine still exhibits a similar operating range when the efficiency is higher than 93%.

4. Conclusions

A novel MAS-HPM machine is proposed to achieve a higher torque capability and a wider high-efficiency operation range for EV applications in this paper. The basic electromagnetic characteristics of the proposed MAS-HPM machine and the benchmark BMW i3 IPM machine are comprehensively investigated and compared by FEA. Due to the MAS effect, the difference between the optimal current angles of maximizing the magnet and reluctance torques is reduced. In addition, it was found that the back-EMF and total torque of the proposed MAS-HPM machine can be effectively improved, compared with the conventional BMW i3 IPM machine. Moreover, the proposed machine shows lower peak mechanical stress, better field-weakening capability, higher peak efficiency and comparable high-efficiency operating range, which confirms the effectiveness of the proposed MAS design concept for performance improvement. However, due to higher harmonics, the proposed MAS-HPM configuration has higher cogging torque and iron losses in a high speed operating range.

Author Contributions: Conceptualization, Y.L. and H.Y.; methodology, Y.L. and H.Y.; software, Y.L. and W.W.; validation, Y.L., H.Y. and H.L., formal analysis, Y.L. and H.Y.; investigation, Y.L. and H.Y.; resources, Y.L. and H.Y.; data curation, Y.L. and H.Y.; writing—original draft preparation, Y.L. and H.Y.; writing—review and editing, Y.L., H.Y., H.L. and S.F.; visualization, Y.L. and H.Y.; supervision, H.Y., H.L., and S.F.; project administration, H.Y.; funding acquisition, H.Y. and H.L.

References

1. Zhu, Z.Q.; Howe, D. Electrical Machines and Drives for Electric, Hybrid, and Fuel Cell Vehicles. *Proc. IEEE* **2007**, *95*, 745–765. [CrossRef]
2. Burress, T.; Rogers, S.A.; Ozpineci, B. *FY 2016 Aannual Progress Report for Electric Drive Technologies Program*; Oak Ridge National Laboratory Department of Energy: Oak Ridge, TN, USA, 2016; pp. 196–207. Available online: https://www.energy.gov/sites/prod/files/2017/08/f36/FY16%20EDT%20Annual%20Report_FINAL (accessed on 5 February 2019).
3. Chalmers, B.J.; Musaba, L.; Gosden, D.F. Variable-Frequency Synchronous Motor Drives for Electric Vehicles. *IEEE Trans. Ind. Appl.* **1996**, *32*, 896–903. [CrossRef]
4. Chalmers, B.J.; Akmese, R.; Musaba, L. Design and Field-weakening Pperformance of Permanent Magnet/Reluctance Motor with Two-part Rotor. *IEE Proc.-Electr. Power Appl.* **1998**, *145*, 133–139. [CrossRef]
5. Chen, X.; Gu, C.; He, X.; Shao, H. Experimental Research on the ALA+SPM Hybrid Rotor Machine. In Proceedings of the International Conference on Electrical Machines and Systems, Beijing, China, 20–23 August 2011; pp. 1–4.
6. Yang, H.; Li, Y.; Lin, H.; Zhu, Z.Q.; Lyu, S.; Wang, H.; Fang, S.; Huang, Y. Novel Reluctance Axis Shifted Machines with Hybrid Rotors. In Proceedings of the Energy Conversion Congress and Exposition (ECCE), Cincinnati, OH, USA, 1–5 October 2017; pp. 2362–2367.
7. Liu, G.; Xu, G.; Zhao, W.; Du, X.; Chen, Q. Improvement of Torque Capability of Permanent-Magnet Motor by Using Hybrid Rotor Configuration. *IEEE Trans. Energy Convers.* **2017**, *32*, 953–962. [CrossRef]
8. Li, Y.; Bobba, D.; Sarlioglu, B. Design and Optimization of a Novel Dual-Rotor Hybrid PM Machine for Traction Application. *IEEE Trans. Ind. Electron.* **2018**, *65*, 1762–1771. [CrossRef]
9. Zhao, W.; Chen, D.; Lipo, T.A.; Kwon, B. Performance Improvement of Ferrite-Assisted Synchronous Reluctance Machines Using Asymmetrical Rotor Configurations. *IEEE Trans. Magn.* **2015**, *51*, 8108504. [CrossRef]
10. Randi, S.A.; Astier, S. Parameters of Salient Pole Synchronous Motor Drives with Two-Part Rotor to Achieve a Given Constant Power Speed Range. In Proceedings of the IEEE 32nd Annual Power Electronics Specialists Conference, Vancouver, BC, Canada, 17–21 June 2001; pp. 1673–1678.
11. Winzer, P.; Doppelbauer, M. Theoretical Analysis of Synchronous Machines with Displaced Reluctance Axis. In Proceedings of the International Conference on Electrical Machines (ICEM), Berlin, Germany, 2–5 September 2014; pp. 641–647.
12. Winzer, P.; Doppelbauer, M. Comparison of Synchronous Machine Designs with Displaced Reluctance Axis Considering Losses and Iron Saturation. In Proceedings of the IEEE International Electric Machines & Drives Conference (IEMDC), Coeur d'Alene, ID, USA, 10–13 May 2015; pp. 1801–1807.

13. Alsawalhi, J.Y.; Sudhoff, S.D. Effects of Positioning of Permanent Magnet Axis Relative to Reluctance Axis in Permanent Magnet Synchronous Machines. In Proceedings of the IEEE Power and Energy Conference at Illinois (PECI), Champaign, IL, USA, 20–21 February 2015; pp. 1–8.

14. Alsawalhi, J.Y.; Sudhoff, S.D. Design Optimization of Asymmetric Salient Permanent Magnet Synchronous Machines. *IEEE Trans. Energy Convers.* **2016**, *31*, 1315–1324. [CrossRef]

15. Zhu, X.; Zhang, X.W.C.; Wang, L.; Wu, W. Design and analysis of a spoke-type hybrid permanent magnet motor for electric vehicles. *IEEE Trans. Magn.* **2017**, *53*, 8208604. [CrossRef]

16. Jeong, C.L.; Hur, J. Design technique for PMSM with hybrid type permanent magnet. In Proceedings of the IEEE International Electric Machines and Drives Conference (IEMDC), Miami, FL, USA, 21–24 May 2017; pp. 1–6.

17. Li, G.J.; Zhu, Z.Q. Hybrid excited switched flux permanent magnet machines with hybrid magnets. In Proceedings of the 8th International Conference on Power Electronics, Machines and Drives (PEMD 2016), Glasgow, UK, 19–21 April 2016; pp. 1–6.

18. Yang, H.; Lin, H.; Zhu, Z.Q.; Wang, D.; Fang, S.; Huang, Y. A variable-flux hybrid-PM switched-flux memory machine for EV/HEV applications. *IEEE Trans. Ind. Appl.* **2016**, *52*, 2203–2214. [CrossRef]

19. Yang, H.; Zhu, Z.Q.; Lin, H.; Zhan, H.L.; Hua, H.; Zhuang, E.; Fang, S.; Huang, Y. Hybrid-excited switched-flux hybrid magnet memory machines. *IEEE Trans. Magn.* **2016**, *52*, 8202215. [CrossRef]

20. Yang, H.; Zhu, Z.Q.; Lin, H.; Wu, D.; Hua, H.; Fang, S.; Huang, Y. Novel high-performance switched flux hybrid magnet memory machines with reduced rare-earth magnets. *IEEE Trans. Ind. Appl.* **2016**, *52*, 3901–3915. [CrossRef]

21. Yang, H.; Zhu, Z.Q.; Lin, H.; Fang, S.; Huang, Y. Synthesis of hybrid magnet memory machines having separate stators for traction applications. *IEEE Trans. Veh. Technol.* **2018**, *67*, 183–195. [CrossRef]

22. Yang, H.; Lin, H.; Zhu, Z.Q.; Fang, S.; Huang, Y. A dual-consequent-pole Vernier memory machine. *Energies* **2016**, *9*, 134. [CrossRef]

23. Jahns, T.M.; Kliman, G.B.; Neumann, T.W. Interior permanent-magnet synchronous motors for adjustable-speed drives. *IEEE Trans. Ind. Appl.* **1986**, *IA-22*, 738–747. [CrossRef]

24. Gieras, J.F. *Permanent Magnet Motor Technology: Design and Application*, 3rd ed.; CRC Press: Boca Raton, FL, USA, 2009.

25. Chu, W.Q.; Zhu, Z.Q. Average torque separation in permanent magnet synchronous machines using frozen permeability. *IEEE Trans. Magn.* **2013**, *49*, 1202–1210. [CrossRef]

26. Yamazaki, K.; Abe, A. Loss investigated of interior permanent-magnet motors considering carrier harmonics and magnet eddy currents. *IEEE Trans. Ind. Appl.* **2009**, *45*, 659–665. [CrossRef]

Mitigation Method of Slot Harmonic Cogging Torque Considering Unevenly Magnetized Permanent Magnets in PMSM

Chaelim Jeong [1], Dongho Lee [2] and Jin Hur [3],*

[1]   Department of Industrial Engineering, University of Padova, 35131 Padova, Italy; cofla827@gmail.com
[2]   Sungshin Precision Global (SPG) Co., Ltd., Incheon 21633, Korea; leedh38126@gmail.com
[3]   Department of Electrical Engineering, Incheon National University, Incheon 22012, Korea
*   Correspondence: jinhur@inu.ac.kr.

Abstract: This paper presents a mitigation method of slot harmonic cogging torque considering unevenly magnetized magnets in a permanent magnet synchronous motor. In previous studies, it has been confirmed that non-uniformly magnetized permanent magnets cause an unexpected increase of cogging torque because of additional slot harmonic components. However, these studies did not offer a countermeasure against it. First, in this study, the relationship between the residual magnetic flux density of the permanent magnet and the cogging torque is derived from the basic form of the Maxwell stress tensor equation. Second, the principle of the slot harmonic cogging torque generation is explained qualitatively, and the mitigation method of the slot harmonic component is proposed. Finally, the proposed method is verified with the finite element analysis and experimental results.

Keywords: cogging torque; permanent magnet machine; torque ripple; uneven magnets

1. Introduction

The cogging torque is one of the most representative components of torque ripple in the permanent magnet synchronous motor (PMSM). Therefore, studies on the reduction method for the cogging torque have been actively carried out to minimize the torque ripple [1–11]. Those studies on cogging torques have been mainly focused on reducing the cogging effect by modulating the combination of the pole and slot number, the pole arc, the shape of the core, the skew angle, the notching, etc. In general, the results of such studies are based on a simple theoretical analysis, so it is assumed that the magnetic components of the motor are ideal. However, since there are many possible manufacturing errors such as eccentricity, machining error, and unevenly magnetized magnets, motors always contain non-ideal components. As a result of those errors, the measured cogging torque of the actual motor may be very different from what is expected in the simulation [12]. This phenomenon can be a critical issue to those applications that need precision control of the motor and are sensitive to noise and vibration.

For this reason, several studies that take manufacturing errors into consideration have emerged [13–19]. In [13–16], analytical solutions of cogging torque are studied by considering the magnet imperfections, rotor eccentricity, geometrical variation, and magnetizing fixture. In addition, [17] mathematically investigated the cogging torque caused by the simultaneous existence of eccentricities and the uneven magnetization. Those studies have focused on analysis methods of cogging torque by considering manufacturing errors, and they reported that those errors generate additional harmonic components. In [18], they show that the unevenly magnetized permanent magnet (PM) can have a negative impact on applying the cogging torque reduction method (teeth curvature modulation method), leading to additional slot harmonic cogging torque. In [19], it is confirmed that the main contributors, which have the greatest effect on the cogging torque distortion, are the inner radius tolerance of the stator and the tolerance of PM remanence (unevenly magnetized magnets), among many other manufacturing errors. In addition to the aforementioned studies, there are a few studies that have analyzed motor performance in consideration of manufacturing errors, but those studies only handle the phenomena analysis caused by them, and there is a lack of research on the mitigation countermeasures.

Unlike most studies that have only analyzed the effects of manufacturing errors on cogging torque, this paper proposes a method to counteract the influence of unevenly magnetized magnets, which are one of the main contributors of cogging torque distortion [19]. Here, unevenly magnetized magnets mean that each magnet has different magnetic strength. This study is carried out in the following order. First, the relationship between the remanence of each PM and the cogging torque is derived from the basic form of the Maxwell stress tensor equation. Second, the principle of slot harmonic cogging torque generation is explained qualitatively. Based on this principle, a new mitigation method of slot harmonic cogging torque is proposed. This method involves a series of processes that select the position of the PMs, taking into account the remanence deviations of each PM. Finally, the proposed method is verified with a finite element analysis (FEA) and experimental results. Here, note that this study assumed that each magnet is pre-magnetized before the assembly. Therefore, the proposed method is more appropriate for small quantity customized production than mass production. Moreover, based on the principle of the slot harmonic component mitigation condition, it is possible to adjust that the manufacturing tolerance of the magnetization yoke, leading to the alleviation of the influence of the uneven magnetization.

2. Analysis of Cogging Torque in PMSM from a Macroscopic Perspective

Before examining the process of the slot harmonic cogging torque generation caused by the unevenness in magnetic strengths of each pole, we first analyzed the generation of cogging torque from a macroscopic perspective.

2.1. The Relation between the Electromagnetic Force and the Remanence of Magnet

The cogging torque refers to a torque caused by an electromagnetic force generated when the PM's magnetic flux passing through the air gap between a rotor and a stator is concentrated in a path that has a relatively small magnetic reluctance. Therefore, in order to analyze the magnitude of the cogging torque, it is necessary to understand the electromagnetic force. In many studies, according to Maxwell stress tensor theory, the electromagnetic force of the tangential component is defined as follows in a single rotor position [13,20–22]:

$$F_t = \frac{R}{\mu_0} \int_0^{2\pi} B_{rgap}(\phi_r)B_{tgap}(\phi_r)d\phi_r \tag{1}$$

where F_t is the tangential force density, μ_0 is the permeability of free space, R is the radius for which the Maxwell stress tensor is calculated, Φ_r is the space angle at single rotor position, and B_{rgap} and B_{tgap} represent radial and tangential components of magnetic flux density, respectively. Here, if the saturation phenomenon of the magnetic material is ignored, B_{rgap} and B_{tgap} are always proportional to the remanence of the PM (B_r), and the following relationship holds:

$$F_t \propto B_r^2 \tag{2}$$

2.2. The Cogging Torque Caused by Single Pole

The cogging torque according to the electromagnetic force described above can be defined as follows:

$$T_{cog}(\theta,l) = R \int_0^{L_{stk}} F_t(\theta,l)dl \tag{3}$$

where θ is rotation angle of the rotor, L_{stk} is stack length, and l is axial length. Here, if the electromagnetic force is uniformly distributed in the stacking direction, the above equation can be simplified as the following equation:

$$T_{cog}(\theta) = RL_{stk}F_t(\theta) \tag{4}$$

To understand the period of cogging torque and the interaction of harmonic components, we first assumed an example model with one magnet (pole), as in Figure 1. Here, the pole-arc of the PM is 24° and the stator has 12 teeth. In this case, since the influences of eccentricity and shape error are deviated from the subject of this study, the cogging torque can be expressed in the following form of the Fourier series [23]:

$$T_{cog}(\theta) = RL_{stk}\sum_{k=1}^{\infty} F_{tk}\sin(kS\theta) = RL_{stk}\sum_{k=1}^{\infty} F_{tk}\sin(k\theta_s) \tag{5}$$

where θ_s is a slot periodic angle that is calculated by multiplying θ (mechanical angle) with the slot number, F_{tk} is Fourier coefficient (amplitude of force) of the 'k'th harmonic component, and S is the number of the slots.

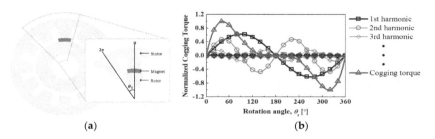

(a)   (b)

Figure 1. (a) The geometry of the example model with one pole and twelve teeth; (b) cogging torque harmonic component due to one pole rotation (θ_s).

The cogging torque result of the example model of Figure 1a is shown in Figure 1b with the harmonic component. Here, the magnitude of each harmonic component is normalized to the peak value of the total cogging torque. Since the first harmonic component is the largest, the total cogging torque has the same period as the first harmonic component.

2.3. The Cogging Torque Caused by Multi Poles

Now suppose that we add some magnets to the example model. Figure 2a–c have 8 poles, 10 poles, and 14 poles, respectively. Each added pole has the same remanence, pole-arc, and thickness. In this case, the cogging torque generated in each adjacent pole has a phase difference by a pole pitch. Further, due to the phase difference, the cogging torques generated by each pole interfere with each other. Using the property from Equation (5), the cogging torque caused by multi-poles can be expressed as follows using superposition technique [24,25].

$$\begin{aligned}
T_{cog}(\theta_s) &= T_{p1} + T_{p2} + T_{p3} + \cdots + T_{pP} \\
&= RL_{stk}\Bigg\{ \sum_{k=1}^{\infty} F_{t1_k}\sin(k\theta_s) + \sum_{k=1}^{\infty} F_{t2_k}\sin\left[k\left(\theta_s + (n_2-1)\frac{2\pi S}{P}\right)\right] \\
&+ \sum_{k=1}^{\infty} F_{t3_k}\sin\left[k\left(\theta_s + (n_3-1)\frac{2\pi S}{P}\right)\right] + \cdots + \sum_{k=1}^{\infty} F_{tP_k}\sin\left[k\left(\theta_s + (n_P-1)\frac{2\pi S}{P}\right)\right] \Bigg\}
\end{aligned} \tag{6}$$

where T_{pP} is the cogging torque that is generated by each pole, F_{tPk} is the 'k'th harmonic component of the tangential magnetic force density generated by each pole, $n_{2\ldots P}$ are the order of the poles, P is the number of the poles, and S is the number of the slots. Here, the rotation angle of the rotor is expressed in the slot periodic angle.

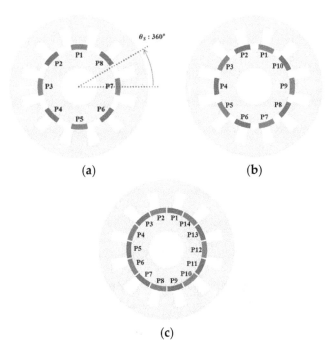

(a) (b)

(c)

Figure 2. The geometry of the example model with multi poles and twelve teeth for (**a**) 8p/12s, (**b**) 10p/12s, and (**c**) 14p/12s.

Figure 3 shows the cogging torque of each example model case that resulted in the mutual interference of cogging torque for each pole. As can be seen from the figure, depending on the phase difference of each pole, the cogging torque can be increased or decreased by overlapping. Also, the harmonic component of cogging torque is demonstrated in Figure 4. Here, if those harmonic orders are calculated based on the mechanical angle (θ), the most dominant harmonic component has the order of least common multiple (LCM) of the number of slots and the number of poles. This is because the fundamental frequency of the cogging torque is calculated by using the LCM [26]. For example, since the LCM of an 8-pole/12-slot motor is 24, the 24th harmonic becomes as the most dominant harmonic component, as shown in Figure 4. This is a well-known fact of the cogging torque period and has been confirmed once again through this analysis.

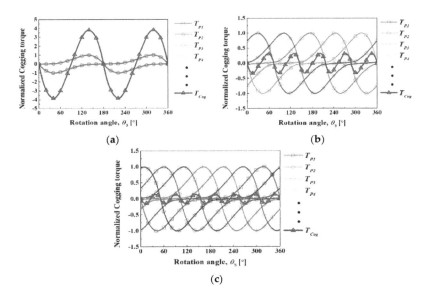

Figure 3. Cogging torque caused by multi poles rotation based on θ_S: (**a**) cogging torque of 8p/12s, (**b**) cogging torque of 10p/12s, and (**c**) cogging torque of 14p/12s.

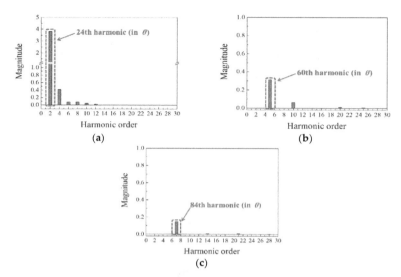

Figure 4. Harmonic spectra of cogging torque caused by multi poles rotation based on θ_s: (**a**) harmonics of 8p/12s, (**b**) harmonics of 10p/12s, and (**c**) harmonics of 14p/12s.

3. Mitigation Method of Slot Harmonic Cogging Torque Component

As the name implies, the slot harmonic component refers to a harmonic component that has the same number of cycles as the number of slots when the rotor rotates 360° (based on θ). Hence, in Equation (6), the first order harmonic component can be defined as the slot harmonic component (because of $\theta_s = 12\theta$). Then, the slot harmonic cogging torque can be expressed as follows:

$$
\begin{aligned}
T_{slot}(\theta_s) = RL_{stk}\Big\{ & F_{t1_1}\sin(\theta_s) + F_{t2_1}\sin\Big[\theta_s + (n_2 - 1)\tfrac{2\pi S}{P}\Big] \\
& + F_{t3_1}\sin\Big[\theta_s + (n_3 - 1)\tfrac{2\pi S}{P}\Big] + \cdots + F_{tP_1}\sin\Big[\theta_s + (n_P - 1)\tfrac{2\pi S}{P}\Big]\Big\}
\end{aligned}
\tag{7}
$$

As can be seen from the result of Figure 4, when the motor is under the ideal condition and all the magnetic forces of each pole are equal to each other, the slot harmonic component does not exist in all cases. This is because the first harmonic component of each pole is canceled out by the phase difference from the other pole. This is easy to understand the phenomenon with the 8-pole/12-slot and 10-pole/12-slot models.

Substituting the number of poles of 8 and the number of slots of 12 into Equation (7), it is summarized as follows:

$$
\begin{aligned}
T_{slot}(\theta_s) = RL_{stk}\Big\{ & F_{t1_1}\sin(\theta_s) + F_{t2_1}\sin\Big[\theta_s + (2 - 1)\tfrac{2\pi\cdot12}{8}\Big] \\
& + F_{t3_1}\sin\Big[\theta_s + (3 - 1)\tfrac{2\pi\cdot12}{8}\Big] + \cdots + F_{t8_1}\sin\Big[\theta_s + (8 - 1)\tfrac{2\pi\cdot12}{8}\Big]\Big\} \\
= RL_{stk}\Big[& \big(F_{t1_1} + F_{t3_1} + F_{t5_1} + F_{t7_1}\big)\sin(\theta_S) + \big(F_{t2_1} + F_{t4_1} + F_{t6_1} + F_{t8_1}\big)\sin(\theta_S + \pi)\Big].
\end{aligned}
\tag{8}
$$

As a result, in the motor of the 8-pole/12-slot, the cogging torque produced by each pole had two phases, and the phase difference was π. Therefore, under ideal conditions, the slot harmonic torque (first harmonic torque of each pole) becomes zero because each pole produces the same amount of magnetic force. Here, through the above equation, the removal condition of the slot harmonic can be more clearly expressed as follows:

$$
F_{t1_1} + F_{t3_1} + F_{t5_1} + F_{t7_1} = F_{t2_1} + F_{t4_1} + F_{t6_1} + F_{t8_1}
\tag{9}
$$

According to this condition, the slot harmonic component is likely to be canceled even if the density of each magnetic force does not exactly coincide. That is, even if the magnetic strength of each pole (the remanence of each magnet) is different, the slot harmonic component may be removed.

In this paper, since the influence of shape error and eccentricity is not considered, the above condition can be rearranged as follows using the relation of (2):

$$B_{r1}^2 + B_{r3}^2 + B_{r5}^2 + B_{r7}^2 = B_{r2}^2 + B_{r4}^2 + B_{r6}^2 + B_{r8}^2 \tag{10}$$

where $B_{r1 \ldots rP}$ are the remanence of the magnet in each pole. Under this condition, the slot harmonic component can be mitigated, and all harmonics except the LCM harmonic component will be also mitigated by superposition because the phase difference is equal to the pole pitch.

As in the example of the 8-pole/12-slot motor, by substituting the number of poles of 10 and the number of slots of 12 into Equation (7), it can be summarized as follows:

$$T_{slot}(\theta_s) = RL_{stk}[(F_{t1_1} + F_{t6_1})\sin(\theta_s) + (F_{t2_1} + F_{t7_1})\sin(\theta_s + \tfrac{12}{5}\pi) \\ + (F_{t3_1} + F_{t8_1})\sin(\theta_s + \tfrac{24}{5}\pi) + (F_{t4_1} + F_{t9_1})\sin(\theta_s + \tfrac{36}{5}\pi) \\ + (F_{t5_1} + F_{t10_1})\sin(\theta_s + \tfrac{48}{5}\pi)]. \tag{11}$$

Under ideal conditions, the slot harmonic torque (first harmonic torque of each pole) becomes zero because each pole produces the same amount of magnetic force. As a result, the removal condition of the slot harmonic can be expressed as follows:

$$F_{t1_1} + F_{t6_1} = F_{t2_1} + F_{t7_1} = F_{t3_1} + F_{t8_1} = F_{t4_1} + F_{t9_1} = F_{t5_1} + F_{t10_1} \tag{12}$$

$$B_{r1}^2 + B_{r6}^2 = B_{r2}^2 + B_{r7}^2 = B_{r3}^2 + B_{r8}^2 = B_{r4}^2 + B_{r9}^2 = B_{r5}^2 + B_{r10}^2. \tag{13}$$

Through the above examples, it is verified that the slot harmonic cogging torque can be zero, when the motor is under the ideal condition and all the magnetic forces of each pole are equal to each other regardless of the odd or even number of magnet set.

Looking at the process of deriving this condition, consequently, it is important to find poles with the same cogging torque phase so that the sum of the remanence of the magnets is equal to the poles with different phases. By using the equation below, the distance of pole (N), which has the same torque phase with the first (reference) pole, can be calculated in the pole number. This can be simply derived with the number of poles and the number of slots, and by taking into account the relationship between pole pitch and slot pitch.

$$N = \left| \frac{P}{(S - P)} \right| \tag{14}$$

According to Equation (14), N of the 14-pole/12-slot is assigned "7." Therefore, in the case of the 14-pole/12-slot, the torque phases of the first pole and the eighth pole are the same. Therefore, the slot harmonic torque removal condition for each pole/slot combination is derived as follows:

$$B_{r1}^2 + B_{r8}^2 = B_{r2}^2 + B_{r9}^2 = B_{r3}^2 + B_{r10}^2 = B_{r4}^2 + B_{r11}^2 \\ = B_{r5}^2 + B_{r12}^2 = B_{r6}^2 + B_{r13}^2 = B_{r7}^2 + B_{r14}^2. \tag{15}$$

In addition to the above condition, in order to make the magnetic flux connected to the winding uniform according to the polarity, the remanence summation of the entire magnets located at the N pole must be equal to that of the S pole. Therefore, the following condition should be also met:

$$B_N(B_{r1} + B_{r3} + B_{r5} + \cdots) = B_S(B_{r2} + B_{r4} + B_{r6} + \cdots) \tag{16}$$

where B_N and B_S are total remanences of the north and south poles.

Considering these conditions when assembling magnets and rotors, the generation of the slot harmonic cogging torque components will be minimized. Hence, the work flow chart of the slot harmonic component mitigation method is shown in Figure 5. Here, the sum of B_r^2 of the poles with the same cogging torque phase is conveniently referred to as Z. The smaller the difference in Z value for each torque phase is the smaller the slot harmonic size is. In addition, the tolerance is denoted as δ, and it is reasonable to choose this to be larger than the measurement uncertainty of the Gauss value of the magnet in practical.

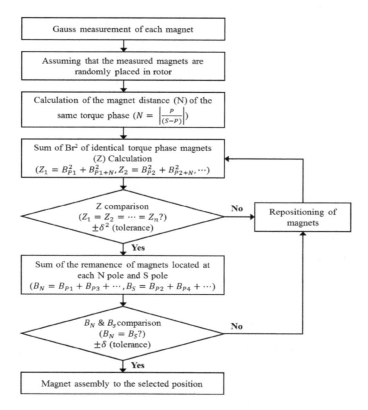

Figure 5. The workflow diagram of a method for mitigating slot harmonic cogging torque before rotor assembly.

4. Verification of the Proposed Method

4.1. Verification Using the Finite Element Analysis (FEA)

In this section, the FEA is conducted to verify the validity of the slot harmonic mitigation method proposed in Figure 5. Two example models are selected for this analysis. The first model is a 6-pole/9-slot interior permanent-magnet motor (IPM), and the other is an 8-pole/12-slot IPM. The geometry with the mesh information and the specification of each model are shown in Figure 6 and Table 1. Here, the saturation point of each core material was adjusted to be lower than the actual property. This is to confirm that the proposed method is still valid under nonlinear material properties.

In order to consider the unevenly magnetized magnet in FEA verification, firstly, the management tolerance on the B_r of the commercial magnets was investigated and is shown in Table 2. Then based on the data, the B_r of each magnet was randomly selected and positioned on the rotor, as shown in Figure 7. The selected B_r results and the magnet position are recorded in Tables 3 and 4. Here, in each table, Case A is each magnet arranged according to its number order, and Case B is where it is arranged according to the proposed method in Figure 5. The change in position of each magnet is shown more clearly in Figure 7. In Tables 3 and 4, it can be seen that the Z comparison and B_N-B_S comparison value are not 'zero,' even in Case B. In fact, since there is very low probability that there can be a magnet arrangement that satisfies this in reality, the tolerances were changed step by step to have the magnet array with the smallest comparison result.

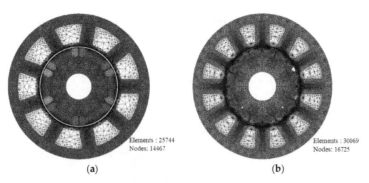

Figure 6. The geometry and mesh information of each FEA model for the verification of the proposed method at **(a)** 6p/9s IPM and **(b)** 8p/12s IPM.

Table 1. Specification of each FEA model for the verification of the proposed method.

Item	6p/9s IPM	8p/12s IPM
Stator outer diameter	100.0 mm	150.0 mm
Rotor outer diameter	54.0 mm	82.0 mm
Stack length	40.0 mm	72 mm
Air gap length	1.0 mm	0.6 mm
Rated power	400 W	5000 W
Rated speed	3500 rpm	2000 rpm
Rated torque	1.1 Nm	23.8 Nm
Rated ph. current	10.3 Arms	120 Arms
Series turn per phase	72	20
Core material	50PN470 (FEM: saturate@1.2T)	50PN470 (FEM: saturate@1.2T)
Magnet material	NMX-36EH	NEOREC 40UH

Table 2. The magnet management tolerance of the manufacturer.

Company	6p/9s IPM	8p/12s IPM
TDK	NEOREC 40UH	1290 ± 30
	NEOREC 40TH	1285 ± 30
	NEOREC 38UX	1250 ± 30
	NEOREC 35NX	1200 ± 30
Hitachi	NMX-43SH	1295 ± 35
	NMX-41SH	1275 ± 35
	NMX-39EH	1235 ± 35
	NMX-36EH	1195 ± 35

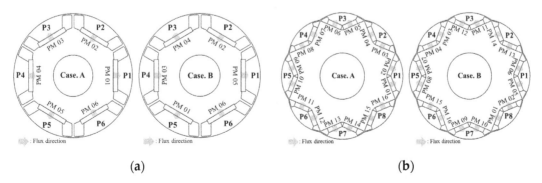

Figure 7. PM position change according to the proposed method for the FEA verification for **(a)** 6p/9s IPM and **(b)** 8p/12s IPM.

Table 3. Random selection of the magnet remanence for the 6p/9s model considering the management tolerance.

Position & Comparison	Case A		Case B	
	PM No.	B_r	PM No.	B_r
P1	PM 01	1.194 T	PM 05	1.226 T
P2	PM 02	1.195 T	PM 02	1.195 T
P3	PM 03	1.198 T	PM 04	1.162 T
P4	PM 04	1.162 T	PM 03	1.198 T
P5	PM 05	1.226 T	PM 01	1.194 T
P6	PM 06	1.172 T	PM 06	1.172 T
Z_1-Z_2	0.212 (T)2		0.042 (T)2	
B_N-B_S	0.089 (T)		0.017 (T)	

Table 4. Random selection of the magnet remanence of 8p/12s model considering the management tolerance.

Position & Comparison	Case A		Case B	
	PM No.	B_r	PM No.	B_r
P1	PM 01	1.265 T	PM 05	1.269 T
	PM 02	1.275 T	PM 06	1.278 T
P2	PM 03	1.290 T	PM 13	1.287 T
	PM 04	1.291 T	PM 14	1.285 T
P3	PM 05	1.269 T	PM 11	1.309 T
	PM 06	1.278 T	PM 12	1.309 T
P4	PM 07	1.294 T	PM 03	1.290 T
	PM 08	1.295 T	PM 04	1.291 T
P5	PM 09	1.278 T	PM 07	1.294 T
	PM 10	1.290 T	PM 08	1.295 T
P6	PM 11	1.309 T	PM 15	1.310 T
	PM 12	1.309 T	PM 16	1.315 T
P7	PM 13	1.287 T	PM 09	1.278 T
	PM 14	1.285 T	PM 10	1.290 T
P8	PM 15	1.310 T	PM 01	1.265 T
	PM 16	1.315 T	PM 02	1.275 T
Z_1-Z_2	−0.483 (T)2		0.009 (T)2	
B_N-B_S	−0.187 (T)		0.004 (T)	

Now, the validity of the proposed method can be verified by examining the variation of the cogging torque harmonic component in each case. The FEA results of the cogging torque harmonic component are shown in Figure 8. Figure 8a shows the result of the 6-pole/9-slot model, and Figure 8b shows the result of the 8-pole/12-slot model. In both models, it can be seen that the slot harmonic component of Case B is much smaller than Case A. Therefore, the proposed method was effective in mitigating the slot harmonic component of cogging torque. Furthermore, it can be seen that the permeability of the core is in a somewhat non-linear region by observing the flux density distribution in Figure 9. Hence, although we ignored the saturation when deriving the method, the result of Figure 8 proves that the proposed method is still effective under the non-linear material characteristics of the ferromagnetic.

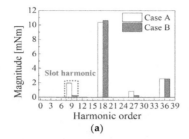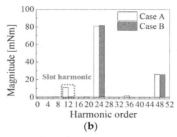

Figure 8. FEA results of cogging torque harmonic component according to each case: (**a**) 6p/9s IPM and (**b**) 8p/12s IPM.

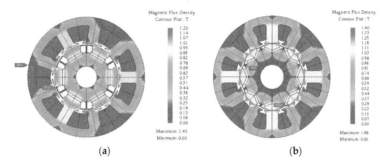

(a)   (b)

Figure 9. Contour plot of magnetic flux density under no load condition for (**a**) 6p/9s IPM and (**b**) 8p/12s IPM.

4.2. Verification with Experimentation

For the experimental verification, both models in Figure 6 were manufactured, one of each. Figure 10 is a picture of the produced motor. Then, the experiment process was performed as follows.

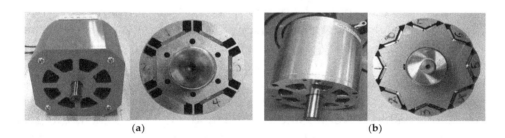

(a)   (b)

Figure 10. The manufactured motors for the experiment: (**a**) 6p/9s IPM and (**b**) 8p/12s IPM.

1. The surface Gauss value of each magnet was measured (with ATM 1000, SCMI) in the space, excluding the magnetic substance. Figure 11 shows a picture of the measurement, and the results are written in Tables 5 and 6. The Gauss average value was calculated from the seven measurement points per each magnet, and the measurement uncertainty was calculated by repeating the measurement five times.

2. The position of each magnet was set according to the proposed method. The results are shown in Tables 5 and 6 and in Figure 12 (Case B).

3. These magnets were alternately assembled to the rotor according to the case of each model shown in Figure 12, and the cogging torque according to each case was measured (with ATM-5KA, SUGAWARA) as shown in Figure 13.

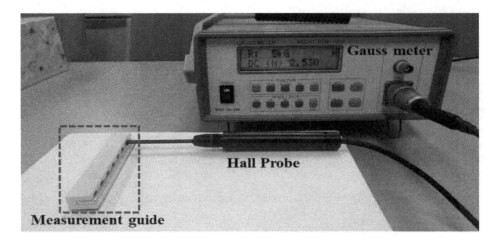

Figure 11. Gauss measurement of the magnet surface.

Table 5. Gauss measurement of each magnet surface of the 6p/9s model and the changes in magnet position according to the proposed method.

Position & Comparison	Case A		Case B	
	PM No.	Gauss Avg.	PM No.	Gauss Avg.
P1	PM 01	194.3 mT	PM 04	198.4 mT
P2	PM 02	198.3 mT	PM 02	198.3 mT
P3	PM 03	196.7 mT	PM 05	196.9 mT
P4	PM 04	198.4 mT	PM 06	199.1 mT
P5	PM 05	196.9 mT	PM 03	196.7 mT
P6	PM 06	199.1 mT	PM 01	194.3 mT
Uncertainty	$\pm 0.2\%$			
Z_1-Z_2	−3113.3 (mT)2		106.9 (mT)2	
B_N-B_S	−7.9 (mT)		0.3 (mT)	

The results of cogging torque measurements are demonstrated in Figure 14. Figure 14a shows the result of the 6-pole/9-slot motor. Case A had a cogging torque of 56.7 mNm$_{pk-pk}$, and Case B had 55.1 mNm$_{pk-pk}$. In Figure 14b the 8-pole/12-slot motor showed 227.3 mNm$_{pk-pk}$ for Case A and 214.8 mNm$_{pk-pk}$ for Case B. As a result, although the shapes of the stator and rotor of the analyzed motors were already optimized for reducing cogging torque, the cogging torque could be improved more by using the proposed method. The main cause of this cogging difference between Case A and B is due to the slot harmonic component of Case B being smaller than Case A, as can be seen in the FFT result of each cogging torque in Figure 15. Consequently, the validity of the proposed method was confirmed again by the experimental results. Overall, since this method only affects the position of each magnet before assembly, it can be compatible with the conventional cogging torque reduction methods using the teeth curvature and rotor shape modulation.

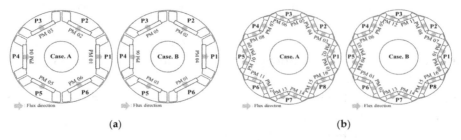

Figure 12. PM position change according to the proposed method for the experimental verification for (**a**) 6p/9s IPM and (**b**) 8p/12s IPM.

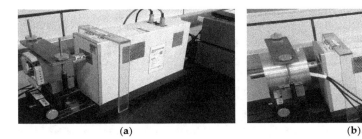

Figure 13. Cogging torque measurement of each motor for (**a**) 6p/9s IPM and (**b**) 8p/12s IPM.

Table 6. Gauss measurement of each magnet surface of the 8p/12s model and the changes in magnet position according to the proposed method.

Position & Comparison	Case A		Case B	
	PM No.	Gauss Avg.	PM No.	Gauss Avg.
P1	PM 01	227.1 mT	PM 09	225.5 mT
	PM 02	228.7 mT	PM 10	226.2 mT
P2	PM 03	233.8 mT	PM 07	231.1 mT
	PM 04	234.0 mT	PM 08	231.5 mT
P3	PM 05	226.3 mT	PM 11	230.6 mT
	PM 06	226.8 mT	PM 12	230.9 mT
P4	PM 07	231.2 mT	PM 05	226.3 mT
	PM 08	231.5 mT	PM 06	226.8 mT
P5	PM 09	225.5 mT	PM 03	233.9 mT
	PM 10	226.2 mT	PM 04	234.0 mT
P6	PM 11	230.6 mT	PM 01	227.1 mT
	PM 12	230.9 mT	PM 02	228.7 mT
P7	PM 13	229.0 mT	PM 13	229.0 mT
	PM 14	229.9 mT	PM 14	229.9 mT
P8	PM 15	234.2 mT	PM 15	234.2 mT
	PM 16	234.3 mT	PM 16	234.2 mT
Uncertainty	$\pm0.1\%$			
Z_1-Z_2	$-10{,}065.1$ (mT)2		133.9 (mT)2	
B_N-B_S	-21.9 (mT)		0.3 (mT)	

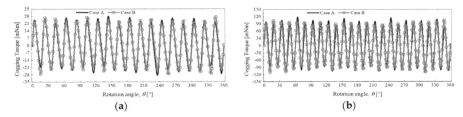

Figure 14. The measured cogging torque of each motor for (**a**) 6p/9s IPM and (**b**) 8p/12s IPM.

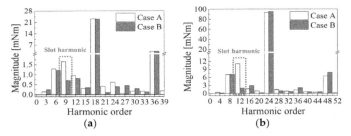

Figure 15. Cogging torque measurement of each motor for (**a**) 6p/9s IPM and (**b**) 8p/12s IPM.

5. Discussion

The reduction effect cannot be clearly seen in the peak-peak comparison of cogging torque in Figure 14. This is because the LCM component is much larger than the slot harmonic component in both cases. In this case, although the slot harmonic component was reduced, as shown in Figure 15, by the proposed method, the effect is not seen much. If the proposed method is applied to a model that is sensitive to the slot harmonic component, the cogging torque can be effectively mitigated, compared with the results of this paper. In other words, the proposed method has a different effect on the mitigation of cogging torque depending on which harmonic component is dominant.

Additionally, since there are some methods to measure the B_r or flux density of PM, the real application for applying the proposed method can be manufactured. Among the measurement methods, the simplest example is using Helmholtz coil. As mentioned in the introduction, the proposed method is more appropriate for small quantity customized production than mass production because the Gauss value of each magnet should be measured before the assembly. In the case of mass production, it is possible that if the manufacturing tolerance of the magnetization yoke is adjusted based on the principle of the slot harmonic component mitigation condition, that the influence of the uneven magnetization can be alleviated.

As described above, there are some limitations to the proposed method. However, it is meaningful that we have dealt with the method to compensate manufacturing tolerance (Uneven PM) that has not been covered in the meantime. Furthermore, this method can prevent an increase in cogging torque caused by unevenly magnetized PMs of motors with a high number of poles. Since small scale customized manufacturing process, which adopts the method of the pre-magnetization of magnets before assembly, cannot adjust and compensate for the unevenness of the PMs, by using the proposed method, it will be possible to ensure the cogging performance of a manufactured motor.

6. Conclusions

In this study, a mitigation method of slot harmonic cogging torque caused by the unevenly magnetized magnet was proposed. This method was drawn through the qualitative analysis of the cogging torque from a macroscopic perspective. As shown in Figure 5, the main process of this method is arranging each magnet according to the non-slot harmonic condition described in Section 3. The validity of the proposed method was verified using FEA and experimentation. Here, the verification was performed by comparing the harmonic components of the cogging torque with and without the proposed method. In this process, it was confirmed that this method is sufficiently effective, even when considering the non-linear material characteristics of the ferromagnetic.

Author Contributions: Conceptualization, C.J. and D.L.; methodology, C.J.; software, D.L.; validation, C.J., and D.L.; formal analysis, C.J.; investigation, C.J. and D.L.; resources, C.J.; data curation, C.J.; writing—original draft preparation, C.J. and D.L.; writing—review and editing, C.J. and J.H.; visualization, C.J.; supervision, J.H.; project administration, J.H.

Acknowledgments: This work was supported by the Incheon National University under Research Grant 2019-0254 (Corresponding author: Jin Hur).

References

1.   Li, G.J.; Ren, B.; Zhu, Z.Q.; Li, Y.X.; Ma, J. Cogging torque mitigation of modular permanent magnet machines. *IEEE Trans. Magn.* **2016**, *52*, 1–10. [CrossRef]
2.   Park, Y.U.; Cho, J.H.; Kim, D.K. Cogging torque reduction of single-phase brushless DC motor with a tapered air-gap using optimizing notch size and position. *IEEE Trans. Ind. Appl.* **2015**, *51*, 4455–4463. [CrossRef]

3. Xue, Z.; Li, H.; Zhou, Y.; Ren, N.; Wen, W. Analytical prediction and optimization of cogging torque in surface-mounted permanent magnet machines with modified particle swarm optimization. *IEEE Trans. Ind. Electron.* **2017**, *64*, 9795–9805. [CrossRef]

4. Ren, W.; Xu, Q.; Li, Q.; Zhou, L. Reduction of cogging torque and torque ripple in interior PM Machines with asymmetrical V-type rotor design. *IEEE Trans. Magn.* **2016**, *52*, 1–5. [CrossRef]

5. Dosiek, L.; Pillay, P. Cogging torque reduction in permanent magnet machines. *IEEE Trans. Ind. Appl.* **2007**, *43*, 1565–1571. [CrossRef]

6. Kim, K. A novel method for minimization of cogging torque and torque ripple for interior permanent magnet synchronous motor. *IEEE Trans. Magn.* **2014**, *50*, 793–796. [CrossRef]

7. Wang, D.; Wang, X.; Jung, S. Cogging torque minimization and torque ripple suppression in surface-mounted permanent magnet synchronous machines using different magnet widths. *IEEE Trans. Magn.* **2013**, *49*, 2295–2298. [CrossRef]

8. Wanjiku, J.; Khan, M.A.; Barendse, P.S.; Pilay, P. Influence of slot openings and tooth profile on cogging torque in axial-flux pm machines. *IEEE Trans. Ind. Electron.* **2015**, *62*, 7578–7589. [CrossRef]

9. Kwon, J.; Lee, J.; Zhao, W.; Kwon, B. Flux-switching permanent magnet machine with phase-group concentrated-coil windings and cogging torque reduction technique. *Energies* **2018**, *11*, 2758. [CrossRef]

10. Hwang, M.; Lee, H.; Cha, H. Analysis of torque ripple and cogging torque reduction in electric vehicle traction platform applying rotor notched design. *Energies* **2018**, *11*, 3053. [CrossRef]

11. Dini, P.; Saponara, S. Cogging torque reduction in brushless motors by a nonlinear control technique. *Energies* **2019**, *12*, 2224. [CrossRef]

12. Kim, J.M.; Yoon, M.H.; Hong, J.P.; Kim, S.I. Analysis of cogging torque caused by manufacturing tolerances of surface-mounted permanent magnet synchronous motor for electric power steering. *IEEE Trans. Electr. Power Appl.* **2016**, *10*, 691–696. [CrossRef]

13. Ortega, A.J.P.; Paul, S.; Islam, R.; Xu, L. Analytical model for predicting effects of manufacturing variations on cogging torque in surface-mounted permanent magnet motors. *IEEE Trans. Magn.* **2016**, *52*, 3050–3061. [CrossRef]

14. Qian, H.; Guo, H.; Wu, Z.; Ding, X. Analytical solution for cogging torque in surface-mounted permanent-magnet motors with magnet imperfections and rotor eccentricity. *IEEE Trans. Magn.* **2014**, *50*, 1–15. [CrossRef]

15. Zhou, Y.; Li, H.; Meng, G.; Zhou, S.; Cao, Q. Analytical calculation of magnetic field and cogging torque in surface-mounted permanent-magnet machines accounting for any eccentric rotor shape. *IEEE Trans. Ind. Electron.* **2015**, *62*, 3438–3447. [CrossRef]

16. Lee, C.J.; Jang, G.H. Development of a new magnetizing fixture for the permanent magnet brushless dc motors to reduce the cogging torque. *IEEE Trans. Magn.* **2011**, *47*, 2410–2413. [CrossRef]

17. Song, J.Y.; Kang, K.J.; Kang, C.H.; Jang, G.H. Cogging torque and unbalanced magnetic pull due to simultaneous existence of dynamic and static eccentricities and uneven magnetization in permanent magnet motors. *IEEE Trans. Magn.* **2014**, *50*, 1–9. [CrossRef]

18. Sung, S.J.; Park, S.J.; Jang, G.H. Cogging torque of brushless DC motors due to the interaction between the uneven magnetization of a permanent magnet and teeth curvature. *IEEE Trans. Magn.* **2011**, *47*, 1923–1928. [CrossRef]

19. Ou, J.; Liu, Y.; Qu, R.; Doppelbauer, M. Experimental and theoretical research on cogging torque of PM synchronous motors considering manufacturing tolerances. *IEEE Trans. Ind. Electron.* **2018**, *65*, 3772–3783. [CrossRef]

20. Ionel, D.M.; Popescu, M.; McGilp, M.I.; Miller, T.J.E.; Dellinger, S.J. Assessment of torque components in brushless permanent-magnet machines through numerical analysis of the electromagnetic field. *IEEE Trans. Electr. Power Appl.* **2005**, *41*, 1149–1158. [CrossRef]

21. Zhu, W.; Pekarek, S.; Fahimi, B.; Deken, B.J. Investigation of force generation in a permanent magnet synchronous machine. *IEEE Trans. Energy Convers.* **2007**, *22*, 557–565. [CrossRef]

22. Meessen, K.J.; Paulides, J.J.H.; Lomonova, E.A. Force calculations in 3-D cylindrical structures using fourier analysis and the Maxwell stress tensor. *IEEE Trans. Magn.* **2013**, *49*, 536–545. [CrossRef]

23. Gao, J.; Wang, G.; Liu, X.; Zhang, W.; Huang, S.; Li, H. Cogging torque reduction by elementary-cogging-unit shift for permanent magnet machines. *IEEE Trans. Magn.* **2017**, *53*, 1–5. [CrossRef]

24.   Zhu, Z.Q.; Ruangsinchaiwanich, S.; Howe, D. Synthesis of cogging-torque waveform from analysis of a single stator slot. *IEEE Trans. Ind. Appl.* **2006**, *42*, 650–657. [CrossRef]
25.   Zhu, Z.Q.; Ruangsinchaiwanich, S.; Chen, Y.; Howe, D. Evaluation of superposition technique for calculating cogging torque in permanent-magnet brushless machines. *IEEE Trans. Magn.* **2006**, *42*, 1597–1603. [CrossRef]
26.   Gieras, J.F. *Electrical Machines: Fundamentals of Electromechanical Energy Conversion*; CRC Press: Boca Raton, FL, USA, 2016.

9

Determining the Position of the Brushless DC Motor Rotor

Krzysztof Kolano

Faculty of Electric Drives and Machines, Lublin University of Technology, 20-618 Lublin, Poland;
k.kolano@pollub.pl

Abstract: In brushless direct current (or BLDC) motors with more than one pole pair, the status of standard shaft position sensors assumes the same distribution several times for its full mechanical rotation. As a result, a simple analysis of the signals reflecting their state does not allow any determination of the mechanical position of the shaft of such a machine. This paper presents a new method for determining the mechanical position of a BLDC motor rotor with a number of pole pairs greater than one. In contrast to the methods used so far, it allows us to determine the mechanical position using only the standard position sensors in which most BLDC motors are equipped. The paper describes a method of determining the mechanical position of the rotor by analyzing the distribution of errors resulting from the accuracy proposed by the BLDC motor's Hall sensor system. Imprecise indications of the rotor position, resulting from the limited accuracy of the production process, offer a possibility of an indirect determination of the rotor's angular position of such a machine.

Keywords: rotor position; BLDC motor; sensor misalignment

1. Introduction

Compared to conventional DC brush motors, BLDC brushless motors using shaft position sensors are becoming increasingly popular in many applications, due to their relatively low cost, high performance and high reliability. The operating issues associated with the control of such machines have been extensively studied under the assumption of the correct operation of the shaft position sensor system, namely Hall sensors, and under the assumption of the symmetry of the signals generated by them [1–4].

Some industrial applications require that the position of the rotor is clearly defined and the information about the rotor is used by the control system during the operation cycle of the device. An example of this is the machine spindle, which must be properly positioned in relation to the cutter hopper, in order for the tools to be automatically picked and replaced.

A common solution to this problem of determining the rotor position is to equip the drive system with additional elements, namely, shaft position sensors, whose resolution is matched to the desired accuracy of its positioning. All kinds of encoders of the motor shaft's absolute position are used for this purpose [5]: absolute position encoders, and incremental encoders with an additional index "I" signal generated once per mechanical rotation.

The presence of Hall sensors, mounted in the motor, led the author to use them as an element defining the mechanical angle of the motor shaft. Determining the position of the BLDC motor rotor with an accuracy of 60 mechanical degrees for motors with one pair of poles is possible according to the formula:

$$\theta_e = p \cdot \theta_m \tag{1}$$

(where p equals the number of pole pairs), and it is determined directly by the state of the shaft position sensors, as the change in the electrical angle δ_e is equal to the change in the mechanical angle δ_m of the motor shaft position – Figure 1a.

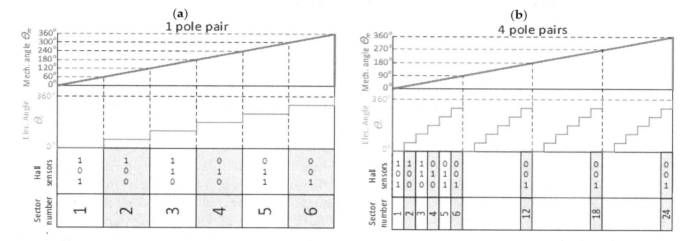

Figure 1. Relationships between the values of mechanical angle θ_m, electrical angle θ_e and the state of Hall sensors for brushless direct current (or BLDC) motors for: **(a)** 1; **(b)** 4 pairs of poles.

However, in machines with the number of $p > 1$ pole pairs ($p = 4$ for most BLDC motors that are present on the market), equipped only with standard Hall sensors, it is not possible to determine the mechanical position of the shaft solely on the basis of the analysis of their state, because the Hall sequence repeats p times per one mechanical revolution, as shown in Figure 1b.

Until now, the solution to this problem has been to install additional elements determining the position of the motor shaft. Unfortunately, this makes it necessary to mechanically modify the typical BLDC motors available on the market.

This article proposes a method that allows us to determine the mechanical position of the rotor with an accuracy equal to $60/p$ mechanical degrees for BLDC motors with the number of pole pairs greater than one, without the need for additional sensors. This accuracy value is sufficient for rotor positioning in many industrial applications.

2. Errors in the Positioning of Motor Sensors

The disadvantages resulting from the real, different from ideal, arrangement of shaft position sensors are known, and described in detail in the literature [6–13]. This problem is caused by the relatively low accuracy of the motor shaft position determination system during mass production. Researchers' efforts focus on compensating for sensor placement errors by assuming perfect symmetry of the rotating magnetic element (Figure 1 $\delta_{mr} = 0$). This makes it possible to treat the multipolar machine as a machine with one pair of poles [14–16]. Defining the errors in the measurement of the speed and position of the motor shaft, resulting only from the incorrect arrangement of sensors, is a large simplification in the analysis of the operation of the real system for determining the position of the BLDC motor rotor. If the error concerned only the sensor location accuracy (δ_{ms}– Figure 2), the speed measurement errors could be uniquely determined for each combination of Hall sensor states. In fact, these errors result to the same extent from the accuracy of sensor placement (δ_{ms}), as from the accuracy of making the magnetic ring rotating coaxially with the shaft (δ_{mr}); see Figure 2. The angular velocity δ_r of the shaft, measured by the microprocessor system as a value inversely proportional to the time t_s of changing the state of the position sensor values, amounts to:

$$\omega_r = \frac{\pi \pm \delta_{ms} \pm \delta_{mr}}{p \cdot t_s} \tag{2}$$

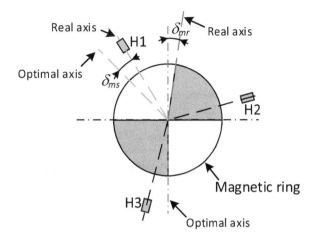

Figure 2. Visualisation of sensor (δ_{ms}) and magnet (δ_{mr}) misalignment in a BLDC shaft sensing system.

If non-zero errors δ_{ms} or δ_{mr} are assumed, the speed values for the individual sectors may vary despite the shaft rotating at a constant speed. This difference can be determined by comparing the measured value with a reference value, measured e.g., by an external measuring system.

In extreme cases, assuming that both the sensor system and the magnetic ring are made with limited accuracy, an error in shaft speed measurement can be calculated for each sector of the BLDC motor (Figure 3), e.g., for a motor with four pole pairs, the number of sectors is 24 per full mechanical rotation of the shaft (Figure 3b).

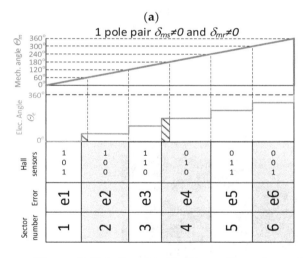
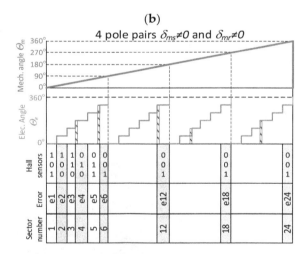

Figure 3. Relationships between the values of mechanical angle θ_m, electrical angle θ_e and the state of Hall sensors for brushless direct current (or BLDC) motors for: (**a**) 1; (**b**) 4 pairs of poles for $\delta_{ms} \neq 0$ and $\delta_{mr} \neq 0$.

For a motor with one pair of poles six unique error values can be assigned to a specific state of Hall sensors (Figure 3a), while for a motor with the number of pole pairs greater than one, these errors cannot be assigned to a specific state of sensors (Figure 3b).

Errors δ_{ms} and δ_{mr} result in switching the keys of the motor stator winding controller in a suboptimal position of the rotor, which in turn leads to the electromagnetic torque pulsation of the motor, increases the amplitude of its current (Figure 4) and the noise level of the motor operation. Boosting the amplitude of the current results in an increase in electrical losses and electromagnetic interference, and forces the constructors of the controller to use in it transistors with a higher rated current, so that they do not suffer thermal and dynamic damage during long-term operation [17].

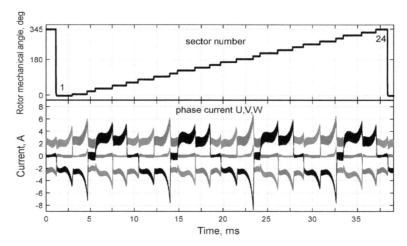

Figure 4. Measurement of phase currents of a BLDC motor (four pole pairs, nominal speed = 3000 rpm/min, nominal power = 380 W) during full mechanical rotation.

3. Laboratory Measurements Taken on Real Motors

In order to confirm the universality of this phenomenon in motors available on the market, it was decided to purchase four types of BLDC motors from four different manufacturers, and to analyze the performance of their shaft positioning systems. For this purpose, an experiment was developed to compare the actual ω_{ref} speed measured by the incremental encoder, with the speed measured using the signal generated by the motor's Hall sensors in an open loop drive system. The measuring functions were executed by the dSpace GmbH (Paderborn, Germany) dSpace MicroLabBox system supported by the Tektronix (Beaverton, Oregon United States) Digital Phosphor Oscilloscope 5054-B (Figure 5). As a measure of the error in determining the speed in individual sectors e_{sec}, a percentage ratio of the ω_{sec} calculated speed to the actual ω_{ref} measured by the reference element, i.e. the incremental encoder, was proposed. Additionally, the values of phase currents of the BLDC motor were measured during the tests.

Figure 5. Research set-up.

One of the aims of the experiment was to determine the influence of sensor distribution error on the pulsation of the actual rotational speed. It was predicted that significant oscillations of the motor phase current amplitude, caused by the premature or delayed activation of the controller transistors, may significantly increase this pulsation. During laboratory tests, it was found that in a wide range of speeds (from 10% to 100% of the n_n rated speed), and for wide ranges of load torque changes (from idle to rated load), the actual speed pulsation is imperceptible (Figure 6), despite significant (up to 80%; Figure 4) phase current fluctuations of the motor.

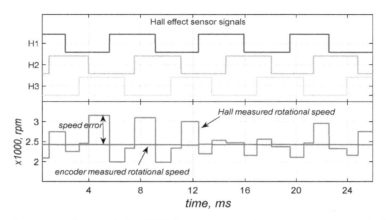

Figure 6. Actual and measured speed of a BLDC motor with factory nonsymmetrical Hall sensor array, for a BLDC motor with four pairs of poles.

The measurements showed that the accuracy of speed determination in the experiment, assumed at 0.5%, was not exceeded. This means that, regardless of errors in the BLDC motor position measurement system, we can assume that its average steady state rotational speed can be considered constant, regardless of the phase current amplitude fluctuations of the motor. This is a very important conclusion, allowing the thesis that, in the steady state the average actual speed for individual motor sectors is equal to the average speed measured during the full mechanical rotation of the shaft. The actual average speed of a full revolution can be measured correctly by any single Hall sensor. This results in the possibility of not using an additional reference speed measurement system (i.e. an additional encoder) in favor of using averaged full revolution speed, which is independent of the sensor placement error [12].

Figure 7 shows the calculated speed error values for the individual e_{sec} sectors for four different BLDC motors with powers ranging from 60 W to 1500 W. All motors had four pairs of poles each, for which 24 sectors of speed measurement can be defined per one mechanical revolution. Each of them corresponds to a mechanical rotation angle equal to $\pi/(3 \cdot p)$, which gives an angle value for each sector equal to 15 mechanical degrees.

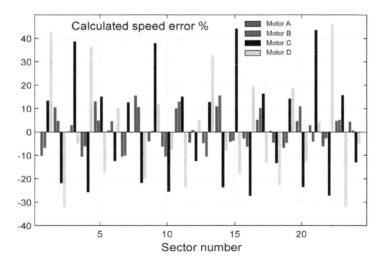

Figure 7. Rotational speed computation error "e_{sec}" in particular sectors of the tested motors.

It can be seen that the values of these errors are significant, and strongly influence the accuracy of motor speed calculation. It is worth emphasizing that all of the tested motors were characterized by a significant error in the arrangement of e_e sensors, which ranged from a few to over 25° of the electric angle, with respect to the optimal location according to the formula $e_e = e_{sec} \cdot 60°$.

4. Determination of the Absolute Position of the BLDC Motor Rotor

In the course of the work described above, errors in determining the rotational speed for individual sectors of the motors, tested at different rotational speeds, were determined. It was shown that for a wide range of rotational speeds, these errors are almost constant, and what is worth underlining, unique for the specific motor used in the test. This is illustrated in Figure 8 for the motor marked as "D".

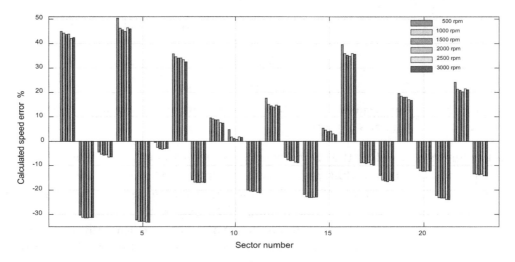

Figure 8. Rotational speed computation errors determined at different rotational speeds of motor D (four pole pairs, nominal speed = 3000 rpm/min, nominal power = 380W).

Figure 9 shows the error distribution of the speed determination for motor "D". Using a standard Hall sensor system causes that during the full rotation of the shaft, the sequence of sensor states is repeated depending on the number of pole pairs of the motor (here, four times).

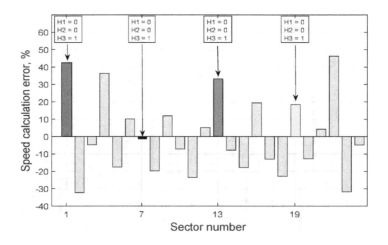

Figure 9. Distribution of errors in determining the e_{sec} speed of the BLDC motor, for motor "D", determined at 2000 rpm.

It can be seen that the value of the speed determination errors for individual sectors varies greatly, creating a sequence unique for a specific motor. This makes it possible to identify the position of the rotor using information about the states of the sensors, with the support of an algorithm that looks for a specific pattern of error distribution. Determination of the rotor position can be done in relation to the reference value of the e_{ref} error distribution, stored in the electrically erasable programmable read-only - EEPROM memory of the motor control device.

The sequence of errors is repeated with the frequency corresponding to the time of full rotation of the shaft by the motor. This sequence can start from different values, which depends only on the initial position of the motor shaft at the moment of starting the drive system.

To simplify the procedure for determining the motor shaft position, only the error distributions corresponding to the same combination of the Hall sensor states can be analyzed, which in the case of the motor under test reduces the number of errors analysed from 24 to only 4.

Figure 10 shows possible error distributions for the selected sensor states (H1 = 0, H2 = 0, H3 = 1). For a motor with four pairs of poles, four different sequences of assigning the error distribution to the absolute position of the motor rotor are possible (depending on the initial position of the rotor after power-up).

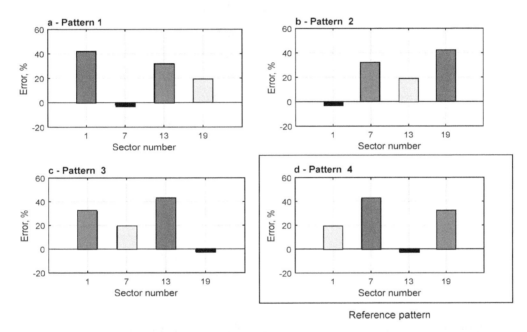

Figure 10. Possible error distributions for determining the rotor speed for the Hall sensor state for different initial rotor positions: (**a**–**c**); (**d**) for H1 = 0; H2 = 0; H3 = 1; (**d**) reference distribution stored in the memory of the motor controller.

5. Example Implementation

The procedure for determining the absolute position of the rotor is, in the simplest possible implementation, to search for the largest e_{sec} error for the same selected sequence of their state, and to assign its sector number to the sector number of the maximum error of the reference distribution. For example, sectors 1, 7, 13 and 19 (Figure 10) correspond to the sequence (H1 = 0; H2 = 0; H3 = 1). After determining the speed calculation errors, the algorithm compares the results obtained with the reference values stored in the controller's non-volatile EEPROM memory. If the reference error values stored for specific sensor signals reached the maximum value, e.g. for Sector 7, then after searching the newly obtained error distribution, and finding the maximum error value, e.g., for sector 13, one should change the sector number in which the greatest error of speed measurement was found. In this particular case, sector number 13 will be changed by the algorithm determining the position of the shaft to number 7, and the remaining sector numbers will be automatically assigned to the following sectors of Figure 11 by means of software incrementation.

The operation of this algorithm allowed to unambiguously assign the BLDC motor sector number to the angle of the mechanical rotor position by analyzing the reference pattern of error distribution, stored e.g., in the EEPROM memory.

When the differences between the errors of individual sectors are small, i.e., when the shaft position detection system was made precisely at the factory, it is possible to expand the algorithm of searching for similarity between measured values and reference values stored in the EEPROM memory. The minimum error value and even all errors in all sectors can also be analysed.

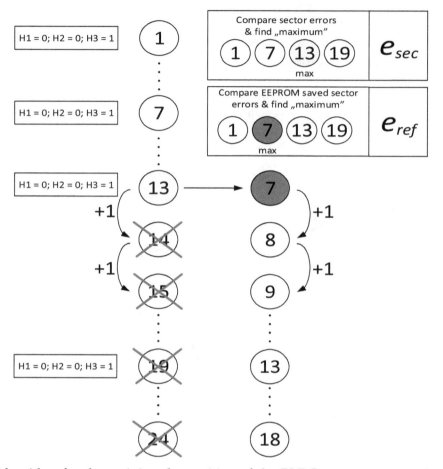

Figure 11. Algorithm for determining the position of the BLDC motor rotor on the basis of the comparison of e_{sec} errors with the reference value of the error distribution e_{ref}, for a given state of the Hall sensors (H1 = 0; H2 = 0; H3 = 1).

6. Conclusions

During the research, four BLDC motors with four pairs of poles each were analysed. Thanks to the proposed method, it was possible to find the mechanical position of a multi-pole BLDC motor, despite the fact that the sequence of the Hall sensor states changes in the same sequence p-times per mechanical revolution. What is very important is that theoretically, if the shaft position monitoring system was made very accurately at the factory, it would not be possible to achieve motor diagnostics results characteristic enough to assign a specific position of the motor rotor to them. Hence, the conclusion that applying the method described in the article can correct a factory's inaccurate fixing of motor components, and result in the possibility of using such machines in drive systems, where the absolute position of the shaft must be determined for the correct operation of the whole system. Another interesting issue is the possibility of identifying a specific electrical machine by analyzing the distribution of errors in speed measurement, using its shaft position sensors. This gives the possibility of the quick, simple and cost-free detection of the replacement of the drive motor, even if it is characterized by the same nominal data. Determination of the angular position of the BLDC motor shaft on the basis of the analysis of the error spectrum allows the application of advanced algorithms for the correction of commutation errors, which result from errors in the position of sensors in multi-pole BLDC motors [3]. This contributes to the reduction of the noise emitted by the drive as well as the use of semiconductor connectors with a lower rated current for the controller.

References

1. Giulii Capponi, F.; De Donato, G.; Del Ferraro, L.; Honorati, O.; Harke, M.C.; Lorenz, R.D. AC brushless drive with low-rersolution hall-effect sensors for surface-mounted PM machines. *IEEE Trans. Ind. Appl.* **2006**, *42*, 526–535. [CrossRef]
2. Hui, T.S.; Basu, K.P.; Subbiah, V. Permanent magnet brushless motor control techniques. In Proceedings of the National Power Engineering Conference, Bangi, Malaysia, 15–16 December 2003; pp. 133–138.
3. Kolano, K. Improved Sensor Control Method for BLDC Motors. *IEEE Access* **2019**. [CrossRef]
4. Pillay, P.; Krishnan, R. Modeling, simulation, and analysis of permanent-magnet motor drives, Part II: The brushless DC motor drive. *IEEE Trans. Ind. Appl.* **1989**, *25*, 274–279. [CrossRef]
5. Santolaria, J.; Conte, J.; Pueo, M.; Carlos, J. Rotation error modeling and identification for robot kinematic calibration by circle point method. *Metrol. Meas. Syst.* **2014**, *21*, 85–98. [CrossRef]
6. Alaeinovin, P.; Jatskevich, J. Filtering of Hall-Sensor Signals for Improved Operation of Brushless DC Motors. *IEEE Trans. Energy Convers.* **2012**, *27*, 547–549. [CrossRef]
7. Alaeinovin, P.; Chiniforoosh, S.; Jatskevich, J. Evaluating misalignment of hall sensors in brushless DC motors. In Proceedings of the 2008 IEEE Canada Electric Power Conference, Vancouver, BC, Canada, 6–7 October 2008; pp. 1–6.
8. Alaeinovin, P.; Jatskevich, J. Hall-sensor signals filtering for improved operation of brushless DC motors. In Proceedings of the 2011 IEEE International Symposium on Industrial Electronics, Gdansk, Poland, 27–30 June 2011; pp. 613–618.
9. Baszyński, M.; Piróg, S. A Novel Speed Measurement Method for a High-Speed BLDC Motor Based on the Signals From the Rotor Position Sensor. *IEEE Trans. Ind. Inform.* **2014**, *10*, 84–91. [CrossRef]
10. Choi, J.H.; Park, J.S.; Gu, B.G.; Kim, J.H.; Won, C.Y. Position estimation and control of BLDC motor for VVA module with unbalanced hall sensors. In Proceedings of the 2012 IEEE International Conference on Power and Energy (PECon), Kota Kinabalu, Malaysia, 2–5 December 2012; pp. 390–395.
11. Park, J.W.; Kim, J.H.; Kim, J.M. Position Correction Method for Misaligned Hall-Effect Sensor of BLDC Motor using BACK-EMF Estimation. *Trans. Korean Inst. Power Electron.* **2012**, *17*, 246–251. [CrossRef]
12. Kolano, K. Calculation of the brushless dc motor shaft speed with allowances for incorrect alignment of sensors. *Metrol. Meas. Syst.* **2015**, *22*, 393–402. [CrossRef]
13. Samoylenko, N.; Han, Q.; Jatskevich, J. Dynamic performance of brushless DC motors with unbalanced hall sensors. *IEEE Trans. Energy Convers.* **2008**, *23*, 752–763. [CrossRef]
14. Lim, J.S.; Lee, J.K.; Seol, H.S.; Kang, D.W.; Lee, J.; Go, S.C. Position Signal Compensation Control Technique of Hall Sensor Generated by Uneven Magnetic Flux Density. *IEEE Trans. Appl. Supercond.* **2008**, *28*, 1–4. [CrossRef]
15. Nerat, M.; Vrančić, D. A Novel Fast-Filtering Method for Rotational Speed of the BLDC Motor Drive Applied to Valve Actuator. *IEEE/ASME Trans. Mechatron.* **2016**, *21*, 1479–1486. [CrossRef]
16. Park, D.H.; Nguyen, A.T.; Lee, D.C.; Lee, H.G. Compensation of misalignment effect of hall sensors for BLDC motor drives. In Proceedings of the 2017 IEEE 3rd International Future Energy Electronics Conference and ECCE Asia (IFEEC 2017—ECCE Asia), Kaohsiung, Taiwan, 3–7 June 2017; pp. 1659–1664. [CrossRef]
17. Zieliński, D.; Fatyga, K. Comparison of main control strategies for DC/DC stage of bidirectional vehicle charger. In Proceedings of the 2017 International Symposium on Electrical Machines (SME), Naleczow, Poland, 18–21 June 2017. [CrossRef]

10

New Modulation Technique to Mitigate Common Mode Voltage Effects in Star-Connected Five-Phase AC Drives

Markel Fernandez [1,*], Andres Sierra-Gonzalez [2], Endika Robles [1], Iñigo Kortabarria [1], Edorta Ibarra [1] and Jose Luis Martin [1]

[1] Derpartment of Electronic Technology, University of the Basque Country (UPV/EHU), Plaza Ingeniero Torres Quevedo 1, 48013 Bilbao, Spain; endika.robles@ehu.eus (E.R.); inigo.kortabarria@ehu.eus (I.K.); edorta.ibarra@ehu.eus (E.I.); joseluis.martin@ehu.eus (J.L.M.)

[2] Tecnalia Research and Innovation, C. Mikeletegi 7, 20009 Donostia, Spain; andres.sierra@tecnalia.com

*   Correspondence: markel.fernandez@ehu.eus

Abstract: Star-connected multiphase AC drives are being considered for electromovility applications such as electromechanical actuators (EMA), where high power density and fault tolerance is demanded. As for three-phase systems, common-mode voltage (CMV) is an issue for multiphase drives. CMV leads to shaft voltages between rotor and stator windings, generating bearing currents which accelerate bearing degradation and produce high electromagnetic interferences (EMI). CMV effects can be mitigated by using appropriate modulation techniques. Thus, this work proposes a new Hybrid PWM algorithm that effectively reduces CMV in five-phase AC electric drives, improving their reliability. All the mathematical background required to understand the proposal, i.e., vector transformations, vector sequences and calculation of analytical expressions for duty cycle determination are detailed. Additionally, practical details that simplify the implementation of the proposal in an FPGA are also included. This technique, HAZSL5M5-PWM, extends the linear range of the AZSL5M5-PWM modulation, providing a full linear range. Simulation results obtained in an accurate multiphase EMA model are provided, showing the validity of the proposed modulation approach.

Keywords: multiphase electric drives; CMV; modulation techniques; PWM

1. Introduction

AC electric drives are used in a wide variety of industrial applications such as in compressors [1], in electric vehicle propulsion systems [2,3] and in more electric aircraft (MEA) [4], among others. Although three-phase systems dominate the AC drive market, multiphase solutions are gaining popularity [5–7]. Multiphase systems are preferable for applications where high fault tolerance is required [8], such as for MEA applications, where electromechanical actuators (EMA) for control surfaces, fuel pumps, landing gears, environmental control systems and starter-generators need to be operated [9–11]. Apart from their intrinsic fault tolerance, other benefits of multiphase drives include a reduced current per phase (reducing copper losses and increasing efficiency) [4], noise and electromagnetic interference (EMI) minimization [12,13], higher power density and lower torque ripple [14], making them attractive for transport electrification. Among the multiphase topologies available in the scientific literature, star-connected five-phase technologies (Figure 1) can be highlighted, as they provide a good trade-off between system complexity and fault tolerance [15,16]. Specifically, multiphase permanent magnet synchronous machines (PMSM) are being considered for aircrafts due to their superior power density [17,18].

In general, AC electric drives can experience issues due to the common-mode voltage (CMV) [19] and common mode currents (CMC) [20]. CMV variations are generated by the commutation of the power converter devices, producing EMI [21] and bearing currents that can compromise the integrity of the electric machine [22]. Such voltage variations create new capacitive paths through the motor bearings, leading to premature aging. Capacitive currents, electrostatic discharge machine (EDM) currents, circulating currents and rotor-to-ground currents can flow through the bearings [23,24] (Figure 2), and their harmful effects depend on the type of bearing, size of the machine and how the machine is used.

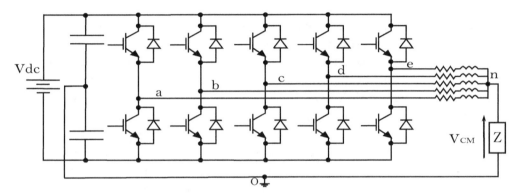

Figure 1. CMV in a five-phase power system.

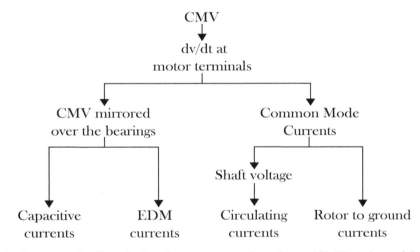

Figure 2. Cause-and-effect chain of common mode voltage (CMV) (adapted from [25]).

Operating at higher switching frequencies can entail more severe CMV related issues (additional EMI generation and larger number of *dv/dt*) [26]. As considerable efforts are being carried out to widespread the usage of wide bandgap (WBG) devices in AC drives, much higher switching frequencies are expected in the future, making the investigation on CMV mitigation a popular topic [13,19,20]. Thus, a wide variety of solutions have been proposed in the literature. Such solutions can be classified as passive or active. Passive solutions are those which mitigate or eliminate the harmful effects generated by CMV, while active solutions are intended to reduce or totally avoid CMV generation. Among passive solutions, Faraday shielding [27,28], ceramic and hybrid bearings [22,29], shielded cables [23,30] and shaft grounding rings [27,31] can be highlighted. On the other hand, modulation techniques and new inverter topologies such as multilevel inverters [32], single-phase transformerless inverters [33,34] and three-phase inverters [35] among others [36] are the most common active solutions. Among all these solutions, modulation algorithms can be considered for CMV reduction in star-connected five-phase AC drives due to their ease of implementation, low cost, and because no additional hardware is needed.

In [37], the authors initially proposed a CMV reduction modulation technique for five-phase inverters, named AZSL5M5-PWM. However, the proposal has been only considered for passive loads (star-connected RL loads) and solely validated in open-loop. From the obtained results, it has been concluded that the linear range of the original AZSL5M5-PWM is limited, which can prevent the utilization of this technique in electric drives where operation close to the base speed (without entering in field weakening region) is desirable, as is the case in most EMA systems. Thus, a hybrid AZSL5M5-PWM technique (HAZSL5M5-PWM) that provides the same linear range as conventional space vector PWM (SV-PWM) is proposed in this work, and its performance is evaluated in an EMA system.

This manuscript is organized as follows. First of all, conventional SV-PWM for star-connected five-phase power systems is presented, where the harmonic projection of the stator voltages into their corresponding orthogonal subspaces by means of Clarke transformation is mathematically justified. After that and considering the third harmonic elimination constraint, it is shown how CMV variations are generated in the multiphase drive. Secondly, the most relevant reduced common-mode voltage PWM (RCMV-PWM) modulation techniques are briefly described, focusing on their limitations. After that, the proposed Hybrid AZSL5M5-PWM modulation technique is presented providing the required tools for duty cycle calculation, and validated by means of simulation. The target of the proposed modulation technique is to effectively reduce CMV in star connected multiphase systems, while the hybridization is performed to cover the whole operation range of the drive. Open-loop and detailed five-phase EMA simulations are conducted to perform the validation, where not only CMV reduction is verified, but other figures such as total harmonic distortion (THD) and efficiency are evaluated in order to demonstrate that the achieved CMV reduction does not significantly penalize other relevant drive figures.

2. Influence of the SV-PWM Technique in the CMV of a Star-Connected Five-Phase AC Drive

SV-PWM is one of the most used modulation techniques in three-phase and multiphase power systems thanks to its easy digital implementation and optimum DC bus voltage utilization. As a star-connected five-phase system has four degrees of freedom, stator voltages and currents can be represented into two separated two-dimensional planes, α-β and x-y, and one homopolar component by means of the following amplitude invariant Clarke transformation [38]:

$$\begin{bmatrix} v_\alpha \\ v_\beta \\ v_x \\ v_y \\ v_0 \end{bmatrix} = \frac{2}{5} \begin{bmatrix} 1 & cos(2\pi/5) & cos(4\pi/5) & cos(6\pi/5) & cos(8\pi/5) \\ 0 & sin(2\pi/5) & sin(4\pi/5) & sin(6\pi/5) & sin(8\pi/5) \\ 1 & cos(4\pi/5) & cos(8\pi/5) & cos(12\pi/5) & cos(16\pi/5) \\ 0 & sin(4\pi/5) & sin(8\pi/5) & sin(12\pi/5) & sin(16\pi/5) \\ \frac{1}{2} & \frac{1}{2} & \frac{1}{2} & \frac{1}{2} & \frac{1}{2} \end{bmatrix} \begin{bmatrix} v_a \\ v_b \\ v_c \\ v_d \\ v_e \end{bmatrix}. \tag{1}$$

The Clarke transformation allows us to decouple the 5-dimensional voltage vector in the $abcde$ reference frame into three orthogonal subspaces (α-β, x-y and 0). For a surface-mounted permanent magnet synchronous machine (SM-PMSM), this decoupling is done through the diagonalization of the inductance matrix $\mathbf{L}$ (2).

$$\mathbf{L} = \begin{bmatrix} L_{11} & L_{12} & L_{13} & L_{14} & L_{15} \\ L_{21} & L_{22} & L_{23} & L_{24} & L_{25} \\ L_{31} & L_{32} & L_{33} & L_{34} & L_{35} \\ L_{41} & L_{42} & L_{43} & L_{44} & L_{45} \\ L_{51} & L_{52} & L_{53} & L_{54} & L_{55} \end{bmatrix}. \tag{2}$$

For SM-PMSMs, the elements of $\mathbf{L}$ can be considered invariant with respect to the rotor angular position, as the surface placed magnets have a permeability near that the one of the air. Therefore, a SM-PMSM behaves like a non-salient pole synchronous machine [39]. As the windings in each phase are manufactured identically, the mutual inductances between any pair of phases separated with the same electrical angle are equal, i.e., $L_{12} = L_{15} = L_{21} = L_{23} = L_{51} = L_{jk}$ (if $|j - k| = 1$) or $L_{13} = L_{31} = L_{25} = L_{52} = L_{jk}$ (if $|j - k| = 2$). Similarly, all the self-inductances are equal ($L_{11} = L_{22} = \cdots = L_{55}$). This type of matrix is known as a circulant matrix, and it has some special properties [40]. For example, it guarantees that **L** is orthogonally diagonalizable by a transformation represented by a 5×5 real matrix [41,42].

The circulant matrices are diagonalized by the Fourier Matrix [40,43]. Therefore, the Clarke transformation decomposes the 5-dimensional vectors according to their harmonic components. In the α-β sub-space, the $h = 5(l - 1) \pm 1$ harmonic components are projected while, in the x-y sub-space, the $h = 5(l - 1) \pm 3$ ones are projected, being $l \in \{1, 3, 5, ...\}$ [41,44]. The harmonic components of order $h = 5l$ are projected into the zero-sequence or homopolar sub-space. In Table 1, the odd harmonics associated with each sub-space according to the Clarke transformation of (1) are presented for a 5-phase machine.

Table 1. Sub-space harmonics mapping for a five-phase machine.

Sub-Space	Harmonics
$\alpha - \beta$	$h = 1, 9, 11, 19...$
$x - y$	$h = 3, 7, 13, 17...$
zero-sequence	$h = 0, 5, 15, 25...$

The number of possible switching states or space vectors is 2^5, where 30 are active vectors and two are zero vectors (Figure 3). Active vectors can be classified depending on their magnitude as:

- Large vectors, where $|V_l| = 4/5 V_{DC} \cos(\pi/5)$, which correspond to the outer decagon of Figure 3.
- Medium vectors, where $|V_m| = 2/5 V_{DC}$, which correspond to the middle decagon of Figure 3.
- Small vectors, where $|V_s| = 4/5 V_{DC} \cos(2\pi/5)$, which correspond to the inner decagon of Figure 3.

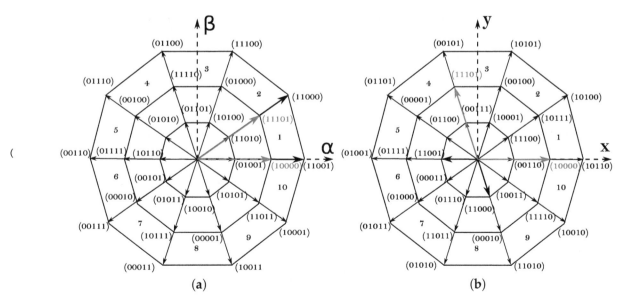

Figure 3. Five-phase SV-PWM: α-β and x-y planes with their corresponding space vectors and switching states (a '1' in a switching state represents that the top switch of a given phase is activated, while a '0' represents that its complementary switch is activated). (**a**) α-β vector plane. (**b**) x-y vector plane.

For an n-phase system, $n - 1$ active vectors must be applied at each commutation period in order to achieve a sinusoidal output [45]. Thus, four active vectors must be used in a star-connected five-phase system. Although various possible active vector combinations are possible to produce a given output voltage vector, the most common alternative consists on using two large and two medium adjacent vectors. As an illustrative example, Figure 3 shows the vectors used to synthesize a given reference voltage vector located in the first sector of the α-β plane, being the following the application sequence that minimizes switching losses: 00000, 10000, 11000, 11001, 11101, 11111, 11101, 11001, 11000, 10000, 00000. If the third harmonic component needs to be eliminated, the application-time ratio between medium and large vectors must satisfy (3), as with this ratio the sum of the applied vectors in the x-y plane is zero (Figure 3) [46].

$$\frac{t_{large}}{t_{medium}} = 1.618. \tag{3}$$

As a result, the maximum achievable output voltage following this modulation approach is:

$$V_{o_{max}} = \frac{4}{5}\cos\left(\frac{\pi}{5}\right)\cos\left(\frac{\pi}{10}\right) = 0.6155 V_{DC}, \tag{4}$$

and the CMV generated in the five-phase system is:

$$V_{CM}(t) = \frac{1}{5}\left[V_{a0}(t) + V_{b0}(t) + V_{c0}(t) + V_{d0}(t) + V_{e0}(t)\right]. \tag{5}$$

When using SV-PWM and applying the vector sequence that corresponds to the first sector of the α-β plane, the CMV waveform of Figure 4 is obtained. From (5), it can be deduced that all large and short vectors generate CMV levels of $\pm 0.3 V_{DC}$, while CMV levels are of $\pm 0.1 V_{DC}$ for medium vectors and of $\pm 0.5 V_{DC}$ for null vectors. When evaluating the impact of the CMV, the difference between the maximum and minimum CMV levels (Δ_{CMV}, Figure 4) must be considered, and the number of CMV variations for each commutation period (N_{CMV}) must also be taken into account.

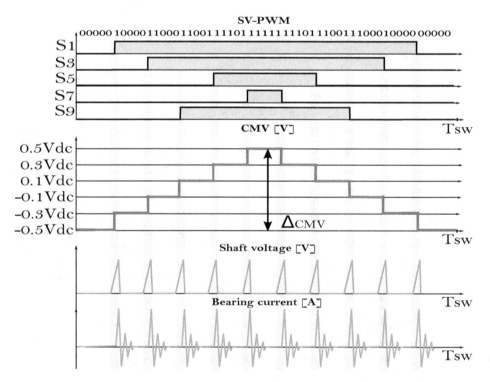

Figure 4. CMV waveform of SV-PWM technique (adapted from [23]).

3. RCMV-PWM Techniques

As zero vectors are responsible for generating the maximum CMV levels (Figure 4), most of the RCMV-PWM techniques avoid the application of these vectors to reduce Δ_{CMV} and N_{CMV}. In [38], an extension of the three-phase active zero state PWM (AZS-PWM) [47] modulation technique to the five-phase scenario is proposed. This technique replaces zero vectors by applying two active vectors with the opposite phase at the same time. In this work, this technique will be named AZSL2M2-PWM as, apart from the active vectors that substitute zero vectors, two large (L2) and two medium (M2) vectors are used at each modulation period. This technique shows a good harmonic performance and DC bus utilization, being its linear range $0 \leq m \leq 1$. However, Δ_{CMV} is not greatly reduced, as only $\pm 0.5 V_{DC}$ CMV levels are avoided (Table 2). Similarly, a modulation algorithm that employs four large active vectors in conjunction with two active vectors with opposite phases (AZSL4-PWM) is proposed in [38]. This technique has the same linear range as SV-PWM and AZSL2M2-PWM, and considerably reduces Δ_{CMV}, as only applies large vectors. Nonetheless, N_{CMV} remains as for SV-PWM (Table 2).

M5-PWM [48] and L5-PWM [49] techniques completely eliminate Δ_{CMV} and N_{CMV} by only using odd or even medium (M5-PWM), or odd or even large (L5-PWM) active vectors. However, this is achieved at the cost of introducing additional power losses, significantly reducing the linear range up to 0.5257 for M5-PWM (Table 2), and generating high harmonic distortion for L5-PWM, making them inappropriate for many industrial applications. Authors in [48,49] also propose variants that use ten medium (M10-PWM) or ten large (L10-PWM) vectors. These techniques enhance the linear range by increasing the available vectors, but do not reduce Δ_{CMV} and N_{CMV} as much as with M5-PWM and L5-PWM (Table 2).

Table 2. Summary of the most relevant features of SV-PWM and RCMV-PWM techniques.

Modulation	Δ_{CMV} [V]	N_{CMV}	Δ_{CMV} Reduction [%]	N_{CMV} Reduction [%]	v_{CM} Waveform	Linear Range
SV-PWM	V_{DC}	10	-	-	0.5V_DC 0.3V_DC 0.1V_DC -0.1V_DC -0.3V_DC -0.5V_DC	$0 \leq m \leq 1$
AZSL2M2-PWM	$0.6 V_{DC}$	6	−40%	−40%	0.3V_DC 0.1V_DC -0.1V_DC -0.3V_DC	$0 \leq m \leq 1$
AZSL4-PWM	$0.2 V_{DC}$	10	−80%	0%	0.1V_DC -0.1V_DC	$0 \leq m \leq 1$
M5-PWM	0	0	−100%	−100%	-0.3V_DC	$0 \leq m \leq 0.5257$
M10-PWM	$0.6 V_{DC}$	2	−40%	−80%	-0 3vD G 0 3vD	$0 \leq m \leq 0.618$
L5-PWM	0	0	−100%	−100%	0.1V_DC	$0 \leq m \leq 0.8507$
L10-PWM	$0.2 V_{DC}$	6	−80%	−40%	0.1V_DC -0.1V_DC	$0 \leq m \leq 1$
AZSL5M5-PWM	$0.4 V_{DC}$	2	−60%	−80%	0.3V_DC 10.- V_DC	$0 \leq m \leq 0.8507$

Among the reviewed techniques, AZSL2M2-PWM and L10-PWM best suit for industrial applications, as they keep the linear range with a reasonable THD while effectively reducing Δ_{CMV}. However, N_{CMV} reduction by means of such modulation algorithms is limited. Thus, a new RCMV-PWM technique that further reduces N_{CMV} while keeping an extended linear range is proposed in the following section.

4. Proposed RCMV-PWM Technique

4.1. *Active Zero State L5M5 PWM Technique (AZSL5M5-PWM)*

The main part of the proposed RCMV-PWM technique, named active zero state L5M5 PWM (AZSL5M5-PWM), is based on the AZSL2M2-PWM technique [38]. However, and unlike AZSL2M2-PWM, the proposed scheme only uses odd or even vectors to further reduce the CMV voltage variations. For example, if odd vectors are only considered, this leads to the sector distribution of Figure 5a. Thus, five medium vectors and five large vectors are exclusively used to synthesize the reference voltage (Figure 5), and CMV varies between $-0.3V_{DC}$ and $0.1V_{DC}$ (if only even vectors are used, CMV varies between $-0.1V_{DC}$ and $0.3V_{DC}$). Consequently, $\Delta_{CMV} = 0.4V_{DC}$ and $N_{CMV} = 2$ (Table 2).

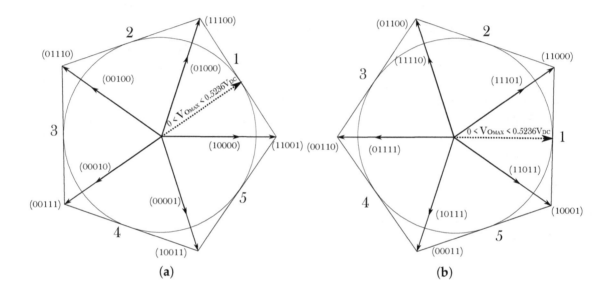

Figure 5. Sector distribution of AZSL5M5-PWM modulation scheme in the α-β plane. (**a**) AZSL5M5-PWM implemented with odd vectors. (**b**) AZSL5M5-PWM implemented with even vectors.

For the sake of simplicity, all the following analyses are conducted considering the AZSL5M5-PWM variant which exclusively uses odd vectors. The procedure is analogous for even vectors.

The application time of each vector can be easily calculated by solving the following system:

$$
\begin{bmatrix} V_\alpha^* \\ V_\beta^* \\ V_x^* \\ V_y^* \end{bmatrix} = \begin{bmatrix} V_{m_{r\alpha}} & V_{l_{l\alpha}} & V_{l_{r\alpha}} & V_{m_{l\alpha}} \\ V_{m_{r\beta}} & V_{l_{l\beta}} & V_{l_{r\beta}} & V_{m_{l\beta}} \\ V_{m_{rx}} & V_{l_{lx}} & V_{l_{rx}} & V_{m_{lx}} \\ V_{m_{ry}} & V_{l_{ly}} & V_{l_{ry}} & V_{m_{ly}} \end{bmatrix} \begin{bmatrix} \delta_1 \\ \delta_2 \\ \delta_3 \\ \delta_4 \end{bmatrix}, \tag{6}
$$

where V_α^*, V_β^*, V_x^* and V_y^* are the reference voltage projections in the α-β and x-y planes, δ_1, δ_2, δ_3 and δ_4 are the duty cycles for each vector, and the 4×4 matrix is composed of the magnitudes of the vectors to be applied in each sector, where V_{m_r} refers to the modulus of the medium vector on the right side of the reference vector (V_{ref}) (Figure 6, ③), V_{m_l} refers to the modulus of the medium vector on the left (Figure 6, ④), and V_{l_l} and V_{l_r} refer to the modulus of the large vectors on both left and right sides, respectively (Figure 6, ② and ①). Consequently, a 4×4 matrix should be defined for each sector. In general, V_x^* and V_y^* are set to zero in order to cancel the voltage third harmonic. From (6), it is possible to explicitly determine the values of the duty cycles δ_1, δ_2, δ_3 and δ_4 with respect to the reference voltages V_α^* and V_β^*:

$$\delta_1 = \alpha_1 V_\alpha^* / V_{DC} + \beta_1 V_\beta^* / V_{DC} = -a_1 \sin\left[\frac{2s\pi}{5}\right] V_\alpha^* / V_{DC} + a_1 \cos\left[\frac{2s\pi}{5}\right] V_\beta^* / V_{DC},$$

$$\delta_2 = \alpha_2 V_\alpha^* / V_{DC} + \beta_2 V_\beta^* / V_{DC} = -a_2 \sin\left[\frac{2(s-1)\pi}{5}\right] V_\alpha^* / V_{DC} + a_2 \cos\left[\frac{2(s-1)\pi}{5}\right] V_\beta^* / V_{DC},$$

$$\delta_3 = \alpha_3 V_\alpha^* / V_{DC} + \beta_3 V_\beta^* / V_{DC} = a_2 \sin\left[\frac{2s\pi}{5}\right] V_\alpha^* / V_{DC} - a_2 \cos\left[\frac{2s\pi}{5}\right] V_\beta^* / V_{DC},$$ (7)

$$\delta_4 = \alpha_4 V_\alpha^* / V_{DC} + \beta_4 V_\beta^* / V_{DC} = a_1 \sin\left[\frac{2(s-1)\pi}{5}\right] V_\alpha^* / V_{DC} - a_1 \cos\left[\frac{2(s-1)\pi}{5}\right] V_\beta^* / V_{DC},$$

where being $s = \{1, 2...5\}$ the corresponding sector in the $\alpha\beta$ plane, and being $a_1 = (-5 + \sqrt{5})/\sqrt{2(5+\sqrt{5})}$ and $a_2 = \sqrt{(10/(5+\sqrt{5})}$.

Regarding the practical implementation of the proposed technique, the 2×4 matrix $\mathbf{M}_s$ can be defined as in (8), where the elements of such matrix can be precalculated for each sector and stored into look-up tables (LUT). In this way, the computational burden and implementation complexity of the algorithm are greatly reduced. Table 3 summarizes the values of $\mathbf{M}_s$ for each sector $s = \{1, 2...5\}$.

$$\mathbf{M_s} = \begin{bmatrix} \alpha_1 & \beta_1 \\ \alpha_2 & \beta_2 \\ \alpha_3 & \beta_3 \\ \alpha_4 & \beta_4 \end{bmatrix}. $$ (8)

Table 3. Values of $\mathbf{M}_s$ depending on the $\alpha\beta$ plane sector.

Sector 1 ($\mathbf{M}_1$)		Sector 2 ($\mathbf{M}_2$)		Sector 3 ($\mathbf{M}_3$)		Sector 4 ($\mathbf{M}_4$)		Sector 5 ($\mathbf{M}_5$)	
0.691	−0.224	0.427	0.588	−0.427	0.588	−0.691	−0.224	0	−0.726
0	1.176	−1.118	0.363	−0.691	−0.951	0.691	−0.951	1.118	0.363
1.118	−0.363	0.691	0.951	−0.691	0.951	−1.118	−0.363	0	−1.176
0	0.726	−0.691	0.224	−0.427	−0.588	0.427	−0.588	0.691	0.224

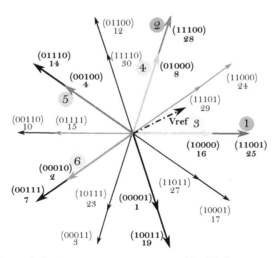

Figure 6. AZSL5M5-PWM modulation vector sequence (①–⑥) for sector 1 when only odd vectors are applied.

The main difference of the AZSL5M5-PWM technique over the AZSL2M2-PWM one is that, there are no strictly phase-opposite vector pairs, as only odd/even vectors can be used (Figure 6). To solve this problem, three active vectors (two medium and one large) are used to replace a zero vector.

First, the large vector on the right side of the sector (Figure 6, ①) is applied during $t_0/3$, and each medium vector (Figure 6, ⑤ and ⑥) is applied during $t_0/3$.

Since the applied vector sequence has a great impact on the CMV and on the switching losses, a sequence with minimum commutations has been chosen for each sector. For instance, when the reference voltage vector lays in sector 1, the next vector sequence is applied: 11001, 11100, 10000, 01000, 00100, 00010, 01000, 10000, 11100 and 11001 (Figure 6). Odd vector variant AZSL5M5-PWM vector sequences depending on the reference voltage sector are given in Table 4. It is important to note that, for AZSL5M5-PWM, more than one commutation is produced at each vector change.

Table 4. AZSL5M5-PWM vector sequences (odd vectors).

AZSL5M5-PWM	Vector Sequence
Sector 1	11001 11100 10000 01000 00100 00010 01000 10000 11100 11001
Sector 2	11100 01110 01000 00100 00010 00001 00100 01000 01110 11100
Sector 3	01110 00111 00100 00010 00001 10000 00010 00100 00111 01110
Sector 4	00111 10011 00010 00001 10000 01000 00001 00010 10011 00111
Sector 5	10011 11001 00001 10000 01000 00100 10000 00001 11001 10011

As the α-β plane is divided into five sectors instead of ten (Figure 5), the linear range of AZSL5M5-PWM is slightly reduced, being the maximum achievable output voltage:

$$V_{o_{MAX}} = \frac{4}{5} \cos\left(\frac{\pi}{5}\right) \cos\left(\frac{\pi}{5}\right) = 0.5236 V_{DC}. \tag{9}$$

For applications where achieving full linear range is mandatory, an hybrid modulation that extends AZSL5M5-PWM's linear range is proposed in the following.

4.2. Hybridization of the Proposed Modulation Algorithm

Three operation areas have been differentiated in Figure 7 to carry out the hybrid modulation algorithm and extend the linear range of AZSL5M5-PWM. The AZSL5M5-PWM hybrid variant (HASZL5M5-PWM) that uses odd vectors (white area) has been chosen as the main modulation scheme. When V_{ref} steps over the boundaries of the white pentagon, two things may occur. On the one hand, V_{ref} might remain within the shadowed boundaries. In such case, AZSL5M5-PWM with even vectors would be applied. On the other hand, if V_{ref} is out of the limits of AZSL5M5-PWM with even or odd vectors, SV-PWM can be used to fulfill the remaining area, marked with lines. This modification extends the linear range of AZSL5M5-PWM up to 26.8%. Further variants with greater CMV reduction could also be considered if a full range RCMV-PWM technique, such as AZSL2M2-PWM, is used instead of SV-PWM.

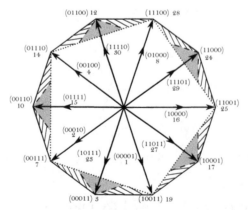

Figure 7. HAZSL5M5-PWM. White: AZSL5M5-PWM with odd vectors; shadowed: AZSL5M5-PWM with even vectors; lines: SV-PWM.

5. Simulation Results

In order to validate the proposal, two simulation platforms have been implemented in Matlab/Simulink. On the one hand, an open-loop model has been created to evaluate the HAZSL5M5-PWM technique and compare its performance with other existing ones regardless of the influence of a control algorithm. On the other, a detailed five-phase EMA model has been implemented to evaluate the proposal in the context of a variable speed AC drive. Without losing generality and in order not to significantly increase the computational burden of the model, ideal switch models that do not consider switching transients nor dead-time effects have been adopted in both simulation platforms. The obtained results and their discussion are provided in the following.

5.1. Open-Loop Model Simulation Results

Figure 8 shows the open-loop model block diagram. SimPowerSystem blocks have been used to model the power elements. The battery has been modeled as an ideal DC voltage source. The power-converter block includes a two-level five-phase voltage source inverter, where each switching device includes a detailed loss and thermal model, allowing an accurate estimation of inverter losses. In this work, a loss model of the International Rectifier AUIRGPS4067D1 IGBT has been implemented for each switch, whose main parameters are detailed in Table 5. The loss and thermal model follows the same approach as the one presented by the authors in [50]. The analytical approach used in this work to estimate the instantaneous conduction and switching losses is commonly used by the scientific community [51] and by the industry [52]. On the other hand, the adopted 1D thermal modeling approach has been verified by the authors in [53], where it has been compared to 3D finite element method (FEM) simulation, obtaining almost the same results. Finally, a passive star-connected five-phase RL load has been included. The most significant parameters of the open-loop model are collected in Table 6.

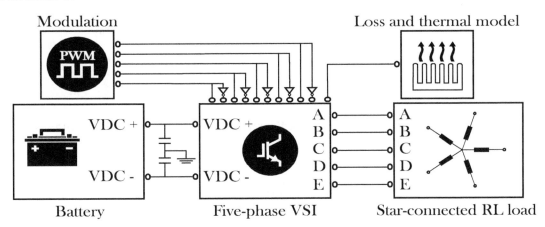

Figure 8. Block diagram of the constituting parts of the open-loop Simulink model.

Table 5. Most relevant parameters of the simulated International Rectifier AUIRGPS4067D1 IGBT.

Parameter	Value	Unit
Nominal current per switch	120	A
Maximum blocking voltage	600	V
Typical IGBT collector-emitter voltage	1.7	V
Typical diode forward voltage	1.7	V
Typical IGBT turn-on switching loss	8.2	mJ
Typical IGBT turn-off switching loss	2.9	mJ
Typical diode reverse recovery	2.4	mJ
IGBT thermal resistance	0.2	°C/W
Diode thermal resistance	0.25	°C/W
Allowable junction temperature	−55 to 175	°C

Table 6. The most significant parameters of the open-loop simulation platform.

Variable	Symbol	Value	Unit
Load resistance	R_{Load}	0.001	Ω
Load inductance	L_{Load}	1	mH
Battery voltage	V_{DC}	320	V
Modulator frequency	f_{mod}	50	Hz
Switching frequency	f_{sw}	10000	Hz

Figure 9 shows the THD and the efficiencies obtained for the proposed algorithm and for other techniques for all the linear range. As it was expected, RCMV-PWM modulations show greater harmonic content when compared to SV-PWM due to the use of phase-opposite vectors. However, for high modulation index values, all the studied modulations produce a similar THD. On the other hand, while AZSL2M2-PWM and SV-PWM have similar efficiencies, the HAZSL5M5-PWM has a slightly lower efficiency, which increases for low modulation indexes.

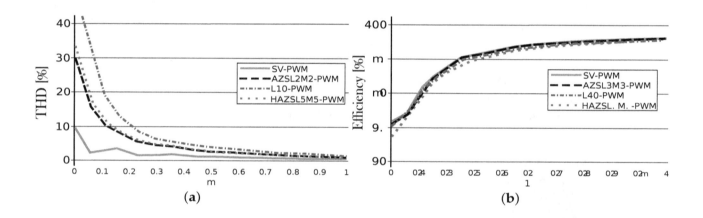

Figure 9. Total harmonic distortion (THD) and efficiency of studied modulation techniques for static operation points. (a) THD. (b) Efficiency.

Power losses can be seen in more detail in Figure 10. As mentioned before, HAZSL5M5-PWM requires more commutations at each vector change which entails an increase of switching losses (Figure 10a). However, conduction losses are almost equal in all modulations (Figure 10b). On the other hand, Figure 10c shows the load power as a function of the modulation index.

Regarding CMV mitigation, the proposed HAZSL5M5-PWM technique reduces Δ_{CMV} and N_{CMV} by a 60% and 80%, respectively, when $m \leq 0.8507$. These percentages are reduced while m gets close to 1. The worst case scenario, when modulation index is 1, AZSL5M5-PWM is active 29.78% of the simulated time while SV-PWM is active the 70.22% of the simulated time. In such a case, the Δ_{CMV} is reduced by 17.86% and N_{CMV} is reduced by 23.82%. In addition, when applying the operation condition equivalent to maximum torque ($T_{em_{max}} = 26$ Nm) and maximum speed ($\omega_{max} = 105$ rpm) that allows this particular application (modulation index = 0.96), AZSL5M5-PWM is active 49.8% of the simulated time while SV-PWM is active the 50.2% of the simulated time, reducing the Δ_{CMV} by 29.88% and N_{CMV} by 39.84%. So, even when the most torque and speed values are considered, HAZSL5M5-PWM reduces the N_{CMV} as much as AZSL2M2-PWM and L10-PWM techniques.

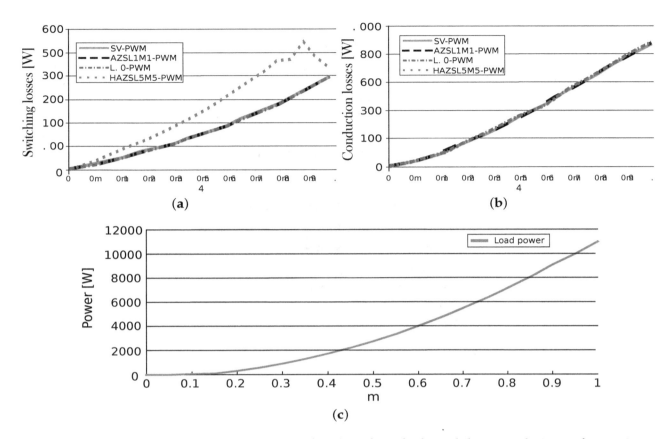

Figure 10. Distribution of power losses for the of studied modulation techniques for static operation points. (**a**) Switching losses. (**b**) Conduction losses. (**c**) Load power.

5.2. EMA Model Simulation Results

Figure 11 shows the block diagram of the implemented EMA model where, as in the open-loop model, the same power loss model based on the International Rectifier AUIRGPS4067D1 IGBT has been implemented. The EMA incorporates a star-connected five-phase PMSM, which third harmonic back-EMF component is negligible. Table 7 shows the main parameters of the simulated EMA.

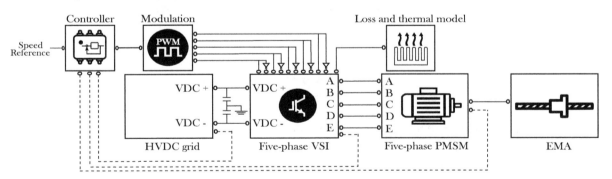

Figure 11. General block diagram of the electromechanical actuator (EMA) simulation platform.

The five-phase PMSM stator voltages are given by:

$$\mathbf{V} = \mathbf{R}\mathbf{I} + \mathbf{L}\frac{d\mathbf{I}}{dt} + \frac{d\mathbf{\Psi}_{PM}}{dt}, \tag{10}$$

where $\mathbf{V}$ and $\mathbf{I}$ are five-dimensional vectors whose element (v_j and i_j, $j \in [a,b,...,e]$) are the per-phase voltages and currents, respectively. $\mathbf{R}$ is a 5×5 diagonal matrix, where each diagonal element represents the phase resistance. $\mathbf{L}$ is the 5×5 stator inductance matrix, where each element L_{ij} ($i,j \in [a,b,...,e]$) represent the self- ($i = j$) and mutual-inductances ($i \neq j$) between phases i

and j. Being the value of the mutual-inductances very low and magnetic saturation phenomena negligible, mutual-inductances have been considered zero and self-inductances have been considered constant in the implemented electric machine model. A perfectly balanced stator has been considered. These assumptions have been done without losing generality for the evaluation of the proposed modulation algorithm as, if any non-ideality is present in the electric drive, torque and speed loops are responsible of their compensation, while the proposed algorithm synthesizes the commanded reference voltages and minimizes CMV. The term $\mathbf{\Psi}_{PM}$ is the five-dimensional flux linkage vector ($\mathbf{\Psi}_{PM} = [\Psi_{PMa}, \Psi_{PMb}, \ldots, \Psi_{PMe}]^T$) produced due to the permanent magnets.

Table 7. Most significant parameters of the simulated EMA.

Parameter	Symbol	Value	Unit
Rated power	P_{nom}	1.51	kW
Rated torque	T_{nom}	12.1	Nm
Rated speed	ω_{nom}	1200	RPM
Pole-pair number	N_p	9	−
Stator resistance	R_s	1.5	Ω
Stator self-inductance	L_s	9.6	mH
PM flux linkage	Ψ_{PM}	0.13	Wb
HVDC grid voltage	V_{DC}	270	V
Switching frequency	f_{sw}	10000	Hz

The torque produced by the motor is given by:

$$T_{em} = \mathbf{I}^T \frac{d\mathbf{\Psi}_{PM}}{d\theta_m},$$

(11)

where θ_m is the angular mechanical rotor position. The dynamics of the rotational movement are given by:

$$T_{em} - T_l = J\frac{d\omega_m}{dt} + B\omega_m,$$

(12)

where T_l is the load torque produced by the EMA, J is the total inertia moment of the rotating masses, including EMA and motor, ω_m is the rotational speed of the rotor and B is the viscous friction coefficient.

Figure 12 shows the detailed diagram of the controller. The controller consists of two control loops. The outer one regulates the rotational speed of the motor. This loop has a proportional-integral (PI) controller tuned in z. For this application, the damping factor has been set to $\zeta = 0.707$, while the settling-time has been set to $T_s = 50$ ms. The inner loop tracks the current references through a vector controller [54]. Again, $\zeta = 0.707$ for the current regulator, while $T_s = 5$ ms. In this particular case, only two PI controllers are required to control the first harmonic components (i_{d1}, i_{q1}), as there is no third harmonic back-EMF component and the proposed PWM technique intrinsically regulates to zero the third harmonic voltages (V_x^* and V_y^* are imposed to be zero). It must be taken into account that, for this particular control approach, a conventional microcontroller sine-triangle PWM peripheral cannot be used due to the modulation algorithm computational requirements. Thus, an FPGA should be incorporated to implement the modulation algorithm, while implementing the speed and current loops in a fixed-point or floating point DSP.

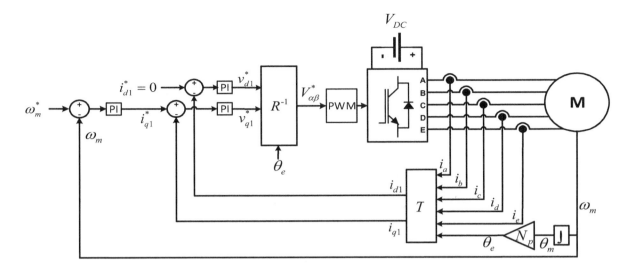

Figure 12. Block diagram of the EMA speed and torque controller.

Therefore, the following *abcde* to *d1-q1* transformation (T) is used in the controller:

$$T = \frac{2}{5} \begin{bmatrix} cos(\theta_e) & cos(\theta_e - 2\pi/5) & cos(\theta_e - 4\pi/5) & cos(\theta_e - 6\pi/5) & cos(\theta_e - 8\pi/5) \\ -sin(\theta_e) & -sin(\theta_e - 2\pi/5) & -sin(\theta_e - 4\pi/5) & -sin(\theta_e - 6\pi/5) & -sin(\theta_e - 8\pi/5) \end{bmatrix}, \qquad (13)$$

this being the matrix product of the transformation in (1) with the following rotational matrix:

$$R = \begin{bmatrix} cos(\theta_e) & sin(\theta_e) & 0 & 0 & 0 \\ -sin(\theta_e) & cos(\theta_e) & 0 & 0 & 0 \end{bmatrix}, \qquad (14)$$

where θ_e is the electrical rotor position of the motor, being $\theta_e = N_p\theta_m$.

Once the inner loop PI controllers provide the voltage references (v^*_{d1}, v^*_{q1}), such references are transformed into the $\alpha\beta$ frame by applying the R^{-1} (pseudo-inverse of R) matrix and fed to the PWM block. Matrix R^{-1} is the classical counter-clockwise rotation transformation [42]:

$$R^{-1} = \begin{bmatrix} cos(\theta_e) & -sin(\theta_e) \\ sin(\theta_e) & cos(\theta_e) \end{bmatrix}, \qquad (15)$$

Several simulations have been performed for various torque and speed conditions that cover the whole operation range of the EMA in order to evaluate the figures of the HAZSL5M5-PWM algorithm compared to other techniques.

Figures 13a–c show the efficiency results and the distribution between conduction and switching losses. Similar results as in open-loop simulations have been obtained for the EMA platform. As expected, switching losses increase when applying HAZSL5M5-PWM. However, such losses are not linear since hybrid AZSL5M5-PWM also includes SV-PWM algorithm and, when high modulation indexes are required, SV-PWM and HAZSL5M5-PWM techniques operate together reducing commutation losses. In terms of overall system efficiency, it is only reduced for about 1% when compared HAZSL5M5-PWM to SV-PWM. However, Δ_{CMV} and N_{CMV} are significantly reduced thanks to the proposed technique. In addition, in this particular application, AZSL5M5-PWM is operating all the time in all the simulated operation points except the one described in the previous section (T_{em} = 26 Nm and ω = 105 rpm). Consequently, the benefits of the AZSL5M5-PWM are fully exploited in the vast majority of the operation points of this application.

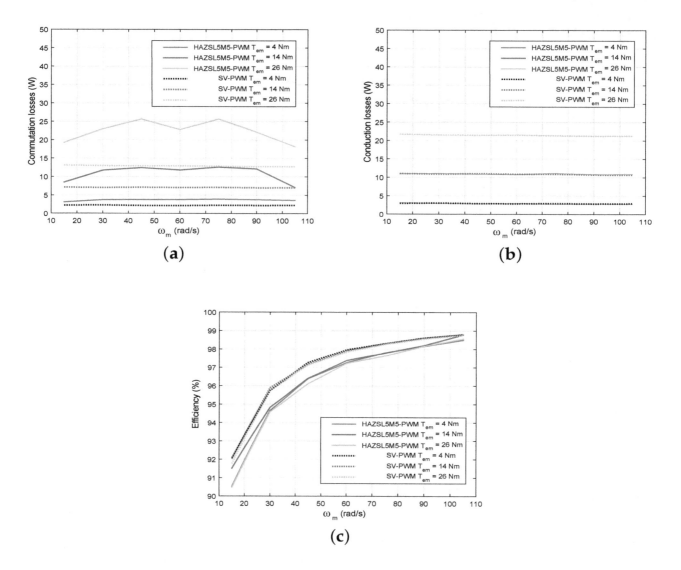

Figure 13. Power losses and efficiency of the proposed HAZSL5M5-PWM algorithm compared to other modulation techniques. (**a**) EMA switching losses for the studied modulation techniques. (**b**) EMA conduction losses for the studied modulation techniques. (**c**) EMA system efficiency with for the studied modulation techniques.

6. Conclusions

This paper introduces the CMV issue in multiphase electric drives and, more precisely, in EMA drives for MEA applications. In this context, the HAZSL5M5-PWM modulation technique is proposed. The basis of this technique merges the usage of phase-opposite vectors and the use of only odd or even active vectors to further reduce the CMV. The THD and efficiency characteristics of the proposed RCMV-PWM algorithm are evaluated and compared with SV-PWM and other RCMV-PWM techniques in an open-loop and in an EMA drive Simulink models. It is shown that the proposed algorithm achieves better THD performance when high modulation indexes are required. In addition, this technique reduces CMV variations (N_{CMV}) up to 80% when compared to SV-PWM, which directly implies a reduction in the leakage currents that affect the bearings. Therefore, the EMA reliability is improved. In exchange, with the proposed modulation the system efficiency slightly decreases. Thus, the investigation of other hybridization alternatives for the HAZSL5M5-PWM modulation technique that also consider THD and efficiency can be considered for future research.

Author Contributions: M.F. concept for the article, writing, proposed modulation technique development, open-loop and closed-loop models simulations and analysis. A.S-G. EMA model development and support on closed-loop model simulations. E.R. IGBT and thermal model development, support on open-loop simulations and review. I.K. support on modulation technique development, conceptual support, review and supervision. E.I. simulation platform development, review and supervision. J.L.M. review. All authors have read and agreed to the published version of the manuscript.

Abbreviations

The following abbreviations are used in this manuscript:

AC	Alternating current
AZS-PWM	Active Zero-State Pulse Width Modulation
AZSL2M2-PWM	Active Zero-State Two Large Two Medium
AZSL4-PWM	Active Zero-State Four Large
AZSL5M5-PWM	Active Zero-State Five Large Five Medium
CMC	Common mode current
CMV	Common mode voltage
DC	Direct current
EDM	Electric discharge machining
EMA	Electromechanical actuator
EMI	Electromagnetic interferences
FEM	Finite element method
GaN	Gallium nitride
HVDC	High-voltage direct current
IGBT	Insulated gate bipolar transistor
L5-PWM	Five Large Pulse Width Modulation
L10-PWM	Ten Large Pulse Width Modulation
M5-PWM	Five Medium Pulse Width Modulation
M10-PWM	Ten Medium Pulse Width Modulation
MEA	More Electric Aircrafts
PI	Proportional Integral
PMSM	Permanent Magnet Synchronous Machine
PWM	Pulse Width Modulation
RCMV-PWM	Reduced Common Mode Voltage Pulse Width Modulation
SiC	Silicon Carbide
SV-PWM	Space Vector Pulse Width Modulation
THD	Total Harmonic Distortion
VSI	Voltage Source Inverter
WBG	Wide Bandgap

References

1. Uzhegov, N.; Smirnov, A.; Park, C.H.; Ahn, J.H.; Heikkinen, J.; Pyrhönen, J. Design Aspects of High-Speed Electrical Machines With Active Magnetic Bearings for Compressor Applications. *IEEE Trans. Ind. Electron.* **2017**, *64*, 8427–8436. [CrossRef]
2. Riba, J.R.; López-Torres, C.; Romeral, L.; Garcia, A. Rare-earth-free propulsion motors for electric vehicles: A technology review. *Renew. Sustain. Energy Rev.* **2016**, *57*, 367–379. [CrossRef]
3. Kumar, M.S.; Revankar, S.T. Development scheme and key technology of an electric vehicle: An overview. *Renew. Sustain. Energy Rev.* **2017**, *70*, 1266–1285. [CrossRef]
4. Riveros, J.A.; Barrero, F.; Levi, E.; Durán, M.J.; Toral, S.; Jones, M. Variable-Speed Five-Phase Induction Motor Drive Based on Predictive Torque Control. *IEEE Trans. Ind. Electron.* **2013**, *60*, 2957–2968. [CrossRef]

5. Negahdari, A.; Yepes, A.G.; Doval-Gandoy, J.; Toliyat, H.A. Efficiency Enhancement of Multiphase Electric Drives at Light-Load Operation Considering Both Converter and Stator Copper Losses. *IEEE Trans. Power Electron.* **2019**, *34*, 1518–1525. [CrossRef]

6. Liu, Z.; Li, Y.; Zheng, Z. A review of drive techniques for multiphase machines. *CES Trans. Electr. Mach. Syst.* **2018**, *2*, 243–251. [CrossRef]

7. Diana, M.; Ruffo, R.; Guglielmi, P. PWM Carrier Displacement in Multi-N-Phase Drives: An Additional Degree of Freedom to Reduce the DC-Link Stress. *Energies* **2018**, *11*, 443. [CrossRef]

8. Zheng, P.; Sui, Y.; Zhao, J.; Tong, C.; Lipo, T.A.; Wang, A. Investigation of a Novel Five-Phase Modular Permanent-Magnet In-Wheel Motor. *IEEE Trans. Magn.* **2011**, *47*, 4084–4087.10.1109/TMAG.2011.2150207. [CrossRef]

9. Cao, W.; Mecrow, B.C.; Atkinson, G.J.; Bennett, J.W.; Atkinson, D.J. Overview of Electric Motor Technologies Used for More Electric Aircraft (MEA). *IEEE Trans. Ind. Electron.* **2012**, *59*, 3523–3531. [CrossRef]

10. Bozhko, S.; Hill, C.I.; Yang, T. More-Electric Aircraft: Systems and Modeling. In *Wiley Encyclopedia of Electrical and Electronics Engineering*; American Cancer Society: New York, NY, USA, 2018; pp. 1–31. [CrossRef]

11. Wheeler, P.W.; Clare, J.C.; Trentin, A.; Bozhko, S. An overview of the more electrical aircraft. *J. Aerosp. Eng.* **2012**, *227*, 578–585. [CrossRef]

12. Deng, Q.; Wang, Z.; Chen, C.; Czarkowski, D.; Kazimierczuk, M.K.; Zhou, H.; Hu, W. Modeling and Control of Inductive Power Transfer System Supplied by Multiphase Phase-Controlled Inverter. *IEEE Trans. Power Electron.* **2019**, *34*, 9303–9315. [CrossRef]

13. Belkhode, S.; Jain, S. Optimized Switching PWM Technique With Common-Mode Current Minimization for Five-Phase Open-End Winding Induction Motor Drives. *IEEE Trans. Power Electron.* **2019**, *34*, 8971–8980. [CrossRef]

14. Liu, Z.; Zheng, Z.; Xu, L.; Wang, K.; Li, Y. Current Balance Control for Symmetrical Multiphase Inverters. *IEEE Trans. Power Electron.* **2016**, *31*, 4005–4012. [CrossRef]

15. Prieto, B. Design and Analysis of Fractional-Slot Concentrated-Winding Multiphase Fault-Tolerant Permanent Magnet Synchronous Machines. Ph.D. Thesis, Tecnum Universidad de Navarra, Donostia, Spain, 2015.

16. Iqbal, A.; Rahman, K.; Abdallah, A.A.; Moin, K.A.S.; Abdellah, K. Current Control of a Five-phase Voltage Source Inverter. In Proceedings of the International Conference on Power Electronics and their Applications (ICPEA), Elazig, Turkey, 27 September 2013.

17. Bennett, J.W.; Atkinson, G.J.; Mecrow, B.C.; Atkinson, D.J. Fault-Tolerant Design Considerations and Control Strategies for Aerospace Drives. *IEEE Trans. Ind. Electron.* **2012**, *59*, 2049–2058. [CrossRef]

18. Bojoi, R.; Cavagnino, A.; Tenconi, A.; Tessarolo, A.; Vaschetto, S. Multiphase electrical machines and drives in the transportation electrification. In Proceedings of the IEEE 1st International Forum on Research and Technologies for Society and Industry Leveraging a better tomorrow (RTSI), Torino, Italy, 18 September 2015. [CrossRef]

19. Takahashi, S.; Ogasawara, S.; Takemoto, M.; Orikawa, K.; Tamate, M. Common-Mode Voltage Attenuation of an Active Common-Mode Filter in a Motor Drive System Fed by a PWM Inverter. *IEEE Trans. Ind. Appl.* **2019**, *55*, 2721–2730. [CrossRef]

20. Karampuri, R.; Jain, S.; Somasekhar, V.T. Common-Mode Current Elimination PWM Strategy Along with Current Ripple Reduction for Open-Winding Five-Phase Induction Motor Drive. *IEEE Trans. Power Electron.* **2019**, *34*, 6659–6668. [CrossRef]

21. Espina, J.; Balcells, J.; Arias, A.; Ortega, C. Common mode EMI model for a direct matrix converter. *IEEE Trans. Ind. Electron.* **2011**, *58*, 5049–5056. [CrossRef]

22. Muetze, A.; Binder, A. Don't lose your bearings. *IEEE Ind. Appl. Mag.* **2006**, *12*, 22–31. [CrossRef]

23. Hadden, T.; Jiang, J.W.; Bilgin, B.; Yang, Y.; Sathyan, A.; Dadkhah, H.; Emadi, A. A Review of Shaft Voltages and Bearing Currents in EV and HEV Motors. In Proceedings of the Industrial Electronics Society (IECON), Florence, Italy, 24 October 2016; pp. 1578–1583.

24. Mütze, A. Thousands of hits: On inverter-induced bearing currents, related work, and the literature. *Elektrotechnik Inf.* **2011**, *128*, 382–388. [CrossRef]

25. Muetze, A. On a New Type of Inverter-Induced Bearing Current in Large Drives With One Journal Bearing. *IEEE Trans. Ind. Appl.* **2010**, *46*, 240–248. [CrossRef]

26. Morya, K.; Gardner, M.C.; Anvari, B.; Liu, L.; Yepes, A.G.; Doval-Gandoy, J.; Toliyat, H.A. Wide Bandgap Devices in AC Electric Drives: Opportunities and Challenges. *IEEE Trans. Transp. Electrif.* **2019**, *5*, 3–20. [CrossRef]

27. Oh, W.; Willwerth, A. Shaft Grounding—A Solution to Motor Bearing Currents. *Am. Soc. Heating Refrig. Air Cond. Eng. Trans.* **2008**, *114*, 246–251.

28. Muetze, A.; Binder, A. Calculation of influence of insulated bearings and insulated inner bearing seats on circulating bearing currents in machines of inverter-based drive systems. *IEEE Trans. Ind. Appl.* **2006**, *42*, 965–972. [CrossRef]

29. Muetze, A. Bearing Currents in Inverter-Fed AC-Motors. Ph.D. Thesis, Der Technischen Universitaet Darmstadt, Darmstadt, Germany, 2004.

30. Schiferl, R.F.; Melfi, M.J. Bearing current remediation options. *IEEE Ind. Appl. Mag.* **2004**, *10*, 40–50. [CrossRef]

31. Muetze, A.; Oh, W. Application of Static Charge Dissipation to Mitigate Electric Discharge Bearing Currents. *IEEE Trans. Ind. Appl.* **2008**, *44*, 135–143. [CrossRef]

32. Nguyen, N.; Nguyen, T.; Lee, H. A reduced siwtching loss PWM strategy to eliminate common-mode voltage in multilevel inverters. *IEEE Trans. Power Electron.* **2015**, *30*, 5425–5438. [CrossRef]

33. Syed, A.; Kalyani, S. Evaluation of single phase transformerless photovoltaic inverters. *Electr. Electron. Eng. Int. J.* **2015**, *4*, 25–39. [CrossRef]

34. Freddy, T.K.S.; Rahim, N.A.; Hew, W.P.; Che, H.S. Comparison and Analysis of Single-Phase Transformerless Grid-Connected PV Inverters. *IEEE Trans. Power Electron.* **2014**, *29*, 5358–5369. [CrossRef]

35. Freddy, T.K.S.; Rahim, N.A.; Hew, W.P.; Che, H.S. Modulation Techniques to Reduce Leakage Current in Three-Phase Transformerless H7 Photovoltaic Inverter. *IEEE Trans. Ind. Electron.* **2015**, *62*, 322–331. [CrossRef]

36. Kouro, S.; Leon, J.I.; Vinnikov, D.; Franquelo, L.G. Grid-Connected Photovoltaic Systems: An Overview of Recent Research and Emerging PV Converter Technology. *IEEE Ind. Electron. Mag.* **2015**, *9*, 47–61. [CrossRef]

37. Fernandez, M.; Robles, E.; Kortabarria, I.; Andreu, J.; Ibarra, E. Novel modulation techniques to reduce the common mode voltage in multiphase inverters. In Proceedings of the IEEE Industrial Electronics Society Conference (IECON), Lisabon, Portugal, 14 October 2019; pp. 1898–1903.

38. Durán, M.J.; Prieto, J.; Barrero, F.; Riveros, J.A.; Guzman, H. Space-Vector PWM with Reduced Common-Mode Voltage for Five-Phase Induction Motor Drives. *IEEE Trans. Ind. Electron.* **2013**, *60*, 4159–4168. [CrossRef]

39. Rahman, A. *Power Electronics and Motor Drives*; CRC Press: Boca Raton, FL, USA, 2016; pp. 5–10.

40. Gray, R. Toeplitz and Circulant Matrices: A Review. *Found. Trends Commun. Inf. Theory* **2006**, *2*, 155–239. [CrossRef]

41. Semail, E; Bouscayrol, A.; Hautier, P.J. Vectorial formalism for analysis and design of polyphase synchronous machines. *Eur. Phys. J. Appl. Phys.* **2003**, *22*, 207–220. [CrossRef]

42. Tang, K.T. *Mathematical Methods for Engineers and Scientists 1*; Springer: Berlin, Germany, 2006.

43. Davis, P. *Circulant Matrices*; John Wiley & Sons Inc.: Hoboken, NJ, USA, 1979.

44. Zhar, H.; Gong, J.; Semail, E.; Sciuller, F. Comparison of Optimized Control Strategies of a High-speed Traction Machine with Five Phases and Bi-Harmonic Electromotive Force. *Energies* **2016**, *9*, 952. [CrossRef]

45. Kelly, W.; Strangas, E.G.; Miller, J.M. Multiphase Space Vector Pulse Width Modulation. *IEEE Power Eng. Rev.* **2002**, *22*, 53–53. [CrossRef]

46. Iqbal, A.; Levi, E. Space vector modulation schemes for a five-phase voltage source inverter. In Proceedings of the European Conference on Power Electronics and Applications (ECPEA), Barcelona, Spain, 10 September 2005; pp. 1–12. [CrossRef]

47. Yen-Shin, L.; Po-Sheng, C.; Hsiang-Kuo, L.; Chou, J. Optimal common-mode voltage reduction PWM technique for inverter control with consideration of the dead-time effects-part II: Applications to IM drives with diode front end. *IEEE Trans. Ind. Appl.* **2004**, *40*, 1613–1620. [CrossRef]

48. Iqbal Alammari, R.; Mosa, M.; Abu-Rub, H. Finite set model predictive current control with reduced and constant common mode voltage for a five-phase voltage source inverter. In Proceedings of the International Symposium on Industrial Electronics (ISIE), Istambul, Turkey, 4 June 2014; pp. 479–484. [CrossRef]

49. Munim, N.A.; Ismail, M.F.; Abidin, A.F.; Haris, H.M. Multi-phase inverter Space Vector Modulation. In Proceedings of the International Power Engineering and Optimization Conference (PEOCO), Langkawi, Malaysia, 22 July 2013; pp. 149–154. [CrossRef]

50. Robles, E.; Fernandez, M.; Ibarra, E.; Andreu, J.; Kortabarria, I. Mitigation of Common Mode Voltage Issues in Electric Vehicle Drive Systems by Means of an Alternative AC-Decoupling Power Converter Topology. *Energies* **2019**, *12*, 3349. [CrossRef]

51. Sadigh, A.K.; Dargahi, V.; Corzine, K. Analytical determination of conduction power loss and investigation of switching power loss for modified flying capacitor multicell converters. *IET Power Electron.* **2016**, *9*, 175–187. [CrossRef]

52. Wintrich, A.; Nicolai, U.; Tursky, W.; Reimann, T. *Application Manual Power Semiconductors*; Semikron: Nürnberg, Germany, 2017.

53. Matallana, A.; Robles, E.; Ibarra, E.; Andreu, J.; Delmonte, N.; Cova, P. A methodology to determine reliability issues in automotive SiC power modules combining 1D and 3D thermal simulations under driving cycle profiles. *Microelectron. Reliab.* **2019**, *102*, 1–9. [CrossRef]

54. Parsa, L.; Toliyat, H. Five-Phase Permanent-Magnet Motor Drives. *IEEE Trans. Ind. Appl.* **2005**, *41*, 30–37. [CrossRef]

11

Wound Synchronous Machine Sensorless Control based on Signal Injection into the Rotor Winding

Jongwon Choi and Kwanghee Nam *

Department of Electrical Engineering, Pohang University of Science and Technology, 77 Cheongam-Ro, Nam-Gu, Pohang 37673, Korea; jongwon@postech.ac.kr
* Correspondence: kwnam@postech.ac.kr.

Abstract: A sensorless position scheme was developed for wound synchronous machines. The demodulation process is fundamentally the same as the conventional signal-injection method. The scheme is different from techniques for permanent-magnet synchronous machines, in that it injects a carrier signal into the field (rotor) winding. The relationship between the high-frequency current responses and the angle estimation error was derived with cross-coupling inductances. Furthermore, we develop a compensation method for the cross-coupling effect, and present several advantages of the proposed method in comparison with signal injection into the stator winding. This method is very robust against magnetic saturation because it does not depend on the saliency of the rotor. Furthermore, the proposed method does not need to check the polarity at a standstill. Experiments were performed to demonstrate the improvement in the compensation of cross-coupling, and the robustness against magnetic saturation with full-load operation.

Keywords: core saturation; cross-coupling inductance; wound synchronous machines (WSM); signal injection; position sensorless; high-frequency model

1. Introduction

Although their speed is not so high, electrical vehicles (EVs) and hybrid EVs (HEVs) are steadily growing their share in the market. Today, permanent-magnet synchronous machines (PMSMs) are widely used in traction applications because of their superior power density and high efficiency. However, the permanent-magnet (PM) materials, typically neodymium (Nd) and dysprosium (Dy), are expensive, and their price fluctuates depending on political situations. Therefore, some research has been directed to developing Nd-free motors. Wound synchronous machines (WSM) is a viable alternative to a PMSM. The main advantage of a WSM is that it has an additional degree of freedom in the field-weakening control because the rotor field can be adjusted [1–3].

There are two types of sensorless angle detection techniques: back-EMF-based and signal-injection methods. The former is based on the relative magnitudes of d and q-axis EMFs, whereas the latter is based on the spatial saliency of rotor. The back-EMF-based methods are reliable and superior in the medium- and high-speed regions [4–10]. However, they exhibit poor performance in the low-speed region owing to lack of the "observability" [11].

On the other hand, signal-injection methods work well in the zero-speed region, even with a full load [12–19]. The injection method does not use the magnetic polarity; it requires the use of a polarity-checking method before starting [14,15]. The signal-injection method is not feasible in some saturation regions where the d and q-axis inductances are close to each other [16,17]. The cross-coupling inductance refers to an incremental inductance developed by the current in the quadrature position. Cross-coupling, being another saturation phenomenon, becomes significant as the load increases. Zhu et al. [18] showed that cross-coupling caused an offset error in the angle estimation and proposed a method eliminating it. However, the cross-coupling inductances change considerably depending on the current magnitudes. Therefore, a lookup table should be used for its compensation method. A group of researchers are working together toward developing specialized motors that are suitable for signal injection [19].

Similar to PMSMs, back-EMF-based sensorless methods were developed for WSMs [20–23]. Boldea et al. [22] proposed an active-flux-based sensorless method, and this paper presents good experimental results in low-speed operation with a heavy load. Amit et al. [23] used a flux observer in the stator flux coordinate. Recently, the sensorless signal injection for WSM was published, whose carrier signal is injected to an estimated d-axis, and the response in estimated q-axis current [24,25]. Griffo et al. [24] applied a signal-injection method to a WSM to start an aircraft engine, and presented full-torque operation from zero to a high speed. Rambetius et al. [25] compared two detection methods when a signal was injected into the stator winding: one from a stator winding and the other from the field (rotor) winding.

Signal injection to the rotor winding of the WSM has been reported in recent years [26–29]. Obviously, detected signals from the (stator) d-q axes differ depending on rotor position. The stator voltage responses [26] and current responses [27–29] are checked to obtain rotor angle information using mutual magnetic coupling between the field coil and the stator coil. Using inverse sine function and q-axis current in the estimated frame, the position estimation algorithm was presented for a sensorless direct torque control WSM and it presented experimental results at zero speed [27]. Rambetius and Piepenbreier [28,29], included cross-saturation effects in the high-frequency model and presented the position estimation method using q-axis current in the estimation synchronous frame and linearization. The model was included in the stator but also in the field dynamics. It is a reasonable approach when a voltage as the carrier signal is injected to rotor winding. However, a 3 × 3 inductance matrix should be handled, and it is pretty complicated and difficult to analyze.

This paper extends the work in [30]. The signal is injected into the rotor winding, and the resulting high-frequency is detected from the stator currents. The effect of the cross-coupling inductances is modeled. Since the field current is modeled as a current source, 2 × 2 inductance matrix can be obtained. It is easy to calculate the inverse matrix and analyze the effect of the cross-coupling inductance on rotor angle estimation. The dq-axes stator flux linkage are obtained by finite-element analysis. Using the flux linkage, the dq-axes self-inductances and the mutual inductances between the stator and rotor are obtained and analyzed. Then, an offset angle caused by the cross-coupling inductances is straightforwardly derived. Using both dq-axes currents in the misaligned frame and inverse tangent function, the rotor angle estimation algorithm is developed without linearization. The offset angle caused by cross-coupling effect is directly compensated, and the stability issue of the compensation method is analyzed. Furthermore, it explains why injection into the rotor winding is more robust than the existing methods. Finally, experimental results verify that the rotor position is obtained accurately at standstill and very low speed.

This paper is organized as follows. In Section 2, the WSM model is derived with coupling inductances. In Section 3, the current responses in misaligned coordinate are derived and a sensorless method is proposed. Some advantages of the field signal-injection method are presented in comparison with stator signal injection in Section 4. Section 5 presents a performance comparison between the injection methods to stator and rotor. In Section 6, the performance of the sensorless method is demonstrated by experiment. Finally, in Section 7, some conclusions are drawn.

2. Modeling of a WSM

A schematic diagram of a WSM is depicted in Figure 1. Please note that a high-frequency signal is injected into the field winding via a slip ring.

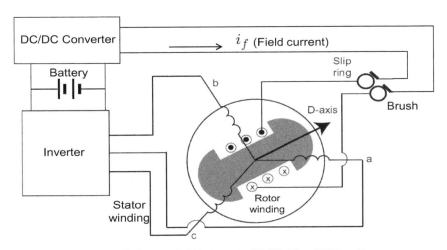

Figure 1. Schematic Diagram of WSM for EV traction.

2.1. WSM Voltage Model

From [31] (in Chapter 11), the voltage equations of WSM are given as

$$v_d^e = r_s i_d^e + \frac{d}{dt}\lambda_d^e - \omega_e \lambda_q^e, \tag{1}$$

$$v_q^e = r_s i_q^e + \frac{d}{dt}\lambda_q^e + \omega_e \lambda_d^e, \tag{2}$$

where v_d^e, v_q^e are the d and q-axis voltages and ω_e is the electrical angular frequency; the superscript e represents the synchronous (rotating) reference frame. Note that $\frac{d}{dt}\lambda_d^e$ and $\frac{d}{dt}\lambda_q^e$ are important terms for sensorless signal-injection methods because the terms are represented as the induced voltage caused by the high-frequency current.

The stator flux linkages of a WSM in the synchronous frame are expressed as $\lambda_d^e(i_d^e, i_q^e, i_f')$, and $\lambda_q^e(i_d^e, i_q^e, i_f')$, where λ_d^e and λ_q^e are non-linear functions [3]; λ_d^e and λ_q^e are the d and q-axes stator flux linkage; i_d^e and i_q^e are the d and q-axes currents; and i_f' is the field current referred to stator. Note that i_f' is represented by $i_f' = \frac{2}{3}\frac{N_s}{N_{fd}}i_f$ [31], where i_f is the field current, N_s and N_{fd} are the number of stator coil and field coil turns. To take account of the cross-coupling effect, the following equations are derived by the chain rule

$$\frac{d\lambda_d^e}{dt} = \underbrace{\frac{\partial\lambda_d^e}{\partial i_d^e}\frac{di_d^e}{dt}}_{\equiv L_{dd}} + \underbrace{\frac{\partial\lambda_d^e}{\partial i_q^e}\frac{di_q^e}{dt}}_{\equiv L_{dq}} + \underbrace{\frac{\partial\lambda_d^e}{\partial i_f'}\frac{di_f'}{dt}}_{\equiv L_{df}} \tag{3}$$

$$\frac{d\lambda_q^e}{dt} = \underbrace{\frac{\partial\lambda_q^e}{\partial i_q^e}\frac{di_q^e}{dt}}_{\equiv L_{qq}} + \underbrace{\frac{\partial\lambda_q^e}{\partial i_d^e}\frac{di_d^e}{dt}}_{\equiv L_{qd}} + \underbrace{\frac{\partial\lambda_q^e}{\partial i_f'}\frac{di_f'}{dt}}_{\equiv L_{qf}}, \tag{4}$$

where L_{dd} and L_{qq} are the self (incremental) inductances, and L_{dq} and L_{qd} are the cross-coupling (incremental) inductances of the stator coil [16]. L_{df} is the mutual (incremental) inductance between the d-axis and the field coils, and L_{qf} is the cross-coupling (incremental) inductance between the stator and field coils. The d-axis inductance can be decomposed as the sum of the mutual and leakage inductances:

$$L_{dd} = L_{df} + L_{ls}, \tag{5}$$

where L_{ls} is the leakage inductance of the stator. Therefore, $L_{dd} \approx L_{df} \gg L_{qf}$.

Substituting (3) and (4) into (1) and (2), the voltage equations are obtained as

$$v_d^e = r_s i_d^e + L_{dd}\frac{di_d^e}{dt} + L_{dq}\frac{di_q^e}{dt} + L_{df}\frac{di_f'}{dt} - \omega_e \lambda_q^e \tag{6}$$

$$v_q^e = r_s i_q^e + L_{qq}\frac{di_q^e}{dt} + L_{qd}\frac{di_d^e}{dt} + L_{qf}\frac{di_f'}{dt} + \omega_e \lambda_d^e. \tag{7}$$

Using (5)–(7), an equivalent circuit can be constructed as shown in Figure 2.

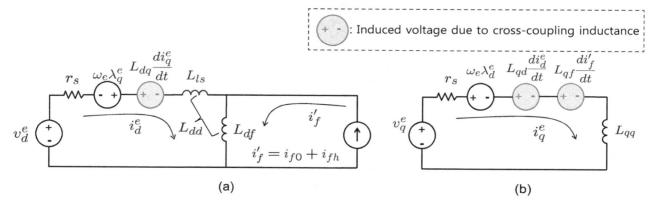

Figure 2. Equivalent circuit of the WSM including cross-coupling inductance: (a) d-axis and (b) q-axis.

Figure 3a shows a FEM model of WSM used in this experiment. The motor has six salient poles, the continuous rated current is 161 Arms, and the maximum power is 65 kW. The other parameters are listed in Table 1. Figure 3b shows plots of the a-, b-, c-phase flux linkages in the stationary frame as the rotor rotates at 500 r/min, which were obtained through finite-element method (FEM) calculations. The $d - q$ axis flux linkages are obtained using

$$\begin{bmatrix} \lambda_d^e \\ \lambda_q^e \end{bmatrix} = \frac{2}{3}\begin{bmatrix} \cos\theta_e & \sin\theta_e \\ -\sin\theta_e & \cos\theta_e \end{bmatrix}\begin{bmatrix} 1 & -\frac{1}{2} & -\frac{1}{2} \\ 0 & \frac{\sqrt{3}}{2} & -\frac{\sqrt{3}}{2} \end{bmatrix}\begin{bmatrix} \lambda_a \\ \lambda_b \\ \lambda_c \end{bmatrix}, \tag{8}$$

where θ_e is the electrical angle. Figure 4 shows the plots of $\lambda_d^e(i_d^e, i_q^e, i_f)$ and $\lambda_q^e(i_d^e, i_q^e, i_f)$ in the (i_d^e, i_q^e) plane when $i_f = 6$ A. Please note that λ_d^e changes more, i.e., the slope becomes steeper as the d-axis current increases negatively. When i_q^e is under 50 A, i_q^e has little effect on λ_d^e. However, when i_q^e is over 100 A, λ_d^e decreases more as i_q^e increases. It is called "cross-coupling phenomenon". Figure 5 shows a contour of $\lambda_d^e(i_d^e, i_q^e, i_f)$ and $\lambda_q^e(i_d^e, i_q^e, i_f)$ when $i_f = 6$ A or $i_f = 6.5$ A. λ_d^e seems to be proportional to i_f. But, in Figure 5b, λ_q^e with $i_f = 6$ A seems to be the same λ_q^e with $i_f = 6.5$ A when i_q^e is under 50 A. But, it is clear that λ_q^e decreases as i_q^e increases when i_q^e is over 100 A.

Table 1. Parameters of a WSM used in the experiments.

Parameter	Value
Maximum power	65 kW
Maximum torque	123 Nm
Maximum current	161 Arms
Numbers of poles (P)	6 poles
Number of slots	36 slot
Back-EMF coefficient	0.121 Wb
Maximum speed	12,000 r/min
Field current (i_f)	6 A
Number of stator coil turns (N_s)	3 turns
Number of field coil turns (N_{fd})	200 turns

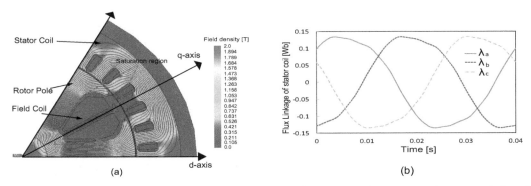

Figure 3. Flux linkage calculation when $i_f = 6$ A: (**a**) Finite-element analysis model and (**b**) flux linkages of stator coils.

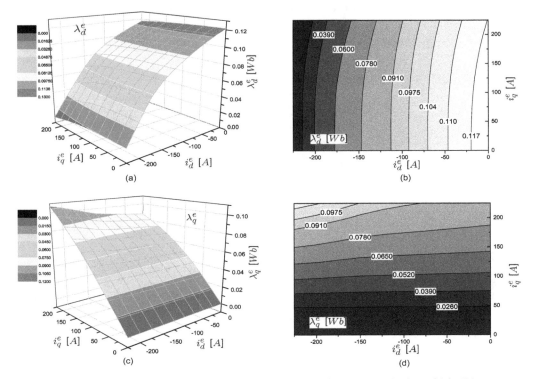

Figure 4. Flux linkages when $i_f = 6$ A (FEM analysis): (**a**) 3-dimensional plot of λ_d^e; (**b**) contour plot of λ_d^e; (**c**) 3-dimensional plot of λ_q^e; and (**d**) contour plot of λ_q^e.

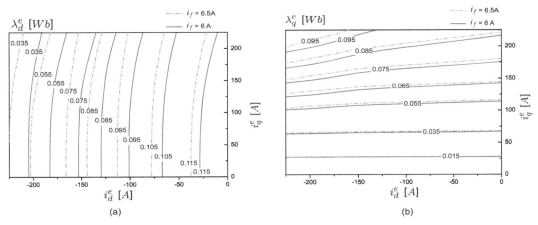

Figure 5. The Influence of the field current on dq-axes stator flux linkage (λ_d^e, λ_q^e): (**a**) Contour plot λ_d^e, and (**b**) contour plot of λ_d^e. (solid and blue line: $i_f = 6$ A, dash dot and magenta line: $i_f = 6.5$ A).

2.2. Incremental Inductance Calculations

As shown in [18,32], the incremental inductances are calculated through numerical differentiation:

$$L_{dq} = \frac{\partial \lambda_d^e}{\partial i_d^e} \approx \frac{\lambda_d^e(i_d^e, i_q^e + \Delta i_q^e, i_f) - \lambda_d^e(i_d^e, i_q^e, i_f)}{\Delta i_q^e}$$

$$L_{df} = \frac{\partial \lambda_d^e}{\partial i_f^e} \approx \frac{\lambda_d^e(i_d^e, i_q^e, i_f + \Delta i_f) - \lambda_d^e(i_d^e, i_q^e, i_f)}{\Delta i_f}.$$

The rest of the inductances are obtained similarly. Figure 6 shows L_{dd}, L_{qq}, and L_{dq} under various current conditions. From Figure 6a, L_{dd} is, in general, independent of the q-axis current when i_q is not so high, whereas it depends strongly on the d-axis current, i.e., L_{dd} increases as i_d^e increases negatively. This is because the rotor core becomes free from the saturation caused by the field current. Specifically, the negative d-axis current induces a field in the opposite direction, i.e., it cancels the rotor field.

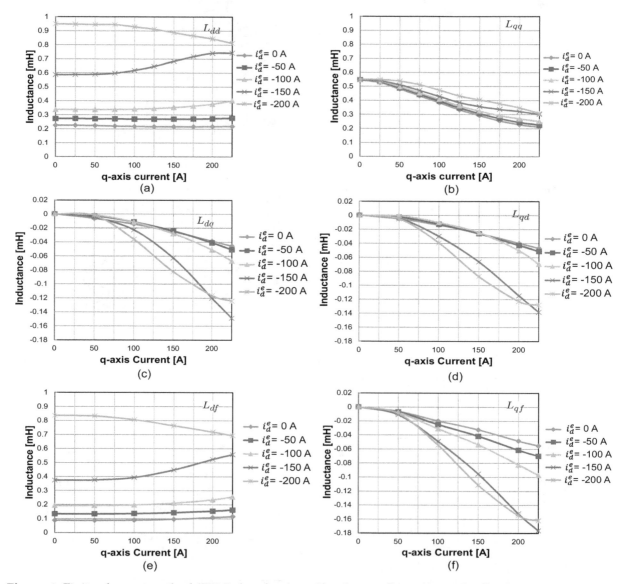

Figure 6. Finite-element method (FEM) data for the self and mutual incremental inductances: (**a**) d-axis self-inductance L_{dd}; (**b**) q-axis self-inductance L_{qq}; (**c**) d-axis and q-axis cross-coupling inductance L_{dq}; (**d**) q-axis and d-axis cross-coupling inductance L_{qd}; (**e**) d-axis and field coil mutual inductance L_{df}; and (**f**) q-axis and field coil cross-coupling inductance L_{qf}.

In addition, also from Figure 6b–d, L_{qq}, L_{dq}, and L_{qd} decrease as i_q^e increases. Figure 6c,f show plots of L_{df} and L_{qf} that show the mutual inductances from the field winding, i.e., the flux variation induced by the field current. Because the high-frequency probing signal is injected into the field coil, L_{df} and L_{qf} play a crucial role in this work. Comparing Figure 6a,c, the shapes of L_{dd} and L_{df} are similar. It is because they are the same except the leakage inductance, as shown in (5). Note from Figure 6c,d,f, the cross-coupling inductances (L_{dq}, L_{qd}, L_{qf}) behave similarly and have negative values. Because of the non-linear magnetic property of core, the core saturation give rise to cross-coupling phenomenon. For instance, when the field current is fixed at 6 A, $\lambda_{d\,(i_d^e=0,i_q^e=50)} = 0.121345$ Wb and $\lambda_{d\,(i_d^e=0,i_q^e=200)} = 0.118252$ Wb were obtained in Figure 4b. Despite the same d-axis current, the d-axis flux linkages decrease as i_q^e increases. This is because the q-axis current saturates the core deeper, and it affects d-axis flux linkages to be reduced. Consequently, the cross-coupling inductance between d and q-axis, L_{dq}, can be negative values, the phenomenon is also exhibited in PMSMs [18]. Correspondingly, the q-axis flux linkages decrease as i_f or i_d^e increases in Figures 4d and 5b. Therefore, L_{qd} and L_{qf} have negative values.

When a signal is injected into the field winding, the current responses are monitored in the stator winding via L_{df} and L_{qf}. The use of L_{qf} is different from PMSM signal-injection methods because it provides another signal path to the q-axis besides the one formed by rotor saliency.

3. High-Frequency Model of a WSM

As shown in Figure 1, the WSM has a separate field controller, which feeds i_f to the field winding via a slip ring and brush. A high-frequency carrier is superposed on the field current. Then, the signal is detected on the stator winding, on which the rotor angle is estimated. The field current controller supplies

$$i_f'(t) = I_{f0} + i_{fh}(t) = I_{f0} - I_h \cos \omega_h t, \tag{9}$$

where I_{f0} and I_h are the amplitudes of the dc and ac components, and ω_h is the angular speed of the carrier. Furthermore, the d and q-axis currents can be separated as

$$i_d^e = i_{d0}^e + i_{dh}^e, \tag{10}$$
$$i_q^e = i_{q0}^e + i_{qh}^e, \tag{11}$$

where i_{d0}^e and i_{q0}^e are dc components current, i_{dh}^e and i_{qh}^e are the high-frequency components. Substituting (9)–(11) into (6) and (7), the voltage equations are written as

$$\begin{bmatrix} v_d^e \\ v_q^e \end{bmatrix} = \begin{bmatrix} r_s & 0 \\ 0 & r_s \end{bmatrix} \begin{bmatrix} i_d^e \\ i_q^e \end{bmatrix} + \begin{bmatrix} L_{dd} & L_{dq} \\ L_{qd} & L_{qq} \end{bmatrix} \frac{d}{dt} \begin{bmatrix} i_{d0} + i_{dh}^e \\ i_{q0} + i_{qh}^e \end{bmatrix}$$
$$+ \begin{bmatrix} -\omega_e \lambda_q^e \\ \omega_e \lambda_d^e \end{bmatrix} + \begin{bmatrix} L_{df} \\ L_{qf} \end{bmatrix} \frac{d}{dt} (I_{f0} - I_h \cos \omega_h t). \tag{12}$$

Please note that i_{dh}^e and i_{qh}^e are induced by the high-frequency part of i_f'. From (12), the terms $r_s i_d^e$, $r_s i_q^e$, $-\omega_e \lambda_q^e$, and $\omega_e \lambda_d^e$ are neglected because ω_h is much larger than ω_e, $\omega_h L_{dd} \gg r_s$, and $\omega_h L_{qq} \gg r_s$. Thus, it is following that

$$\begin{bmatrix} v_d^e \\ v_q^e \end{bmatrix} = \begin{bmatrix} L_{dd} & L_{dq} \\ L_{qd} & L_{qq} \end{bmatrix} \frac{d}{dt} \begin{bmatrix} i_{d0}^e + i_{dh}^e \\ i_{q0}^e + i_{qh}^e \end{bmatrix} + \begin{bmatrix} L_{df} \\ L_{qf} \end{bmatrix} \frac{d}{dt} (I_{f0} - I_h \cos \omega_h t). \tag{13}$$

Please note that L_{dd}, L_{qq}, L_{dq}, L_{qd}, L_{df}, and L_{qf} are the incremental inductance in Section 2.2. It means that the inductances are calculated at a specific operating point, $(i_{d0}^e, i_{q0}^e, I_{f0})$. Therefore, all inductances can be assumed the constant value at the operation point. Based on the superposition law at $(i_{d0}^e, i_{q0}^e, I_{f0})$, the high-frequency part is separated as

$$\begin{bmatrix} 0 \\ 0 \end{bmatrix} = \begin{bmatrix} L_{dd} & L_{dq} \\ L_{qd} & L_{qq} \end{bmatrix} \frac{d}{dt} \begin{bmatrix} i_{dh}^e \\ i_{qh}^e \end{bmatrix} + \omega_h \begin{bmatrix} L_{df} \\ L_{qf} \end{bmatrix} I_h \sin \omega_h t. \tag{14}$$

Thus, the solution to (14) is obtained such that

$$\begin{bmatrix} i_{dh}^e(t) \\ i_{qh}^e(t) \end{bmatrix} = I_h \begin{bmatrix} \alpha_A \\ \alpha_B \end{bmatrix} \cos \omega_h t + \begin{bmatrix} i_{dh}^e(0) \\ i_{qh}^e(0) \end{bmatrix}, \tag{15}$$

where

$$\alpha_A = \frac{L_{qq}L_{df} - L_{dq}L_{qf}}{L_{dd}L_{qq} - L_{dq}L_{qd}}$$

$$\alpha_B = \frac{-L_{qd}L_{df} + L_{dd}L_{qf}}{L_{dd}L_{qq} - L_{dq}L_{qd}}.$$

Please note that (15) is the equation in the synchronous (rotor) frame based on the right angle θ_e. As shown in Figure 7, the angle of the misaligned frame is denoted by $\hat{\theta}_e$, and the angle error is defined as

$$\Delta\theta_e = \hat{\theta}_e - \theta_e, \tag{16}$$

where $\Delta\theta_e \in \mathbb{S} \equiv [-\pi, \pi)$.

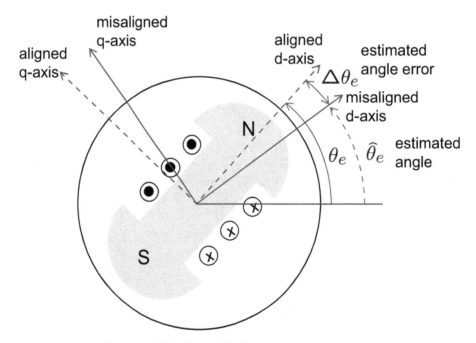

Figure 7. Misaligned dq-frame.

It is assumed that current is measured in a misaligned frame, in which the currents are denoted by $\hat{i}_{dh}^e$ and $\hat{i}_{qh}^e$. The current equation can be transformed to the misaligned frame by the following rotation matrix,

$$\mathbf{R}(\Delta\theta_e) = \begin{bmatrix} \cos(\Delta\theta_e) & \sin(\Delta\theta_e) \\ -\sin(\Delta\theta_e) & \cos(\Delta\theta_e) \end{bmatrix}. \tag{17}$$

By multiplying (17) to (15), we obtain

$$
\begin{bmatrix} \hat{i}^e_{dh} \\ \hat{i}^e_{qh} \end{bmatrix} = I_h \begin{bmatrix} \alpha_A \cos \Delta\theta_e + \alpha_B \sin \Delta\theta_e \\ -\alpha_A \sin \Delta\theta_e + \alpha_B \cos \Delta\theta_e \end{bmatrix} \cos \omega_h t
$$

$$
= I_h \sqrt{\alpha_A^2 + \alpha_B^2} \begin{bmatrix} \cos(\Delta\theta_e - \eta) \\ -\sin(\Delta\theta_e - \eta) \end{bmatrix} \cos \omega_h t, \tag{18}
$$

where

$$
\eta = \tan^{-1}\left(\frac{\alpha_B}{\alpha_A}\right) = \tan^{-1}\left(\frac{-L_{qd}L_{df} + L_{dd}L_{qf}}{L_{qq}L_{df} - L_{dq}L_{qf}}\right). \tag{19}
$$

Please note that η is an angle offset caused by the cross-coupling inductances. Specifically, $\eta = 0$ when $L_{dq} = L_{qd} = L_{qf} = 0$, which means that η is caused by the cross-coupling.

4. Position Estimation Using Signal Injection into the Rotor Winding

Phase currents are measured and transformed into an estimated frame. Then, a band-pass filter (BPF) is applied to $\hat{i}^e_d$ and $\hat{i}^e_q$ to remove the dc components. To perform synchronous rectification, $\cos \omega_h t$ is multiplied with $\hat{i}^e_{dh}$ and $\hat{i}^e_{qh}$. Then, a low pass filter (LPF) is applied to extract the dc signals X and Y [33]:

$$
X \equiv LPF(\hat{i}^e_{dh} \times \cos \omega_h t)
$$

$$
\approx \frac{I_h}{2} \sqrt{\alpha_A^2 + \alpha_B^2} \cos(\Delta\theta_e - \eta) \tag{20}
$$

$$
Y \equiv LPF(\hat{i}^e_{qh} \times \cos \omega_h t)
$$

$$
\approx -\frac{I_h}{2} \sqrt{\alpha_A^2 + \alpha_B^2} \sin(\Delta\theta_e - \eta). \tag{21}
$$

Using the filtered signals, the angle error $\Delta\theta_e$ can be estimated via

$$
\Delta\theta_e = -\tan^{-1}\left(\frac{Y}{X}\right) + \eta. \tag{22}
$$

Figure 8 shows a block diagram of the signal processing illustrating in the above. A lookup table for η is used to compensate the bias depending on the load condition and current angle. Finally, a phase locked loop (PLL) is employed to obtain an estimate $\hat{\theta}_e$.

The bandwidth of the filter has an impact on the estimation bandwidth. To enhance the performance, the bandwidth of PLL and filter should be increased. However, the high-frequency current $(\hat{i}^e_{dh}, \hat{i}^e_{qh})$ can be contaminated with a noise [34]. In practice, the noise limits the bandwidth of the filter and PLL.

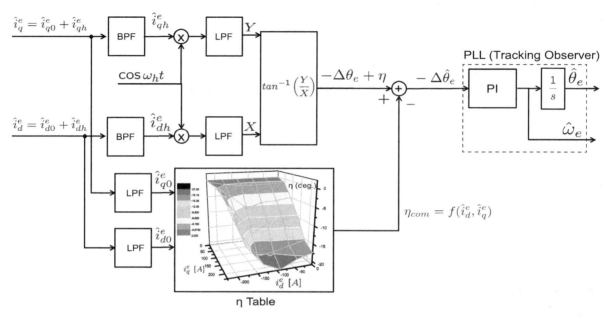

Figure 8. Signal processing block based on the signal-injection method into the field winding.

4.1. The Analysis of Cross-Coupling Offset Angle η

It is clear from (22) that the exact information of η is necessary to obtain an accurate value of $\Delta\theta$. On the other hand, η is a function of L_{dq}, L_{qd}, and L_{qf} in (19), which change nonlinearly with the currents. For practical purposes, it is better to make a lookup table of η over (i_d^e, i_q^e). Figure 9 shows the variations in η when the currents change. In general, the magnitudes of η increases as the core saturation develops. Note also that $|\eta|$ increases with the d-axis current until $i_d^e = -150$ A. However, it has the smallest values when $i_d^e = -200$ A. That situation could be illustrated as follows: The rotor flux generated by the field winding is almost canceled out by the negative d-axis current, $i_d^e = -200$ A. More specifically, note that the ampere-turn of the field winding is $N_{fd}i_f = 200 \times 6$ A-turns, where N_{fd} is the number of turns of the field winding. The ampere-turn of the d-axis stator winding is equal to $\frac{3}{2} \times N_a \times i_a = \frac{3}{2} \times 6 \times \frac{2}{3} \times 200 = 1200$ A-turns, where N_a is the number of turns of a phase winding per pole, and the winding factor is assumed to be unity. That is, they are the same when $i_d^e = -200$ A. Hence, the core is relieved from the saturation induced by the field winding, when $i_d^e = -200$ A. Thus, the non-linear behavior of η is mitigated when $i_d^e = -200$ A.

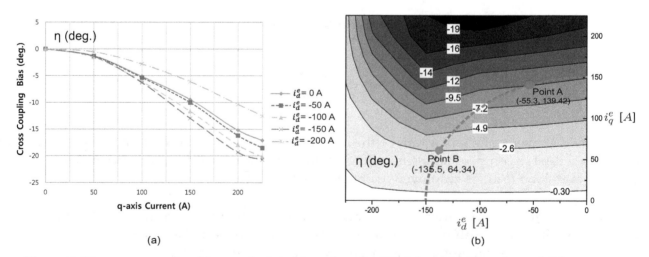

(a)   (b)

Figure 9. The cross-coupling bias η calculated based on the FEM data: (**a**) Plot of η and (**b**) contour plot of η.

4.2. The Compensation Method for Cross-Coupling Offset Angle η

The cross-coupling bias angle relies on the lookup table, which is the function of i_d^e and i_q^e. In practice, the dq-axes currents in the synchronous reference frame (i_d^e, i_q^e) are unknown in sensorless control. Thus, they are replaced by the dq-axes currents in misaligned frame, $(\hat{i}_d^e, \hat{i}_q^e)$. However, the inaccurate compensation for η can be made by the difference between (i_d^e, i_q^e) and $(\hat{i}_d^e, \hat{i}_q^e)$. Therefore, it has an effect on stability of the sensorless observer. To analyze the stability, define the function of η as

$$\eta = f(i_d^e, i_q^e), \tag{23}$$

f is shown in Figure 9. Using (17), the dq-axes currents in misaligned frame are derived as

$$\begin{bmatrix} \hat{i}_d^e \\ \hat{i}_q^e \end{bmatrix} = \begin{bmatrix} \cos \Delta\theta_e & \sin \Delta\theta_e \\ -\sin \Delta\theta_e & \cos \Delta\theta_e \end{bmatrix} \begin{bmatrix} i_d^e \\ i_q^e \end{bmatrix},$$

$$= I_s \begin{bmatrix} -\sin(\beta - \Delta\theta_e) \\ \cos(\beta - \Delta\theta_e) \end{bmatrix}, \tag{24}$$

where $I_s = \sqrt{i_d^{e2} + i_q^{e2}}$ and $\beta = \tan^{-1}(-i_d^e/i_q^e)$. Using (24), the compensation offset angle is obtained by $\eta_{com} = f(\hat{i}_d^e, \hat{i}_q^e)$. In Figure 8, $-\Delta\theta_e$ is the input of the PLL (tracking filter). Subtracting $\tan^{-1}\frac{Y}{X}$ from the compensation angle η_{com}, $-\Delta\hat{\theta}_e$ can be calculated

$$\begin{aligned}
-\Delta\hat{\theta}_e &= -\Delta\theta_e + \eta - \eta_{com} \\
&= -\Delta\theta_e + f(i_d^e, i_q^e) - f(\hat{i}_d^e, \hat{i}_q^e) \\
&= -\Delta\theta_e + f(-I_s \sin\beta, I_s \cos\beta) - f(-I_s \sin(\beta - \Delta\theta_e), I_s \cos(\beta - \Delta\theta_e)) \\
&= -\Delta\theta_e \underbrace{\left[1 + \frac{f(-I_s \sin(\beta - \Delta\theta_e), I_s \cos(\beta - \Delta\theta_e)) - f(-I_s \sin\beta, I_s \cos\beta)}{\Delta\theta_e} \right]}_{\kappa}.
\end{aligned} \tag{25}$$

Please note that κ should be the positive value to ensure the stability of the PLL observer, i.e., $\kappa \geq 0$. If $\kappa < 0$, the estimated position error will be amplified. Therefore, our task is to prove that κ is the positive value in the whole operation region. However, it is difficult for analytical demonstration because f is not mathematically represented and is highly non-linear. Figure 9b shows the contour of the cross-bias angle, η. From (24), $(\hat{i}_d^e, \hat{i}_q^e)$ are rotated by $\Delta\theta_e$ clockwise. It is evident that $f(\hat{i}_d^e, \hat{i}_q^e)$ slightly decreases as $\Delta\theta_e$ increases. It means that κ is maintained over 0, i.e., $\kappa \geq 0$. It is because $\frac{f(\hat{i}_d^e, \hat{i}_q^e) - f(i_d^e, i_q^e)}{\Delta\theta} > -1$. For example, the point B is (i_d^e, i_q^e) and the point A is $(\hat{i}_d^e, \hat{i}_q^e)$ in Figure 9b. β is 64.6° and $\Delta\theta$ is 42.9°. Substituting point A and B into (26), it was obtained as $-\Delta\hat{\theta}_e = -\Delta\theta_e(1 + \frac{-9.5° + 2.6°}{42.9°}) = -0.84 \Delta\theta_e$. Therefore, the convergence of the estimated angle error can be locally guaranteed due to $\kappa \geq 0$. By contrast, PMSMs have a positive offset angle caused by cross-coupling inductance [15,18] when q-axis current is positive. It shows that the offset angle increases as the q-axis current increases. An offset angle compensation method for PMSM was reported [35], and this paper proposed two different estimation angles: the saliency-based angle and the estimation rotor angle (compensated). Consequently, double-synchronous frames should be used, it causes the increasing of the calculation burden. In comparison, the proposed method has only one estimated synchronous frame.

5. Performance Comparison between Rotor and Stator Injection

Sensorless control performance degrades as the load increases because the inductances vary significantly along with core saturation. Specifically, the saliency ratio decreases, i.e., $L_{dd} \approx L_{qq}$ [17,19]. The accuracy of a sensorless method is determined by signal-to-noise ratio, which the saliency ratio affects. In this section, a common method of injecting a signal into the stator is also considered for the purpose of comparison.

Rambetius et al. studied the WSM model by incorporating the effects of the field winding into the stator. For signal injection into the stator, the following inductances should be used [25]:

$$L_{ddt} = L_{dd} - \frac{3}{2} \frac{L_{df}^2}{(L_{df} + L_{lf})}, \tag{26}$$

$$L_{qqt} = L_{qq} - \frac{3}{2} \frac{L_{qf}^2}{(L_{df} + L_{lf})}, \tag{27}$$

$$L_{dqt} = L_{dq} - \frac{3}{2} \frac{L_{df}L_{qf}}{(L_{df} + L_{lf})}, \tag{28}$$

where $L_{dq} = L_{qd}$ is assumed, L_{lf} is the leakage inductance of field, and L_{ddt}, L_{qqt}, and L_{dqt} are substituted for L_{dd}, L_{qq}, and L_{dq}, respectively. Please note that $L_{ddt} \leq L_{dd}$ and $L_{qqt} \leq L_{qq}$, because the field coil acts as a damper winding and reduces the high-frequency component [25]. The angle error is estimated by

$$\Delta\theta_e \approx \frac{\omega_h}{v_h} \frac{L_{ddt}L_{qqt}}{L_{diff}} LPF(\hat{i}_{qh}^e \sin \omega_h t), \tag{29}$$

where v_h is a high-frequency voltage and $L_{diff} = \frac{L_{qqt} - L_{ddt}}{2}$. It is emphasized that L_{diff} plays a crucial role in the estimation [19]. A smaller error is expected for a larger value of L_{diff}.

Figure 10a shows the loci of constant L_{diff} in the current plane along with the maximum torque per ampere (MTPA) line. As mentioned in the above, L_{diff} decreases as i_q^e increases. In other words, the electromagnetic saliency ratio decreases as the load increases.

On the other hand, the proposed method does not depend on the saliency ratio. When the signal is injected into the field winding, L_{df} plays a similar role as L_{diff}. However, its magnitude is less affected by the current magnitudes. Figure 10b shows the loci of constant L_{df} in the current plane. L_{df} increases as i_d^e increases negatively. Also note that L_{df} increases slightly when i_q^e increases. This can be illustrated by a small increase in L_{df} along with i_q^e as shown in Figure 6c. This supports the robust property of the field coil injection method.

According to the saliency-based method, the angle error is recovered from the term $\sin(2\Delta\theta_e)$. Because "2" is multiplied with $\Delta\theta_e$, polarity check should be carried out. Therefore, before starting the saliency-based sensorless algorithm, a polarity-checking procedure needs to be performed. However, the angle error is estimated from $\sin(\Delta\theta_e - \eta)$ with the field current injection method; therefore, no polarity-checking step is necessary.

Another advantage is that the field current injection method does not undermine the PWM duty of an inverter, which should be used for motor operation. Normally with the stator injection method, approximately 50 Vpeak is used for high-frequency injection for a proper SNR in a 300 V dc-link inverter. However, for the field current injection method, a high-frequency signal is synthesized in a separate dc-dc converter.

For implementation, the field current injection method requires a DSP-based dc-dc converter which can produce a high-frequency signal with a dc bias. Two methods are compared in Table 2.

Table 2. Comparison between the injection methods into stator and field windings.

	Signal Injection into Stator Winding	Signal Injection into Field Winding
Signal amplitude	L_{diff}(Saliency)	L_{df}(Mutual)
Polarity check	Necessary	Not necessary
SNR under core saturation	Small	Large
Signal generation	Inverter	dc-dc converter
Stable region ($\Delta\theta_e$)	$<45°$	$<90°$
Implementation	Easy	Medium difficulty

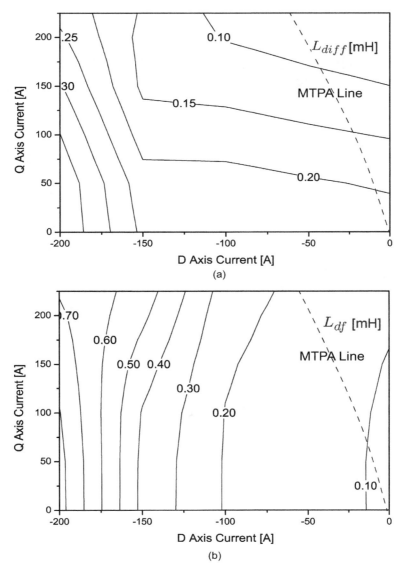

(a)

(b)

Figure 10. (a) L_{diff}, which is important for sensorless control based on signal injection into stator and (b) L_{df}, which is important for sensorless control based on signal injection into rotor.

6. Experimental Results

Figure 11 shows the experimental environment consisting of a dynamometer, a test WSM, an inverter, a dc to dc converter to supply the field current, etc. A zero-voltage switching (ZVS) full-bridge topology was used in the dc to dc converter, and operated at 50 kHz. As shown in Figure 12, the inverter dc-link voltage is 360 V and shared with the dc-dc converter. Practically, the field inductance is very large, and the bandwidth of field winding is very low. Consequently, it is difficult for some WSMs to inject high-frequency signal to the field winding. However, the field current supplier (DC-DC converter) is directly connected to the high voltage dc-link of the inverter. Therefore, it is enough to inject the high-frequency signal into the field winding. A carrier signal of 500 Hz, 90 V (peak-to-peak) was superposed on a dc output, which generated a 25 mA (peak-to-peak) current ripple on the dc component, $i_f = 6$ A. In truth, it cannot be guaranteed that I_h is a constant in the whole operation region due to the inductance variations from (9). Fortunately, the impedance between d-axis and rotor winding was not significantly changed in the MTPA operation. Therefore, it may have a small impact on the position estimation. The dynamometer governed the shaft speed, and the WSM was operated in the torque control mode. A Freescale MPC5554 was used as a processor for the inverter control board, and the inverter switching frequency was 8 kHz. The real angle was monitored using a resolver mounted on the WSM shaft.

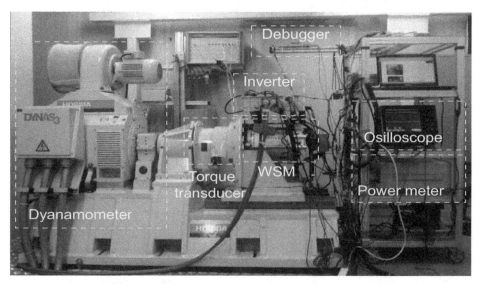

Figure 11. Experiment environment for testing sensorless control: the dynamometer, inverter, debugger, osilloscope, torque transducer, and WSM.

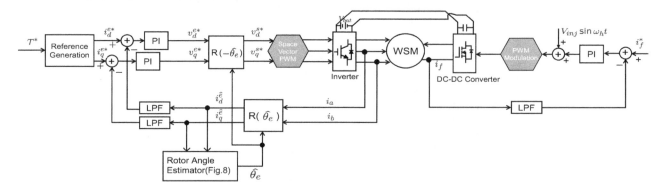

Figure 12. Block diagram of a proposed sensorless control method for a WSM.

The cross-coupling bias angle η is changed depending on the load condition. The bias angle can be directly measured by simple experimental method. The motor was controlled using the real rotor angle from a resolver when the carrier signal is injected into the rotor field. Using the dq-axes currents in the misaligned frame, the estimated angle $\hat{\theta}_e$ can be obtained without the compensation method. Then, the angle offset can be measured by $\eta_{exp} = \hat{\theta}_e - \theta_e$. Figure 13a shows the experimental results of the cross-coupling bias angle. Figure 13b shows η_{exp} and its differences, $\eta_{exp} - \eta$, from the ones computed using the FEM data. Please note that the maximum difference is approximately 6° (ele. degrees).

Figure 14 shows the angle estimate $\hat{\theta}_e$, measured angle θ_e, and their difference $\Delta\theta_e = \hat{\theta}_e - \theta_e$ during the transitions between 0 to 50 r/min and 50 r/min to 0 r/min, when a 28 Nm shaft torque is applied. Please note that the angle error was bounded under $\pm 3.33°$ (mechanical angle).

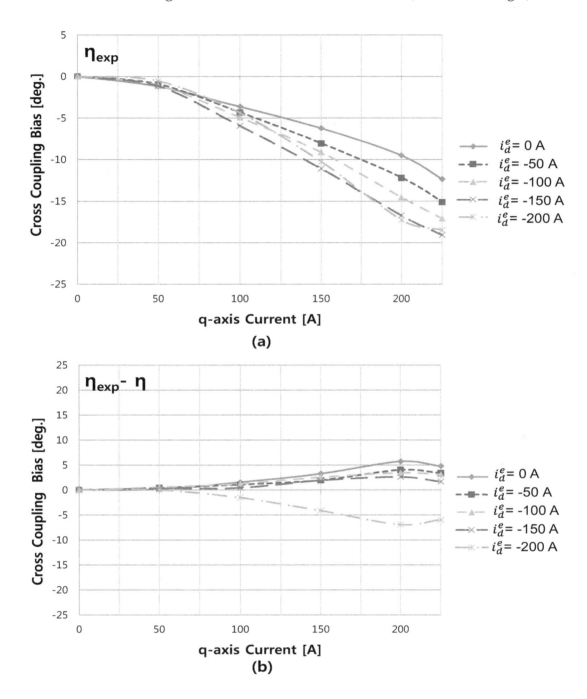

Figure 13. (a) Experimental angle offset due to the cross-coupling inductance and (b) The differences between the angle offset measured η_{exp} and the angle offset calculated η using FEM data.

Figure 15 shows the current response to a step command $i_q^{e*} = 228$ A when $i_d^e = -10$ A when the motor speed was regulated at a standstill by the dynamometer motor, showing that 123 Nm (1 pu) was produced. However, the torque response appears sluggish owing to a strong filter in CAN communication. Please note that the angle error was regulated below $20°$ (ele. degrees) under a rated step torque.

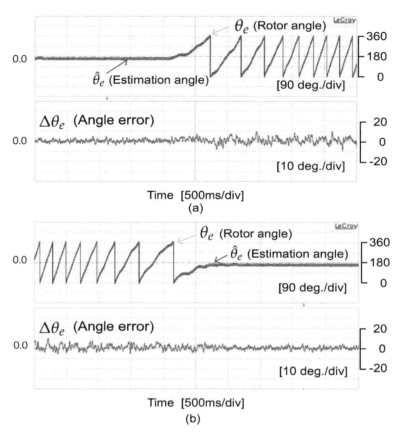

Figure 14. Proposed sensorless control during speed transitions (28 Nm, $i_d^e = -20$ A, $i_q^e = 50$ A): (a) $0 \rightarrow 50$ r/min and (b) $50 \rightarrow 0$ r/min.

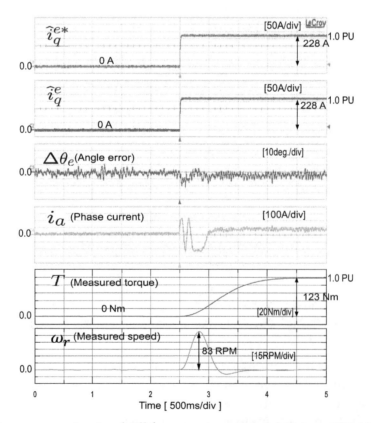

Figure 15. Current responses at a standstill for a q-axis command (0 A $\rightarrow$ 228 A) when $i_d^e = -10$ A: q-axis current command, q-axis current, angle error, phase current, shaft torque, and shaft speed.

Figure 16 shows the angle error trend at a fixed speed (100 r/min) when the q-axis current increases to a rated point without and with compensation for η, which was caused by cross-coupling. The effectiveness of the cross-coupling compensation was demonstrated, in which the bias error (14° ele. degrees) was monitored without the compensation method whereas no bias error was observed with the compensation method.

Figure 17 shows the experimental result using sensorless control based on signal injection into stator winding. The dyanamometer regulated the WSM speed at 50 r/min. For a light load, the estimated angle error $\Delta\theta_e$ is not over 10° (ele. degrees). However, for a heavy load, the estimated angle error is oscillated between −30° and 5° (ele. degrees) owing to magnetic saturation.

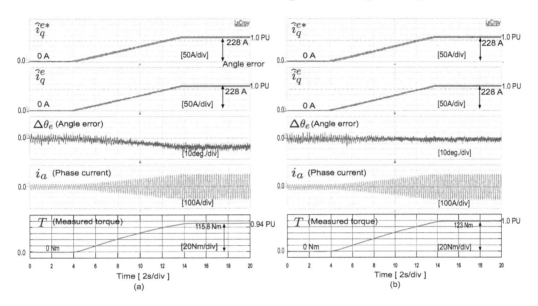

Figure 16. Field current injection method (**a**) without compensation for η caused by cross-coupling and (**b**) with compensation for η: ramp q-axis command (0 A → 228 A), q-axis current response, angle error, phase current, and torque.

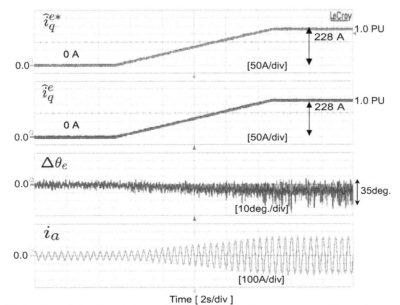

Time [2s/div]

Figure 17. Signal injection into the stator winding without a compensation of cross-coupling effects: response at 50 r/min for a ramp q-axis command (0 A → 228 A) when $i_d^e = -50$ A.

7. Discussion

This study has attempted to investigate sensorless control for a WSM based on signal injection into the field winding. To conclude, we summarize the following contributions of the paper. The sensorless method, which injected a carrier signal into field winding, is not based on the saliency ratio. Therefore, the method does not become unstable caused owing to magnetic saturation phenomenon. In addition, the absolute position angle can be obtained. Both d- and q-axis high-frequency signals were used for angle estimation. Mutual incremental inductances are used to predict and compensate the angle offset caused by the cross-coupling effect. In this work, L_{df} and L_{qf} are significant factors. The algorithm was developed to eliminate the estimation bias caused by the cross-coupling inductance. Experiments were performed for full-torque operation. Furthermore, we obtained an accurate estimated angle in the presence of the cross-coupling effect.

Author Contributions: J.C. and K.N. designed the proposed sensorless algorithm for WSM and J.C. did the experiments and analyzed the data.

Abbreviations

The following abbreviations are used in this manuscript:

v_d^e, v_q^e	d and q-axes voltage in the synchronous frame
i_d^e, i_q^e	d and q-axes current in the synchronous frame
i_f	Field current
$\lambda_a, \lambda_b, \lambda_c$	Phase a, b, c stator flux linkage
λ_d^e, λ_q^e	d and q-axes stator flux linkage in the synchronous frame
r_s	Stator winding resistance
L_{dd}, L_{qq}	d and q-axes self-inductance
L_{dq}, L_{qd}	d and q-axes cross-coupling inductance
L_{df}	Mutual inductance between d-axis and field coil
L_{qf}	Cross-coupling inductance between q-axis and field coil
L_{ls}	Leakage inductance of stator
L_{lf}	Leakage inductance of field
θ_e	Rotor flux angle
$\hat{\theta}_e$	Estimation rotor flux angle
$\Delta\theta_e$	Estimation rotor flux angle error
ω_e, ω_r	Electrical speed and mechanical speed
ω_h	Angular speed of the carrier signal
η	Angle offset caused by the cross-coupling
η_{com}	Compensation for angle offset
$-\Delta\hat{\theta}_e$	The input of the PLL
T	Shaft torque
WSM	Wound Synchronous Machine
EV	Electrical vehicles
HEV	Hybrid EVs
PMSMs	Permanent-magnet synchronous motors
Nd	Neodymium
Dy	Dysprosium
EMF	Electromotive force
MTPA	Maximum torque per ampere
FEM	Finite-element method
LPF	Low pass filter
BPF	Band-pass filter

References

1. Santiago, J.D.; Bernhoff, H.; Ekergard, B.; Eriksson, S.; Ferhatovic, S.; Waters, R.; Leijon, M. Electrical Motor Drivelines in Commercial All-Electric Vehicles: A Review. *IEEE Trans. Veh. Technol.* **2013**, *61*, 475–484. [CrossRef]
2. Kim, Y.; Nam, K. Copper-Loss-Minimizing Field Current Control Scheme for Wound Synchronous Machines. *IEEE Trans. Power Electron.* **2017**, *32*, 1335–1345. [CrossRef]
3. Friedrich, G.; Girardin, A. Integrated starter generator. *IEEE Trans. Ind. Appl. Mag.* **2009**, *15*, 26–34. [CrossRef]
4. Wang, Y.; Wang, X.; Xie, W.; Dou, M. Full-Speed Range Encoderless Control for Salient-Pole PMSM with a Novel Full-Order SMO. *Energies* **2018**, *11*, 2423. [CrossRef]
5. Morimoto, S.; Kawamoto, K.; Sanada, M.; Takeda, Y. Sensorless control strategy for salient-pole PMSM based on extended EMF in rotating reference frame. *IEEE Trans. Ind. Appl.* **2002**, *38*, 1054–1061. [CrossRef]
6. Chen, G.-R.; Yang, S.-C.; Hsu, Y.-L.; Li, K. Position and Speed Estimation of Permanent Magnet Machine Sensorless Drive at High Speed Using an Improved Phase-Locked Loop. *Energies* **2017**, *10*, 1571. [CrossRef]
7. Wang, M.-S.; Tsai, T.-M. Sliding Mode and Neural Network Control of Sensorless PMSM Controlled System for Power Consumption and Performance Improvement. *Energies* **2017**, *10*, 1780. [CrossRef]
8. Genduso, F.; Miceli, R.; Rando, C.; Galluzzo, G.R. Back EMF Sensorless-Control Algorithm for High-Dynamic Performance PMSM. *IEEE Trans. Ind. Electron.* **2010**, *57*, 2092–2100. [CrossRef]
9. Lee, J.; Hong, J.; Nam, K.; Ortega, R.; Praly, L.; Astolfi, A. Sensorless Control of Surface-Mount Permanent-Magnet Synchronous Motors Based on a Nonlinear Observer. *IEEE Trans. Power Electron.* **2010**, *25*, 290–297.
10. Choi, J.; Nam, K.; Bobtsov, A.A.; Pyrkin, A.; Ortega, R. Robust Adaptive Sensorless Control for Permanent-Magnet Synchronous Motors. *IEEE Trans. Power Electron.* **2017**, *32*, 3989–3997. [CrossRef]
11. Koteich, M.; Maloum, A.; Duc, G.; Sandou, G. Observability analysis of sensorless synchronous machine drives. In Proceedings of the 2015 European Control Conference (ECC), Linz, Austria, 15–17 July 2015; pp. 3560–3565.
12. Tian, L.; Zhao, J.; Sun, J. Sensorless Control of Interior Permanent Magnet Synchronous Motor in Low-Speed Region Using Novel Adaptive Filter. *Energies* **2016**, *9*, 1084. [CrossRef]
13. Gabriel, F.; De Belie, F.; Neyt, X.; Lataire, P. High-Frequency Issues Using Rotating Voltage Injections Intended For Position Self-Sensing. *IEEE Trans. Ind. Electron.* **2013**, *60*, 5447–5457. [CrossRef]
14. Wu, X.; Wang, H.; Huang, S.; Huang, K.; Wang, L. Sensorless Speed Control with Initial Rotor Position Estimation for Surface Mounted Permanent Magnet Synchronous Motor Drive in Electric Vehicles. *Energies* **2015**, *8*, 11030–11046. [CrossRef]
15. Gong, L.M.; Zhu, Z.Q. Robust Initial Rotor Position Estimation of Permanent-Magnet Brushless AC Machines With Carrier-Signal-Injection-Based Sensorless Control. *Trans. Ind. Appl.* **2013**, *49*, 2602–2609. [CrossRef]
16. Li, Y.; Zhu, Z.Q.; Howe, D.; Bingham, C.M. Modeling of Cross-Coupling Magnetic Saturation in Signal-Injection-Based Sensorless Control of Permanent-Magnet Brushless AC Motors. *IEEE Trans. Mag.* **2007**, *43*, 2552–2554. [CrossRef]
17. Zhu, Z.Q.; Gong, L.M. Investigation of Effectiveness of Sensorless Operation in Carrier-Signal-Injection-Based Sensorless-Control Methods. *IEEE Trans. Ind. Electron.* **2011**, *58*, 3431–3439. [CrossRef]
18. Li, Y.; Zhu, Z.Q.; Howe, D.; Bingham, C.M.; Stone, D.A. Improved Rotor-Position Estimation by Signal Injection in Brushless AC Motors, Accounting for Cross-Coupling Magnetic Saturation. *IEEE Trans. Ind. Appl.* **2009**, *44*, 1843–1850. [CrossRef]
19. Sergeant, P.; De Belie, J.; Melkebeek, J. Rotor Geometry Design of Interior PMSMs With and Without Flux Barriers for More Accurate Sensorless Control. *IEEE Trans. Ind. Electron.* **2012**, *59*, 2457–2465. [CrossRef]
20. Li, S.; Ge, Q.; Wang, X.; Li, Y. Implementation of Sensorless Control with Improved Flux Integrator for Wound Field Synchronous Motor. In Proceedings of the 2007 2nd IEEE Conference on Industrial Electronics and Applications, Harbin, China, 23–25 May 2007; Volume 59, pp. 1526–1530.
21. Maalouf, A.; Ballois, S.L.; Idekhajine, L.; Monmasson, E.; Midy, J.; Biais, F. Sensorless Control of Brushless Exciter Synchronous Starter Generator Using Extended Kalman Filter. In Proceedings of the 2009 35th Annual Conference of IEEE Industrial Electronics, Porto, Portugal, 3–5 November 2009; Volume 59, pp. 2581–2586.
22. Boldea, I.; Andreescu, G.D.; Rossi, C. Active Flux Based Motion-Sensorless Vector Control of DC-Excited Synchronous Machines. In Proceedings of the 2009 IEEE Energy Conversion Congress and Exposition, San Jose, CA, USA, 20–24 September 2009; pp. 2496–2503.

23. Jain, A.K.; Ranganathan, V.T. Modeling and Field Oriented Control of Salient Pole Wound Field Synchronous Machine in Stator Flux Coordinates. *IEEE Trans. Ind. Electron.* **2011**, *58*, 960–970. [CrossRef]

24. Griffo, A.; Drury, D.; Sawata, T.; Mellor, P.H. Sensorless starting of a wound-field synchronous starter/generator for aerospace applications. *IEEE Trans. Ind. Electron.* **2012**, *59*, 3579–3587. [CrossRef]

25. Rambetius, A.; Ebersberger, S.; Seilmeier, M.; Piepenbreier, B. Carrier Signal Based Sensorless Control of Electrically Excited Synchronous Machines at Standstill and Low Speed Using The Rotor Winding as a receiver. In Proceedings of the 2013 15th European Conference on Power Electronics and Applications (EPE), Lille, France, 2–6 September 2013; pp. 1–10.

26. Deng, X.; Wang, L.; Zhang, J.; Ma, Z. Rotor Position Detection of Synchronous Motor Based on High-frequency Signal Injection into the Rotor. In Proceedings of the 2011 Third International Conference on Measuring Technology and Mechatronics Automation, Shangshai, China, 6–7 January 2011; pp. 195–198.

27. Zhou, Y.; Long, S. Sensorless Direct Torque Control for Electrically Excited Synchronous Motor Based on Injecting High-Frequency Ripple Current Into Rotor Winding. *IEEE Trans. Energy Convers.* **2015**, *30*, 246–253. [CrossRef]

28. Rambetius, A.; Piepenbreier, B. Comparison of carrier signal based approaches for sensorless wound rotor synchronous machines. In Proceedings of the International Symposium on Power Electronics, Electrical Drives, Automation and Motion, Schia, Italy, 18–20 June 2014; pp. 1152–1159.

29. Rambetius, A.; Piepenbreier, B. Sensorless control of wound rotor synchronous machines using the switching of the rotor chopper as a carrier signal. In Proceedings of the International Symposium on Sensorless Control for Electrical Drives and Predictive Control of Electrical Drives and Power Electronics, Munich, Germany, 17–19 October 2013; pp. 1–8.

30. Choi, J.; Jeong, I.; Jung, S.; Nam, K. Sensorless Control for Electrically Energized Synchronous Motor Based on Signal Injection to Field Winding. In Proceedings of the IECON2013—39th Annual Conference of the IEEE Industrial Electronics Society, Vienna, Austria, 10–13 November 2013; pp. 3120–3129.

31. Selmon, G.R. *Electric Machines and Drives*; Addison Welsley: Boston, MA, USA, 1992; ISBN 0-201-57885-9.

32. Stumberger, B.; Stumberger, G.; Dolinar, D.; Hamler, A.; Trlep, M. Evaluation of Saturation and Cross-Magnetization Effects in Interior Permanent-Magnet Synchronous Motor. *IEEE Trans. Ind. Appl.* **2009**, *48*, 1576–1587.

33. Nam, K.H. *AC Motor Control and Electric Vehicle Application*, 1st ed.; CRC Press: Boca Raton, FL, USA, 2010; ISBN 978-1-49-81963-0.

34. Garcia, P.; Briz, F.; Degner, M.W.; Diaz-Reigosa, D. Accuracy, Bandwidth, and Stability Limits of Carrier-Signal-Injection-Based Sensorless Control Methods. *IEEE Trans. Ind. Appl.* **2009**, *43*, 990–1000. [CrossRef]

35. De Kock, H.W.; Kamper, M.J.; Kennel, R.M. Anisotropy Comparison of Reluctance and PM Synchronous Machines for Position Sensorless Control Using HF Carrier Injection. *IEEE Trans. Power Electron.* **2009**, *24*, 1905–1913. [CrossRef]

12

Analytical Modeling and Comparison of Two Consequent-Pole Magnetic-Geared Machines for Hybrid Electric Vehicles

Hang Zhao [1,2], Chunhua Liu [1,2,*], Zaixin Song [1,2] and Jincheng Yu [1,2]

[1] School of Energy and Environment, City University of Hong Kong, 83 Tat Chee Avenue, Kowloon Tong, Hong Kong, China; zhao.hang@my.cityu.edu.hk (H.Z.); zaixin.song@my.cityu.edu.hk (Z.S.); jincheng.yu@my.cityu.edu.hk (J.Y.)

[2] Shenzhen Research Institute, City University of Hong Kong, Nanshan District, Shenzhen 518057, China

* Correspondence: chunliu@cityu.edu.hk.

Abstract: The exact mathematical modeling of electric machines has always been an effective tool for scholars to understand the working principles and structure requirements of novel machine topologies. This paper provides an analytical modeling method—the harmonic modeling method (HMM)—for two types of consequent-pole magnetic-geared machines, namely the single consequent-pole magnetic-geared machine (SCP-MGM) and the dual consequent-pole magnetic-geared machine (DCP-MGM). By dividing the whole machine domain into different ring-like subdomains and solving the Maxwell equations, the magnetic field distribution and electromagnetic parameters of the two machines can be obtained, respectively. The two machines were applied in the propulsion systems of hybrid electric vehicles (HEVs). The electromagnetic performances of two machines under different operating conditions were also compared. It turns out that the DCP-MGM can reach a larger electromagnetic torque compared to that of the SCP-MGM under the same conditions. Finally, the predicted results were verified by the finite element analysis (FEA). A good agreement can be observed between HMM and FEA. Furthermore, HMM can also be applied to the mathematical modeling of other consequent-pole electric machines in further study.

Keywords: harmonic modeling method; magnetic-geared machine; hybrid electric vehicle; magnetic field; electromagnetic performance; analytical modeling

1. Introduction

The last decade has witnessed rapid developments of magnetic gears (MGs) and electric machines that utilize the magnetic-gearing effect, which are also called magnetic-geared machines (MGMs) [1–3]. Ever since their invention in 2001 [4], MGs have become a research hotspot due to their high efficiency and self-protection characteristics [5–7].

The concept of MGMs is derived from MGs. By substituting stator windings with AC current for one rotating permanent magnet (PM) component, MGMs change one mechanical port of MGs into an electrical port. Thus, the two rotating components of the MGM and its stator windings can be regarded as a combination of a magnetic gear and an electric machine [8–10]. Indeed, with the introduction of another rotating component and the ability to alternate the speed ratio and torque ratio between two rotating components, MGMs have broadened the application scenarios of electric machines [11–13]. A good example is that MGMs can serve as the power split component (PSC) in hybrid electric vehicles (HEVs) to realize energy exchange among the internal combustion engine (ICE), wheels, and battery [14–16]. The ICE and electric machine can provide traction for the wheels independently. The electric machine can work as a generator and a motor. When the electric machine serves as a generator, it can absorb power from the ICE or wheels (depending on working modes) to get the battery charged. When the electric machine serves as a motor, the power flows from the battery to the electric machine to drive the wheels. Hence, the ICE can always work at its highest efficiency to save fuel by alternating the working modes of the electric machine. This application scenario has drawn more and more attention as environmental problems become severe [17]. HEVs do offer a chance to alleviate the exhaust gas emission problem caused by fuel vehicles [18]. Moreover, compared to its counterpart, namely the mechanical gearbox with an electric machine, MGMs not only save space, but also improve efficiency and reduce noise and vibration by eliminating the physical contact of two gear sets [19,20].

Just like permanent magnet synchronous machines (PMSMs), MGMs utilize permanent magnets as the magnetic sources instead of using the electrical excitation method. Thus, the carbon brush structure can be eliminated and the durability of electric machines can be enhanced. However, the rare earth elements make the price of PMs extremely expensive [21]. To solve this problem, a consequent-pole structure can be adopted. The consequent-pole structure can not only reduce the flux linkage, but also improve the torque density [22–24]. Two different topologies of consequent-pole MGMs, i.e., single consequent-pole magnetic-geared machines (SCP-MGM) and dual consequent-pole magnetic-geared machines (DCP-MGM) have been proposed [25], but their mathematical modeling has not been well studied.

Although the MGMs offer many new possibilities for electric machines, their magnetic field distribution is much more complex compared to traditional electric machines with one rotor. Many scholars have focused on the magnetic field distribution calculation of MGMs [26–28]. Yet, to the best of author's knowledge, no literature has studied the magnetic field distribution of consequent-pole MGMs. The introduction of soft magnetic material (SMM) to replace the PM part will make the magnetic field distribution of consequent-pole MGMs even more complicated. Research [29] has solved the magnetic field distribution of a PMSM with PMs inserted into the SMM part, but did not consider of the saturation of SMM. Additionally, the subdomain division method [30] is not suitable for MGMs, since too many subdomains increase the calculation time rapidly. Research [31] has proposed a new harmonic modeling method (HMM) to calculate the magnetic distribution of electric machines. By introducing complex Fourier series and a convolution matrix of permeability, HMM can reduce the number of subdomains to within ten. This is because the total number of these ring-like subdomains will not increase with the increase of modulator pieces and slots.

In this paper, two consequent-pole MGMs were studied using HMM. The paper is organized as follows. Section 2 discusses the configurations and operating principles of consequent-pole MGMs. Mathematical models of SCP-MGM and DCP-MGM considering iron saturation are then proposed and elaborated in Section 3. Finally, the effectiveness of proposed HMM is validated by using finite element analysis (FEA) in Section 4.

2. Configurations and Operating Principles of SCP-MGM and DCP-MGM

When the MGM (either SCP-MGM or DCP-MGM) is applied in HEV, its inner rotor can be connected to the ICE, while the outer rotor can be connected to a permanent magnet synchronous machine (PMSM), which will be further connected to the differential to drive the wheels; the battery provides energy to the windings of both the MGM and PMSM via an inverter. The whole system configuration can be seen in Figure 1. The MGM together with the PMSM can be regarded as the E-CVT in a Toyota Prius. They can cooperate with each other according to different working conditions of HEVs [32]. Briefly speaking, either the torque from the ICE or the torque on the outer rotor driven by AC current can be the prime power to drive the HEV, and they can also work together to enhance the output power. Additionally, the battery can be charged under a regenerative braking state. The concept that a PMSM is added after the CP-MGM is derived from that in E-CVT [33]. The PMSM in Figure 1 is used to regulate the performances of the outer rotor. For instance, it can be used to drag the outer rotor of CP-MGM to a synchronous state (the rated rotating speed) at startup state. In addition, it can deliver extra output torque to the outer rotor shaft if the output torque of the CP-MGM cannot meet the requirement. Since this paper mainly focuses on the operating modes of the CP-MGM, it is reasonable to assume that there is no power flow between the PMSM and the outer rotor shaft at the four steady states mentioned in this paper. In fact, power exchange between the PMSM and the wheels would not affect the conclusion obtained in this paper.

The working principle of MGMs is similar to that of magnetic gears. By adopting a modulator layer, the magnetic field distribution can be changed. Assuming that the pole pair number of the original magnetic field is P_i, and the modulator number is Q, then a novel magnetic field will have a component that has $(Q - P_i)$ pole pairs. Thus, the fundamental structural requirement of an MGM is [4]:

$$P_i + P_s = Q \tag{1}$$

where P_s is the pole pair number of stator windings.

Under steady working conditions, the rotating speed of two rotating rotors and the current frequency f within stator windings should then satisfy:

$$P_i \omega_i - Q \omega_o = P_s w_s = 60f \tag{2}$$

where ω_i, ω_o, and ω_s are the rotating speed of the inner rotor, outer rotor, and the equivalent rotating speed of stator windings.

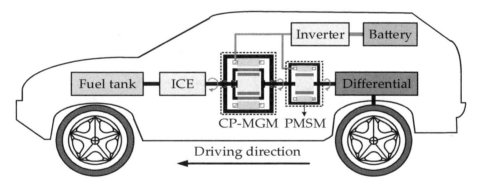

Figure 1. Propulsion system configuration of the consequent-pole magnetic-geared machine (MGM) applied in hybrid electric vehicles (HEVs).

Since the ICE reaches its highest efficiency at the range of ~2000 r/min–3000 r/min, the rotating speed of outer rotor and the current frequency of stator winding must cooperate with the rotating speed of the inner rotor to ensure the highest efficiency of the ICE. However, if the stator windings need to provide energy for the HEV, the rotating speed of the inner rotor must be smaller than that of the outer rotor. Thus, a gearbox must come into service under hybrid mode to reduce the rotating speed of the inner rotor. Therefore, the operation modes of the proposed HEV propulsion system can be divided into four kinds, and their typical operating parameters are shown in Table 1. The rotating speed of the outer rotor is calculated according to the different driving speeds of the HEV, and the current frequency of stator winding is obtained via Equation (2).

Table 1. Operating parameters of MGM under different modes of HEV.

Operation Modes	Rotating Speed of Inner Rotor ω_o	Rotating Speed of Outer Rotor ω_i	Current Frequency f
Pure electric mode	0 r/min	500 r/min	108.3 Hz
Pure mechanical mode	1200 r/min	1015 r/min	0 Hz (DC)
Hybrid mode	1200 r/min	2000 r/min	213.3 Hz
Regenerative braking mode	0 r/min	1000 r/min	216.6 Hz

The topologies of SCP-MGM and DCP-MGM are shown in Figure 2. By substituting SMM for PMs with the same polarity, a consequent-pole structure is obtained. The name "consequent-pole" is due to SMM, and PM appears alternately on the circumferential direction. Although SMM cannot generate a magnetic field itself, it can be easily magnetized to conduct flux lines. Hence, SMM in a consequent-pole structure can be regarded as a magnetic source to some degree. The greatest advantage of using the consequent-pole structure is saving PM material, which is the most expensive material in an electric machine. Both SCP-MGM and DCP-MGM utilize a consequent-pole structure to save PM material. The SMM part in the outer rotor of a DCP-MGM not only works as a consequent-pole structure for the PMs inserted in the outer rotor, it also modulates the magnetic field of the inner rotor. Thus, the P_s-th harmonic component within the DCP-MGM is larger than that of the SCP-MGM. Additionally, the saturation of the DCP-MGM is more severe than that of the SCP-MGM.

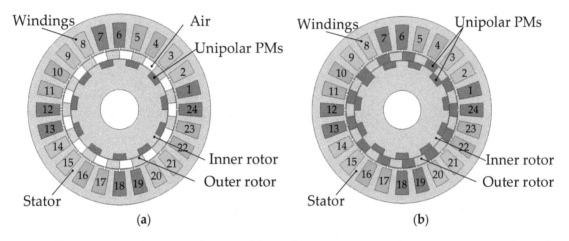

Figure 2. Proposed machine topologies: (**a**) single consequent-pole magnetic-geared machines (SCP-MGMs); (**b**) dual consequent-pole magnetic-geared machines (DCP-MGMs).

3. Mathematical Modeling of SCP-MGMs and DCP-MGMs

3.1. Assumption and Parameter Definition

The machine structure chosen to be studied in this paper was a 24 slot 11 pole-pair SCP-MGM and DCP-MGM, as shown in Figure 2. A few assumptions must be made to simplify the mathematical modeling:

- The geometrical shape of the machine has a radial side and a tangential side;
- The magnetic field distribution is constrained in the 2D plane: the axial component is ignored;
- The machine has infinite axial length, so the end effect is ignored;
- The radial component of the permeability of SMM within a certain region is regarded as a constant;
- Eddy–current effects within SMM and PMs are ignored.

Since there exists a z-direction current within the windings of the studied machines and the machine topology has a circular shape, a vector magnetic potential (VMP) A_z in a polar coordinate is adopted to calculate the magnetic flux density distribution within the machines. The machine structures are then divided into several ring-like regions based on the different material interfaces, as can be seen in Figure 3, where α represents the angle of the inner PM arc, β represents the angle of the slot opening in the modulator, δ is the slot opening angle, and γ is the stator slot angle. The whole machine is divided into ten subdomains: the innermost one (region I) represents the shaft part; region II is the rotor yoke; region III is the inner consequent-pole PM; region IV is the inner air gap; region V is the modulator pieces (it should be noted that for SCP-MGMs, the gap between each two modulator pieces is air, while for DCP-MGM, bipolar PMs are inserted in that gap). Region VI is the outer air gap;

region VII is the stator teeth; region VIII is the stator slots together with windings; region IX is the stator yoke; region X is the outside of the studied machines.

The angular position of the j-th PM part of the inner rotor θ_{PM}, the position of the k-th modulator piece θ_{Mod}, the position of the t-th stator tooth part θ_{tooth}, and the position of the s-th stator slot part θ_{slot} can be defined, respectively, as:

$$\theta_{PM} = \varphi_{in} + \frac{j \cdot 2\pi}{P_i} \tag{3}$$

$$\theta_{Mod} = \frac{k \cdot 2\pi}{Q} + \theta_0 - \frac{\beta}{2} \tag{4}$$

$$\theta_{tooth} = t \cdot 2\pi / P \tag{5}$$

$$\theta_{slot} = s \cdot 2\pi / P \tag{6}$$

where φ_{in}, and θ_0 the initial angular positions of the inner rotor and outer rotor, respectively. Due to the symmetrical structure of the inner rotor and outer rotor, φ_{in}, θ_0 has a range of $[0, 2\pi/P_i]$, $[0, 2\pi/Q]$, respectively. Specifically, φ_{in} is defined as zero when the lower edge of the PM in the inner rotor coincides with the positive direction of the angular axis; θ_0 is defined as zero when the center of the slot of the outer rotor (air in SCP-MGM and PM in DCP-MGM) coincides with the positive direction of the angular axis, as shown in Figure 2.

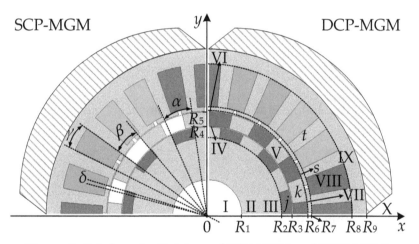

Figure 3. Subdomain divisions of SCP-MGM and DCP-MGM.

The VMP $\vec{A}$, the magnetic flux density $\vec{B}$, the magnetic field strength $\vec{H}$, and the current density distribution $\vec{J}$ in stator windings can be written in vector form as:

$$\vec{A} = A_z(r, \theta) \cdot \vec{u}_z \tag{7}$$

$$\vec{B} = B_r(r, \theta) \cdot \vec{u}_r + B_\theta(r, \theta) \cdot \vec{u}_\theta \tag{8}$$

$$\vec{H} = H_r(r, \theta) \cdot \vec{u}_r + H_\theta(r, \theta) \cdot \vec{u}_\theta \tag{9}$$

$$\vec{J} = J_z(r, \theta) \cdot \vec{u}_z \tag{10}$$

To simplify the solving process, all the parameters related to magnetic field are expressed in terms of complex Fourier series. Thus, the vector amplitude above can be further expressed as:

$$A_z(r, \theta) = \sum_{n=-\infty}^{n=\infty} \hat{A}_{z,n}(r) \cdot e^{-in\theta} \tag{11}$$

$$B_r(r,\theta) = \sum_{n=-\infty}^{n=\infty} \hat{B}_{r,n}(r) \cdot e^{-in\theta} \text{ and } B_\theta(r,\theta) = \sum_{n=-\infty}^{n=\infty} \hat{B}_{\theta,n}(r) \cdot e^{-in\theta} \qquad (12)$$

$$H_r(r,\theta) = \sum_{n=-\infty}^{n=\infty} \hat{H}_{r,n}(r) \cdot e^{-in\theta} \text{ and } H_\theta(r,\theta) = \sum_{n=-\infty}^{n=\infty} \hat{H}_{\theta,n}(r) \cdot e^{-in\theta} \qquad (13)$$

$$J_z(r,\theta) = \sum_{n=-\infty}^{n=\infty} \hat{J}_{z,n}(r) \cdot e^{-in\theta} \qquad (14)$$

where n represents the n-th order coefficient of the corresponding Fourier series. It should be noticed that, in numerical calculation, a reasonable harmonic order N is used to truncate the infinite Fourier series. If N is too small, the Fourier series will have a large error, if N is too large, the calculation time will be rather long.

3.2. Partical Differential Equation Solution

The magnetic field within the machine follows quasistatic Maxwell equations:

$$\nabla \times \vec{H} = \vec{J} \qquad (15)$$

$$\nabla \cdot \vec{B} = 0 \qquad (16)$$

The relationship between $\vec{B}$ and $\vec{A}$ can be further expressed as:

$$\vec{B} = \nabla \times \vec{A} \qquad (17)$$

The radial component and tangential component matrix of magnetic flux density $\vec{B}$ are then obtained in matrix form as [31]:

$$\mathbf{B}_r = \frac{1}{r}\frac{\partial \mathbf{A}_z}{\partial \theta} = -i\frac{1}{r}\mathbf{KA}_z \qquad (18)$$

$$\mathbf{B}_\theta = -\frac{\partial \mathbf{A}_z}{\partial r} \qquad (19)$$

where $\mathbf{K}$ represents the harmonic order coefficient diagonal matrix that is related to N, given by:

$$\mathbf{K} = \begin{bmatrix} -N & \cdots & 0 \\ \vdots & \ddots & \vdots \\ 0 & \cdots & N \end{bmatrix} \qquad (20)$$

Similar to Equations (11)–(14), the relative permeability of each region can also be expressed in a complex Fourier series form:

$$\mu(\theta) = \sum_{n=-\infty}^{n=\infty} \hat{\mu}_n \cdot e^{-in\theta} \qquad (21)$$

Next, based on the relation between $\vec{B}$ and $\vec{H}$, as expressed below:

$$\vec{B} = \mu\vec{H} + \mu_0\vec{M} \qquad (22)$$

where $\vec{M}$ is the magnetization vector. The first item on the right is a product of two Fourier series, which can be rewritten in matrix form by using the Cauchy product theorem:

$$\mathbf{B}_r = \mathbf{\mu}_{r,\text{cov}}\mathbf{H}_r + \mu_0\mathbf{M}_r \qquad (23)$$

$$\mathbf{B}_\theta = \mathbf{\mu}_{\theta,\mathrm{cov}}\mathbf{H}_\theta + \mu_0\mathbf{M}_\theta \tag{24}$$

where $\mu_{r,\mathrm{cov}}$ and $\mu_{\theta,\mathrm{cov}}$ are convolution matrices of the radial and tangential components of permeability, respectively. $\mathbf{M}_r$ and $\mathbf{M}_\theta$ are the radial and tangential components of magnetization intensity, respectively. $\mathbf{M}_r$ and $\mathbf{M}_\theta$ can all written in complex Fourier series. The convolution matrix $\mu_{r,\mathrm{cov}}$ can be defined as:

$$\mathbf{\mu}_{r,\mathrm{cov}} = \begin{bmatrix} \hat{\mu}_0 & \cdots & \hat{\mu}_{-2N} \\ \vdots & \ddots & \vdots \\ \hat{\mu}_{2N} & \cdots & \hat{\mu}_0 \end{bmatrix} \tag{25}$$

where $\hat{\mu}_n$ is the n-th order coefficient of the Fourier series of corresponding μ.

In Equation (24), $\mathbf{H}_\theta$ is continuous at the interface between two regions, but $\mathbf{B}_\theta$ is discontinuous at the interface. Hence, the matrix $\mu_{\theta,\mathrm{cov}}$ cannot be settled using Equation (25). Instead, a fast Fourier factorization is applied to calculate $\mu_{\theta,\mathrm{cov}}$, for the sake of keeping the rate of convergence the same for the left and right side of Equation (24) [34]. $\mu_{\theta,\mathrm{cov}}$ can be given by [31]:

$$\mathbf{\mu}_{\theta,\mathrm{cov}} = \begin{bmatrix} \hat{\mu}_0^{rec} & \cdots & \hat{\mu}_{-2N}^{rec} \\ \vdots & \ddots & \vdots \\ \hat{\mu}_{2N}^{rec} & \cdots & \hat{\mu}_0^{rec} \end{bmatrix}^{-1} \tag{26}$$

From Equation (26), it can be seen that $\mu_{\theta,\mathrm{cov}}$ is acquired by replacing $\hat{\mu}_n$ with the corresponding n-th order Fourier coefficient of $1/\mu_\theta$ for each element, and there is a matrix inversion outside.

The region V in SCP-MGM is used to illustrated the convolution matrix with respect to relative permeability, as shown in Figure 4.

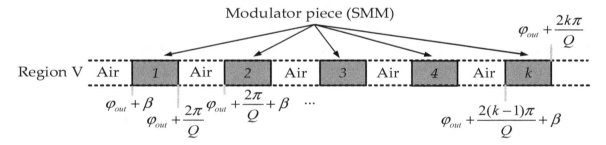

Figure 4. The calculation instance of the convolution matrix with respect to relative permeability.

The relative permeability distribution on the circumferential direction can be expressed as:

$$\mu(\theta) = \begin{cases} \mu_0 & \theta \in [\varphi_{out} + \frac{2(k-1)\pi}{Q}, \varphi_{out} + \frac{2(k-1)\pi}{Q} + \beta) \\ \mu_{iron,k} & \theta \in [\varphi_{out} + \frac{2(k-1)\pi}{Q} + \beta, \varphi_{out} + \frac{2k\pi}{Q}) \end{cases} \tag{27}$$

where $\varphi_{out} = \theta_0 - \frac{\beta}{2}$.

When a Fourier expansion on $[0, 2\pi]$ is applied to Equation (27), the expressions of μ_k and μ_k^{rec} are given by:

$$\hat{\mu}_n = \begin{cases} \sum\limits_{k=1}^{Q} \frac{\mu_0}{2\pi in} e^{in(\frac{2k\pi}{Q}+\theta_0)}\left(e^{\frac{in\beta}{2}} - e^{-\frac{in\beta}{2}}\right) + \sum\limits_{k=1}^{Q} \frac{\mu_{iron,k}}{2\pi in} e^{in(\frac{2k\pi}{Q}+\theta_0)}\left(e^{in(\frac{2\pi}{Q}-\frac{\beta}{2})} - e^{\frac{in\beta}{2}}\right) & n \neq 0 \\ \dfrac{Q\beta\mu_0 + (\frac{2\pi}{Q}-\beta)\sum\limits_{k=1}^{Q}\mu_{iron,k}}{2\pi} & n = 0 \end{cases} \tag{28}$$

$$
\hat{\mu}_n^{rec} = \begin{cases} \sum_{k=1}^{Q} \frac{1}{2\pi in\mu_0} e^{in\left(\frac{2k\pi}{Q}+\theta_0\right)}\left(e^{\frac{in\beta}{2}} - e^{-\frac{in\beta}{2}}\right) + \sum_{k=1}^{Q} \frac{1}{2\pi in\mu_{iron,k}} e^{in\left(\frac{2k\pi}{Q}+\theta_0\right)}\left(e^{in\left(\frac{2\pi}{Q}-\frac{\beta}{2}\right)} - e^{\frac{in\beta}{2}}\right) n \neq 0 \\[2em] \dfrac{\frac{Q\beta}{\mu_0} + \left(\frac{2\pi}{Q}-\beta\right)\sum_{k=1}^{Q}\frac{1}{\mu_{iron,k}}}{2\pi} n = 0 \end{cases}
\tag{29}
$$

The convolution matrices $\mu_{r,cov}$ and $\mu_{\theta,cov}$ can then be obtained by substituting Equations (28) and (29) into (25) and (26).

Combining Equations (18)–(26) together, the VMP satisfies:

$$
\frac{\partial^2 \mathbf{A}_z^k}{\partial r^2} + \frac{1}{r}\frac{\partial \mathbf{A}_z^k}{\partial r} - \frac{\mathbf{V}\mathbf{A}_z^k}{r^2} = -\mu_{\theta,cov}\mathbf{J}_z - \frac{\mu_0}{r}(\mathbf{M}_\theta + i\mathbf{U}\mathbf{M}_r)
\tag{30}
$$

where $\mathbf{V} = \mu_{\theta,cov}\mathbf{K}\mu_{r,covc}^{-1}\mathbf{K}$ and $\mathbf{U} = \mu_{\theta,cov}\mathbf{K}\mu_{r,covc}^{-1}$. The derivation process of Equation (30) is given in Appendix A.

Equation (30) is a Cauchy–Euler differential equation system [35]. The general solution of a single differential equation in Equation (30) is given by:

$$
y = c_1 r^{v^{\frac{1}{2}}} + c_2 r^{-v^{\frac{1}{2}}}
\tag{31}
$$

where c_1, c_2 are unknown coefficients. Similarly, the general solution of the differential equation system in Equation (30) can be written in a matrix form, where the new element (i, j) in matrix $r^{\mathbf{V}}$ is defined as:

$$
r^{\mathbf{V}}(i, j) = r^{\mathbf{V}(i,j)}
\tag{32}
$$

Hence, the complementary solution of Equation (30) $\mathbf{A}_z^k\big|_{com}$ can be written as:

$$
\mathbf{A}_z^k\big|_{com} = r^{\mathbf{V}^{\frac{1}{2}}}\mathbf{C}_1 + r^{-\mathbf{V}^{\frac{1}{2}}}\mathbf{C}_2
\tag{33}
$$

where $\mathbf{C}_1$ and $\mathbf{C}_2$ are unknown coefficient vectors. Matrix $r^{\mathbf{V}^{\frac{1}{2}}}$ can be factorized as:

$$
r^{\mathbf{V}^{\frac{1}{2}}} = \mathbf{P}r^{\lambda}\mathbf{P}^{-1}
\tag{34}
$$

where λ is the eigenvalue matrix of $\mathbf{V}^{1/2}$, and matrix $\mathbf{P}$ is the eigenvector matrix of $\mathbf{V}^{1/2}$. Therefore, Equation (33) can be simplified as:

$$
\mathbf{A}_z^k\big|_{com} = \mathbf{P}r^{\lambda}(\mathbf{P}^{-1}\cdot\mathbf{C}_1) + \mathbf{P}r^{-\lambda}(\mathbf{P}^{-1}\cdot\mathbf{C}_2) = \mathbf{P}r^{\lambda}\mathbf{D} + \mathbf{P}r^{-\lambda}\mathbf{E}
\tag{35}
$$

As for the particular solution of Equation (30), $\mathbf{A}_z^k\big|_{par}$ is given by:

$$
\mathbf{A}_z^k\big|_{par} = r^2\mathbf{F} + r\mathbf{G}
\tag{36}
$$

where $\mathbf{F} = (\mathbf{V} - 4\mathbf{I})^{-1}\mu_{\theta,cov}\mathbf{J}_z$, $\mathbf{G} = \mu_0(\mathbf{V} - \mathbf{I})^{-1}(\mathbf{M}_\theta + i\mathbf{U}\mathbf{M}_r)$.

The general solution of VMP in Equation (30) is the sum of the complementary solution of Equation (35), and particular solution Equation (36):

$$
\mathbf{A}_z^k = \mathbf{A}_z^k\big|_{com} + \mathbf{A}_z^k\big|_{par}
\tag{37}
$$

The expression of VMP in each subdomain is given in Appendix A.

According to the geometrical characteristic of SCP-MGMs and DCP-MGMs, the magnetization intensity only exists in region III and region V, and the current density distribution only exists in region VIII. Their distribution waveforms and Fourier series coefficients can be seen in Table 1. Additionally, for regions I, II, IV, VI, IX, and X, there only exists one material type, so the coefficients within the convolution matrix are a constant. However, the permeability distributions are different in region III, V, VII, and VIII, as shown in Table 2; their coefficients can be obtained by substituting a, b, and c in Table 3 into the following equations:

$$\hat{\mu}_n = \frac{1}{2\pi}\left(\sum_{k=1}^{c}\int_{a}^{a+b}\mu_0 e^{in\theta}d\theta + \sum_{k=1}^{c}\int_{a+b}^{a+2\pi/c}\mu_{iron}(k)e^{in\theta}d\theta\right) \tag{38}$$

$$\hat{\mu}_n^{rec} = \frac{1}{2\pi}\left(\sum_{k=1}^{c}\int_{a}^{a+b}\frac{e^{in\theta}}{\mu_0}d\theta + \sum_{k=1}^{c}\int_{a+b}^{a+2\pi/c}\frac{e^{in\theta}}{\mu_{iron,k}}d\theta\right) \tag{39}$$

Table 2. Mathematical modeling of magnetic sources within SCP-MGMs and DCP-MGMs.

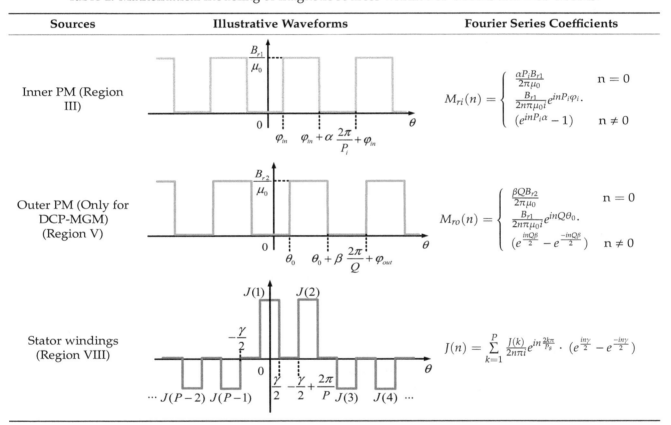

Sources	Illustrative Waveforms	Fourier Series Coefficients
Inner PM (Region III)		$M_{ri}(n) = \begin{cases} \frac{\alpha P_i B_{r1}}{2\pi\mu_0} & n=0 \\ \frac{B_{r1}}{2n\pi\mu_0 i}e^{inP_i\varphi_i}. \\ (e^{inP_i\alpha}-1) & n\neq 0 \end{cases}$
Outer PM (Only for DCP-MGM) (Region V)		$M_{ro}(n) = \begin{cases} \frac{\beta Q B_{r2}}{2\pi\mu_0} & n=0 \\ \frac{B_{r1}}{2n\pi\mu_0 i}e^{inQ\theta_0}. \\ (e^{\frac{inQ\beta}{2}}-e^{\frac{-inQ\beta}{2}}) & n\neq 0 \end{cases}$
Stator windings (Region VIII)		$J(n) = \sum_{k=1}^{P}\frac{J(k)}{2n\pi i}e^{in\frac{2k\pi}{P_s}}\cdot(e^{\frac{in\gamma}{2}}-e^{\frac{-in\gamma}{2}})$

Table 3. Mathematical modeling of the permeability distribution of different regions within SCP-MGMs and DCP-MGMs.

Regions	Illustrative Waveforms	Coefficients
III		$a = \varphi_{in}, b = \alpha, c = P_i$
V		$a = \varphi_{in}, b = \beta, c = Q$
VII		$a = \delta/2, b = 2\pi/P_s - \delta, c = P_s$
VIII		$a = \gamma/2, b = 2\pi/P_s - \gamma, c = P_s$

3.3. Bondary Condition Application

At the interfaces between two different subdomains, the radial component of $\vec{B}$ and the tangential component of $\vec{H}$ should be continuous across the boundary. Due to their ring-like shapes, each subdomain only interfaces with two subdomains at most, and region I and X have only one interface. Hence, all the boundary conditions can be written as:

$$\mathbf{B}_r^k(R_k) = \mathbf{B}_r^{k+1}(R_k) \tag{40}$$

$$\mathbf{H}_\theta^k(R_k) = \mathbf{H}_\theta^{k+1}(R_k) \tag{41}$$

where k represents the k-th subdomain of the proposed machine, and $2 \leq k \leq 9$.

Suppose the subdomain number of the proposed machine is L, and the harmonic order to be calculated is N. By applying the above boundary conditions to each interface of the machines, a system of 2*(L − 1) linear equations with 2*(L − 1) unknowns can be obtained, and each unknown is an (N × 1) vector. The system of linear equations can be further written in matrix form, as below:

$$\mathbf{SX} = \mathbf{T} \tag{42}$$

where the expressions of **S**, **X**, and **T** in this paper are given in Appendix A.

As long as the coefficient matrix **S** is invertible, the unknown vector **X** can be acquired. The numerical solution of Equation (42) can be obtained in the MATLAB software.

3.4. Saturation Consideration of Soft-Magnetic Material

For the SMM in the consequent-pole part, modulator, and stator teeth, the flux line is concentrated, thus, the saturation of the SMM must be considered. The nonlinear B-μ curve of 50JN1300 is given in Figure 4. In HMM, the relative permeability μ is obtained by an iterative method, as shown by the dot lines in Figure 5, where the number "1, 2, 3, 4" means the iteration number. The detailed iterative process is shown in Figure 6. First, an initial value, namely $\mu_0 = 1500$, is assigned to a given region with SMM for the first iteration. In the k-th iteration, the average flux density $\vec{B}$ of a specific region can be obtained by substituting $\overline{\mu}_{i,k}$ and solving the matrix Equation (42). A new average relative permeability $\overline{\mu}_{i,cal}$ in region i can then be acquired. Where i belongs to {III, V, VII, VIII}. The average relative permeability $\mu_{i,k+1}$ for the $(k + 1)$-th iteration in region i is given as:

$$\mu_{i,k+1} = \frac{\overline{\mu}_{i,k} + \overline{\mu}_{i,kcal}}{2} \tag{43}$$

The iteration will stop only if the iteration time n_i exceeds the maximum number of iterations N_i (N_i is set to be 50 here), or the maximum error Δ in all these regions is below the error requirement ξ, Δ is defined as:

$$\Delta = \max\left\{ \frac{|\overline{\mu}_{i,k} - \overline{\mu}_{i,cal}|}{\overline{\mu}_{i,k}} \right\}, \ i \in \{III, V, VII, VIII\} \tag{44}$$

where ξ is set to be 0.05 in this paper. The saturations of SMM in SCP-MGMs and DCP-MGMs are calculated respectively using this method. The relative permeabilities of each region under on-load conditions are shown in Table 4. It can be observed that the saturation of regions V and VII is more severe in DCP-MGMs, since there are PMs inserted in the modulator, thus, there are more flux lines in the modulator and stator teeth.

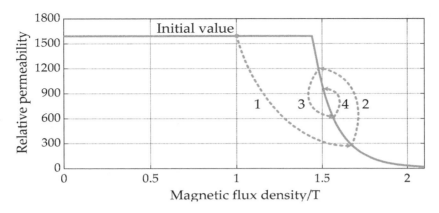

Figure 5. Iterative process illustration.

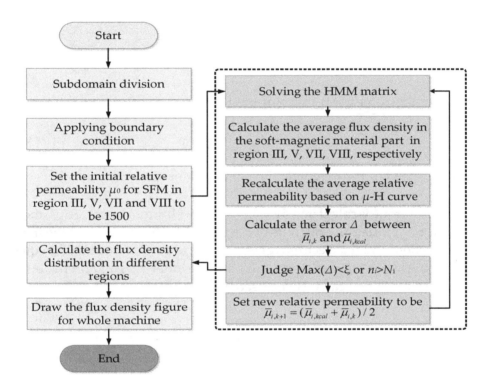

Figure 6. Fast Fourier transform (FFT) of no-load outer-air-gap radial magnetic flux density for SCP-MGMs and DCP-MGMs.

Table 4. Relative permeabilities of different regions within SCP-MGMs and DCP-MGMs.

Regions	SCP-MGM	DCP-MGM
III	27.46	27.92
V	1309.15	1257.49
VII	888.05	838.35
VIII	1013.2	917.24

3.5. Electromagnetic Parameters Calculation

Once the VMP is solved in Equation (42), the related electromagnetic parameters of the two machines can be calculated. The electromagnetic torque of the two rotors of SCP-MGMs and DCP-MGMs can be calculated by using the Maxwell stress tensor.

The flux linkage of each coil side can be given by [36]:

$$\varphi_k = \frac{N_{turn}L}{S_{coil}} \int_{-\frac{\gamma}{2}+\frac{2k\pi}{P}}^{\frac{\gamma}{2}+\frac{2k\pi}{P}} \int_{R_7}^{R_8} A_z^{VIII}(r,\theta)rdrd\theta \tag{45}$$

where N_{turn} is the number of coil turns in each slot, and S_{coil} is the cross section area of a single slot. When all the coils in each phase are in series, the three-phase flux linkage can be written as:

$$\psi = \begin{bmatrix} \psi_A \\ \psi_B \\ \psi_C \end{bmatrix} = \mathbf{C}_{turn}\begin{bmatrix} \varphi_1 & \varphi_2 & \cdots & \varphi_{24} \end{bmatrix} \tag{46}$$

where $\mathbf{C}_{turn}$ is coil-connecting matrix of the proposed machine, the coil number is given in Figure 2, and $\mathbf{C}_{turn}$ can be expressed as:

$$\mathbf{C}_{turn} = \begin{bmatrix} 1 & 0 & 0 & 0 & 0 & -1 & -1 & 0 & 0 & 0 & 0 & 1 & 1 & 0 & 0 & 0 & 0 & -1 & -1 & 0 & 0 & 0 & 0 & 1 \\ 0 & 0 & 0 & 1 & 1 & 0 & 0 & 0 & 0 & -1 & -1 & 0 & 0 & 0 & 0 & 1 & 1 & 0 & 0 & 0 & 0 & -1 & -1 & 0 \\ 0 & -1 & -1 & 0 & 0 & 0 & 0 & 1 & 1 & 0 & 0 & 0 & 0 & -1 & -1 & 0 & 0 & 0 & 0 & 1 & 1 & 0 & 0 & 0 \end{bmatrix} \tag{47}$$

The back electromotive force (EMF) is computed by the derivative of ψ with respect to time:

$$\mathbf{E}_{ABC} = \frac{d\psi}{dt} = \frac{d\psi}{d\theta}\cdot\omega \tag{48}$$

where ω is the rotating speed of the magnetic field in the outer air gap.

The electromagnetic torque is calculated by using a Maxwell stress tensor. Thus, the electromagnetic torque of the inner rotor T_{in} equals the calculus of the Maxwell stress tensor of the inner air gap along the circumferential direction, and the electromagnetic torque of the outer rotor T_{out} equals the algebraic sum of the calculus of the Maxwell stress tensor of both the inner air gap and the outer air gap along the circumferential direction. They can be expressed as:

$$T_{in} = \frac{LR_i^2}{\mu_0}\int_0^{2\pi} B_r^{IV}(R_i,\theta)B_\theta^{IV}(R_i,\theta)d\theta \tag{49}$$

$$T_{out} = \frac{LR_o^2}{\mu_0}\int_0^{2\pi} B_r^{VI}(R_o,\theta)B_\theta^{VI}(R_o,\theta)d\theta - \frac{LR_i^2}{\mu_0}\int_0^{2\pi} B_r^{IV}(R_i,\theta)B_\theta^{IV}(R_i,\theta)d\theta \tag{50}$$

where R_i and R_o are the middle radius of the inner air gap and outer air gap, respectively.

4. Validation and Comparison

4.1. Simulation Environment and Machine Parameters

To make quantitative comparison between the SCP-MGMs and DCP-MGMs, all the geometrical parameters of these two machines should be set as the same, and other parameters, such as the slot filling factor and root mean square value of the winding current should be also set as the same, as shown in Table 5. The analytical prediction of the HMM was carried out using MATLAB, and the FEA model was constructed and run in JMAG software. The FEA model had 35,597 elements and 25,368 nodes; the element size near the air gap was set as 1 mm, to maintain calculation accuracy. It took 4.6 s for the computer to obtain the magnetic field distribution for one step. The computer system configuration was as follows: Processor: Intel Core i7-4790 CPU @ 3.60 GHz; Installed Memory (RAM):

28.0 GB-System type: 64-bit Windows Operating System. Additionally, the mean error percentage ε_1 and maximum difference ε_2 of the two methods was defined as:

$$\varepsilon_1 = \frac{1}{N_c} \sum_{n=1}^{N_c} \left| \frac{V_{FEA,n} - V_{HMM,n}}{V_{FEA,n}} \right| \times 100\% \tag{51}$$

$$\varepsilon_2 = \max\left\{ \left| V_{FEA,n} - V_{HMM,n} \right| \right\}, \; n \in \{1, 2, \ldots, N_c\} \tag{52}$$

where $V_{HMM,n}$ is the n-th value, calculated using the HMM, and $V_{FEA,n}$ is n-th the value obtained by FEA. N_c is the total number of calculated points.

4.2. Comparison between HMM and FEA

Based on the calculation and simulation, a good agreement between the HMM and FEA can be observed in Figures 7–12 for SCP-MGMs and DCP-MGMs. It was also found that the difference between HMM and FEA at a no-load condition was less compared to that at a load condition. This is due to the truncation of the infinite Fourier series; the high-order components take up a greater proportion at a no-load condition, leading to a larger error for HMM. Additionally, the numerical values calculated via MATLAB were constrained by the computational accuracy. Values exceeding the computational accuracy were ignored, which led to errors of the magnetic field calculation. Generally, if the harmonic order and computational accuracy is improved, the error will decrease. However, the computational time increases rapidly with the increase of harmonic order and computational accuracy. Thus, there is a tradeoff between calculation accuracy and calculation time. In this paper, the computational accuracy of MATLAB was set as 32 bits.

The mean error percentage ε_1 and maximum difference ε_2 for the SCP-MGM and DCP-MGM are listed in Table 6.

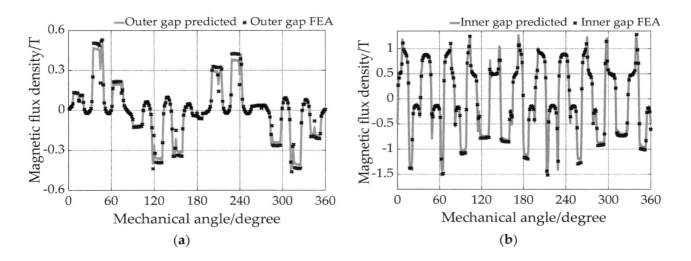

Figure 7. On-load inner and outer air-gap radial magnetic flux density distribution of the SCP-MGM: (**a**) outer air gap; (**b**) inner air gap.

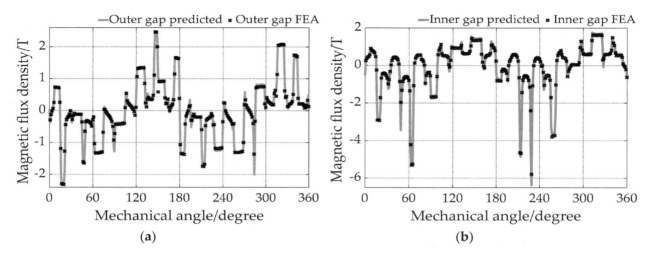

Figure 8. On-load inner and outer air-gap radial magnetic flux density distribution of the SCP-MGM: (**a**) outer air gap; (**b**) inner air gap.

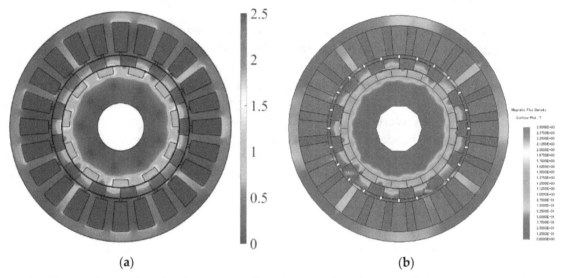

Figure 9. Comparison of on-load magnetic flux density distribution of the SCP-MGM, drawn by harmonic modeling method (HMM) and finite element analysis (FEA): (**a**) HMM; (**b**) FEA.

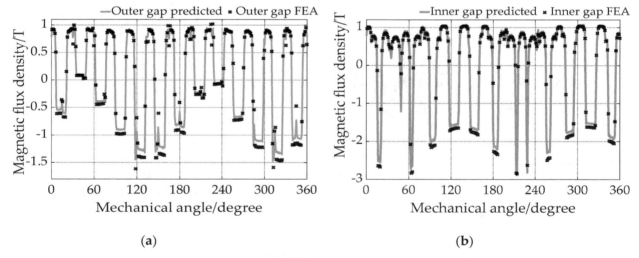

Figure 10. No-load inner and outer air-gap radial magnetic flux density distribution of the DCP-MGM: (**a**) outer air gap; (**b**) inner air gap.

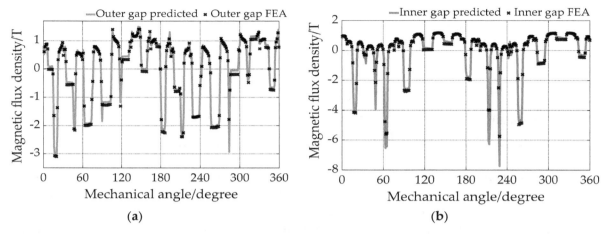

Figure 11. On-load inner and outer air-gap radial magnetic flux density distribution of the DCP-MGM: (**a**) outer air gap; (**b**) inner air gap.

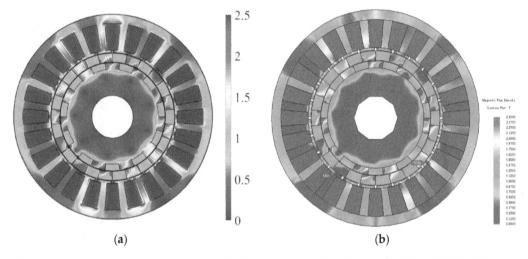

Figure 12. Comparison of on-load magnetic flux density distribution of the DCP-MGM drawn by HMM and FEA: (**a**) HMM; (**b**) FEA.

Table 5. Geometrical parameters of the SCP-MGM and DCP-MGM.

Parameters	Symbols	Values	Units
Number of inner PM pole pairs	P_i	11	-
Number of modulator pieces	Q	13	-
Number of stator slots	P	24	-
Number of stator winding pole pairs	P_s	2	-
Axial length	L	110	mm
Outer radius of shaft	R_1	25	mm
Inner radius of inner PM	R_2	55.5	mm
Outer radius of inner rotor	R_3	63.2	mm
Inner radius of outer rotor	R_4	63.8	mm
Outer radius of outer rotor	R_5	74.4	mm
Inner radius of stator	R_6	75	mm
Radius of stator slot bottom	R_7	78	mm
Outer radius of stator slot	R_8	109	mm
Outer radius of stator	R_9	120	mm
Angle of inner PM arc	α	0.286	rad
Angle of Modulator piece	β	0.242	rad
Angle of slot opening	δ	0.032	rad
Angle of stator slot	γ	0.168	rad
Slot filling factor	Fa	60%	-
Current density in stator windings	I_D	5	A/mm^2

Table 6. Mean error percentage and maximum difference of the magnetic flux density distributions between the FEA and HMM of the SCP-MGM and DCP-MGM.

State	SCP-MGM				DCP-MGM			
	Inner Rotor		Outer Rotor		Inner Rotor		Outer Rotor	
	ε_1	ε_2	ε_1	ε_2	ε_1	ε_2	ε_1	ε_2
No-load	9.3%	0.55 T	15.9%	0.11 T	11.4%	1.06 T	14.4%	0.70 T
On-load	14.2%	1.62 T	14.3%	0.83 T	20.8%	3.11 T	16.7%	1.36 T

A fast Fourier transform (FFT) was executed on the no-load radial component of the magnetic flux density of the outer air gap for both the SCP-MGM and DCP-MGM, as shown in Figure 13. It can be seen that the second harmonic component was much higher for the DCP-MGM, since the inserted PM on the outer rotor produced a second magnetic field after the modulation of the consequent-pole iron part of the inner rotor. Hence, the consequent-pole structure of the inner rotor and outer rotor of the DCP-MGM could modulate the magnetic field generated by its counterpart. The electromagnetic torque of the DCP-MGM was expected to be larger than that of the SCP-MGM under the same working conditions. Additionally, the difference of each frequency component between the HMM and FEA was very small. Specifically, the error percentages of the second harmonic component for the SCP-MGM and the DCP-MGM using HMM and FEA were 4.14% and 2.15%, respectively.

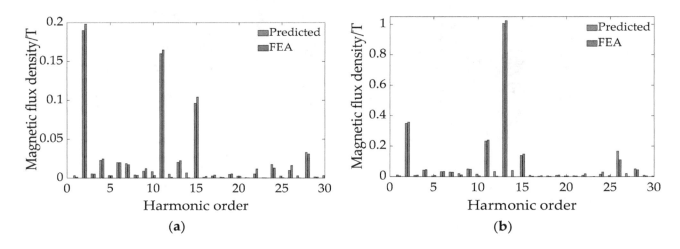

(a)

(b)

Figure 13. FFT of no-load outer-air-gap radial magnetic flux density for the SCP-MGM and the DCP-MGM: (a) SCP-MGM; (b) DCP-MGM.

4.3. Electromagnetic Performance Analysis under Different Working Conditions

The electromagnetic performances of the two MGMs are predicted by HMM and simulated in FEA software under different operating conditions, where the rotating speed of two rotors and the current frequency of stator windings are set to cooperate with the practical driving conditions of the HEC, as shown in Table 1. Since the ICE is connected to the inner rotor to provide the power; the outer rotor is connected to the differential, which is further connected to the wheels; the battery is connected to the stator windings, and there can be two-way power flow between the battery and the stator windings. Thus, the power transmission relation among inner rotor, outer rotor and stator windings is equal to the power transmission relation among ICE, wheels and battery. Assume the anti-clock direction is positive, the torque is positive if it's on the anti-clock direction, otherwise it's negative. The power of a component is defined as a positive one when this component is inputting energy; when a component is outputting energy, its power is defined as a negative one.

4.3.1. Back EMF under No-Load Condition

The amplitude of back EMF was determined by the rotating speed of both the inner rotor and the outer rotor. Figure 14 shows the no-load back EMF of the SCP-MGM and the DCP-MGM under the same operating conditions, namely ω_i = 1200 r/min, ω_o = 1500 r/min. It can be seen that there was an error between the predicted back EMF and FEA result; because the back EMF was calculated as the derivative of flux linkage with respect to time, a small error of flux linkage will be amplified on the back EMF. Additionally, the back EMF of the DCP-MGM was larger than that of the SCP-MGM, due to the flux enhancing effect of the PMs inserted into the modulator. The maximum errors for the SCP-MGM and DCP-MGM were 165 V and 520 V, respectively. The average errors of the SCP-MGM and DCP-MGM were 3.7% and 6.9%, respectively.

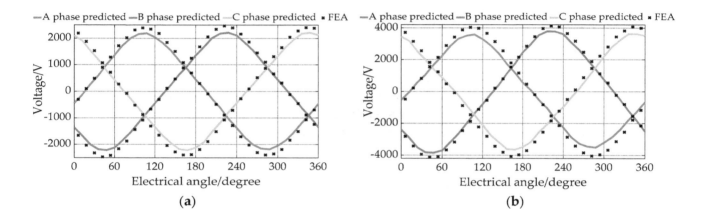

Figure 14. Analytically predicted and FEA simulated back electromotive force (EMF) of the SCP-MGM and DCP-MGM: (**a**) SCP-MGM; (**b**) DCP-MGM.

4.3.2. Pure Electric Mode (Mode 1)

Under this mode, the HEV is at a low driving speed, the inner rotor is locked, and the ICE does not come into service; stator windings only work to provide the power that the HEV needs. Thus, ω_i = 0 r/min, ω_o = 500 r/min. The torque waveforms of the power distribution of the two machines are shown in Figures 15 and 16.

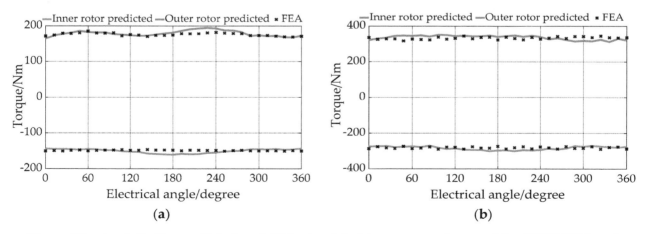

Figure 15. Analytically predicted and FEA simulated torque waveforms of the SCP-MGM and DCP-MGM under pure electric mode: (**a**) SCP-MGM; (**b**) DCP-MGM.

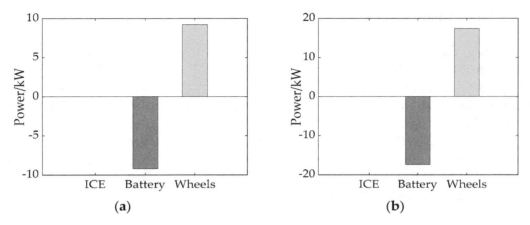

Figure 16. Power distribution among the internal combustion engine (ICE), battery, and wheels under pure electric mode: (**a**) SCP-MGM; (**b**) DCP-MGM.

4.3.3. Pure Mechanical Mode (Mode 2)

Under this mode, the HEV is running at a medium speed, so the battery does not output power anymore and the ICE comes into use. However, to maintain the magnetic field, the stator windings are electrified with DC current [14]. The rotating speeds of two rotors are: $\omega_i = 1200$ r/min, $\omega_o = 1015$ r/min. The torque waveforms of the power distribution of the two machines are shown in Figures 17 and 18.

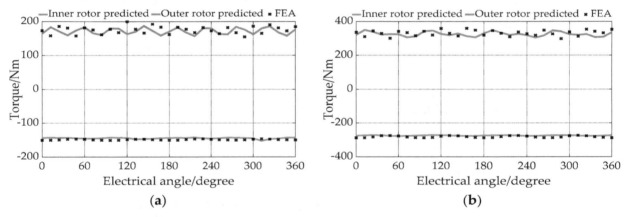

Figure 17. Analytically predicted and FEA simulated torque waveforms of SCP-MGM and DCP-MGM under pure mechanical mode: (**a**) SCP-MGM; (**b**) DCP-MGM.

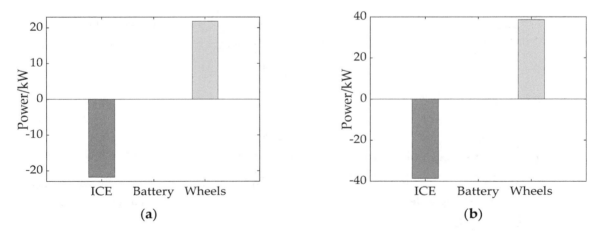

Figure 18. Power distribution among ICE, battery and wheels under pure mechanical mode: (**a**) SCP-MGM; (**b**) DCP-MGM.

4.3.4. Hybrid Mode (Mode 3)

When the HEV needs to further accelerate, the ICE alone is not enough to provide the power that the HEV needs. The SCP-MGM and DCP-MGM can then switch into hybrid mode. In this mode, both ICE and battery provide the energy for the wheels. The rotating speeds of two rotors are: $\omega_i = 1200$ r/min, $\omega_o = 2000$ r/min. The torque waveforms of the power distribution of the two machines are shown in Figures 19 and 20.

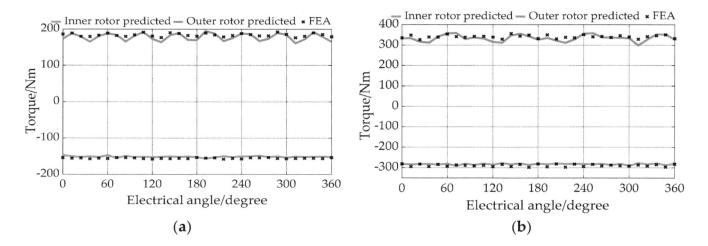

Figure 19. Analytically predicted and FEA simulated electromagnetic torque of SCP-MGM and DCP-MGM under hybrid mode: (**a**) SCP-MGM; (**b**) DCP-MGM.

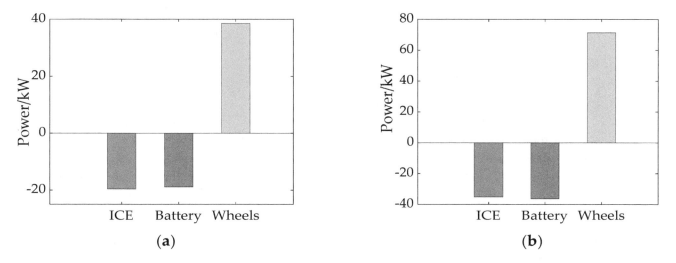

Figure 20. Power distribution among ICE, battery, and wheels under hybrid mode: (**a**) SCP-MGM; (**b**) DCP-MGM.

4.3.5. Regenerative Braking Mode (Mode 4)

When the HEV deaccelerates, the ICE stops working. The magnetic field generated via the stator windings provides a resistance for the wheels, and thus the power flows from the wheels to the battery. At this time, the power flows from the stator windings to the battery. Thus, the battery can be charged under this mode. The rotating speeds of two rotors are: $\omega_i = 0$ r/min, $\omega_o = 1000$ r/min. The torque waveforms of the power distribution of the two machines are shown in Figures 21 and 22.

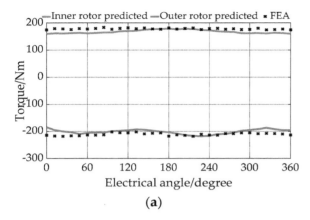 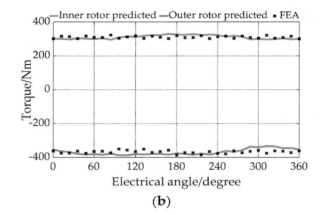

Figure 21. Analytically predicted and FEA simulated back EMF of SCP-MGM and DCP-MGM: (a) SCP-MGM; (b) DCP-MGM.

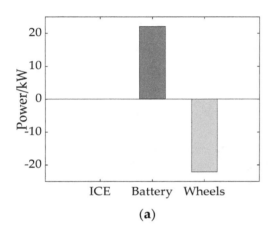 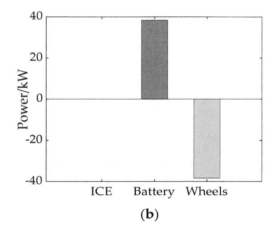

Figure 22. Power distribution among ICE, battery, and wheels under hybrid mode: (a) SCP-MGM; (b) DCP-MGM.

4.3.6. Quantitative comparison between HMM and FEA

Table 7 shows the mean error percentage and maximum difference of the torque waveforms between FEA and HMM of the SCP-MGM and DCP-MGM under different operating modes. It can be observed that the mean error percentage was below 8%, which is acceptable for the electromagnetic torque prediction of SCP-MGMs and DCP-MGMs. The error of torque prediction was mainly caused by the error of the air-gap magnetic flux density prediction on both the radial and tangential direction, since the torque was calculated using the calculus of the product of the radial component and tangential component of air-gap magnetic flux density.

Table 7. Mean error percentage and maximum difference of the torque waveforms between FEA and HMM of the SCP-MGM and DCP-MGM under different operating modes.

| Mode | SCP-MGM | | | | DCP-MGM | | | |
| | Inner Rotor | | Outer Rotor | | Inner Rotor | | Outer Rotor | |
	ε_1	ε_2	ε_1	ε_2	ε_1	ε_2	ε_1	ε_2
1	3.59%	12.1 Nm	2.64%	14.8 Nm	3.55%	22.5 Nm	4.46%	30.3 Nm
2	3.00%	7.85 Nm	7.39%	37.2 Nm	2.57%	14.9 Nm	5.88%	48.6 Nm
3	2.75%	8.9 Nm	3.61%	15.2 Nm	2.27%	15.3 Nm	3.27%	30.3 Nm
4	5.73%	18.1 Nm	5.06%	30.0 Nm	3.50%	28.4 Nm	4.56%	43.0 Nm

From the above comparisons between the SCP-MGM and DCP-MGM under various working conditions, it can be seen that the torque ratio between outer rotor and inner rotor was about 1.18, which is equal to the pole-pair ratio of outer rotor to inner rotor. The ripple rate of the outer rotor was larger than that of the inner rotor, which can be alleviated by adopting a skewed stator when these machines are applied in practical applications. Additionally, the electromagnetic torque of the DCP-MGM was about 1.88 times of that of SCP-MGM. However, the torque per unit weight of the PM of the DCP-MGM was only about 0.72 of that of SCP-MGM. Considering the fact that the consequent–pole structure already saves half the PM material that would otherwise have been used, the total usage of PM material for DCP-MGM is reasonable. Thus, the DCP-MGM is more suitable to be used in the propulsion systems of HEVs.

4.3.7. Discussion of HMM

The greatest advantage of HMM is that the modeling process is simpler compared to the conventional subdomain method [30]. Since it unitizes Fourier series of relative permeability, the boundary conditions become simpler and the number of unknowns becomes less. For instance, the consequent-pole structure leads to a very complex general solution if the subdomain method is used [29]. Additionally, magnetic saturation can be taken into account to an extent in HMM, and a magnetic flux density distribution figure can be obtained. However, the HMM also has some drawbacks. First, HMM can only provide a simplified magnetic saturation model, since the magnetic saturation varies from point to point in some SMM parts. The relative permeability in the radial direction is also regarded as a constant, but, in reality, it could change. To the authors' knowledge, no existing analytical method can really reflect the magnetic saturation at every point within an electric machine. Secondly, the use of convolution sometimes leads to an extremely large or small value, which further leads to a large error in the final result due to the accuracy limit of the numerical calculation.

5. Conclusions

In this paper, an analytical modeling method for consequent-pole MGMs was proposed and elaborated. Two machine topologies, namely the SCP-MGM and DCP-MGM, were analyzed and compared quantitatively using HMM and FEA. A good agreement was achieved for these two methods. Furthermore, these two machines were embedded into the propulsion system of HEVs under different operating conditions. By inserting extra PMs in the modulator, the electromagnetic torque of the DCP-MGM increased greatly compared to its counterpart. Therefore, the DCP-MGM has the potential to be applied in the propulsion systems of HEVs. Additionally, the proposed HMM can also be applied in the mathematical modeling of other consequent-pole electric machines.

Author Contributions: The work presented in this paper is the output of the research projects undertaken by C.L. In specific, H.Z. and C.L. developed the topic. H.Z. carried out the calculation and simulation, analyzed the results, and wrote the paper. Z.S. gave some suggestions on the calculation process. J.Y. helped to carried out a part of the simulation.

Appendix A

The expressions of $\mathbf{H}_r$ and $\mathbf{H}_\theta$ are given by:

$$\mathbf{H}_r = -i\frac{1}{r}\mu_{r,\text{cov}}^{-1}\mathbf{K}\mathbf{A}_z - \mu_0\mu_{r,\text{cov}}^{-1}\mathbf{M}_r \text{ and } \mathbf{H}_\theta = -\mu_{\theta,\text{cov}}^{-1}\frac{\partial \mathbf{A}_z}{\partial r} - \mu_0\mu_{\theta,\text{cov}}^{-1}\mathbf{M}_\theta \tag{A1}$$

In polar coordinate, (15) can be simplified as:

$$\nabla \times \mathbf{H} = \frac{1}{r}\frac{\partial}{\partial r}(r\mathbf{H}_\theta) + i\frac{\mathbf{K}\mathbf{H}_r}{r} = \mathbf{J}_z \tag{A2}$$

By substituting (A1) into (A2), one can obtain that:

$$\frac{\partial^2 \mathbf{A}_z^k}{\partial r^2} + \frac{1}{r}\frac{\partial \mathbf{A}_z^k}{\partial r} - \frac{b_{\theta,\text{cov}}\mathbf{K}\boldsymbol{\mu}_{r,\text{covc}}^{-1}\mathbf{K}\mathbf{A}_z^k}{r^2} = -\mu_{\theta,cov}\mathbf{J}_z - \frac{\mu_0}{r}\left(\mathbf{M}_\theta + i\mu_{\theta,\text{cov}}\mathbf{K}\boldsymbol{\mu}_{r,\text{covc}}^{-1}\mathbf{M}_r\right) \tag{A3}$$

The vector magnetic potential in region I~X are given in Table A1.

Table A1. The Vector Magnetic Potential Expressions.

Region	VMP Expression	Region	VMP Expression
I	$\mathbf{A}_z^I = \left(\frac{r}{R_1}\right)^{|\mathbf{K}|}\mathbf{D}^I$	II	$\mathbf{A}_z^{II} = \left(\frac{r}{R_2}\right)^{|\mathbf{K}|}\mathbf{D}^{II} + \left(\frac{R_1}{r}\right)^{|\mathbf{K}|}\mathbf{E}^{II}$
III	$\mathbf{A}_z^{III} = \mathbf{P}^{III}\left(\frac{r}{R_3}\right)^{\boldsymbol{\lambda}^{III}}\mathbf{D}^{III} + \mathbf{P}^{III}\left(\frac{R_2}{r}\right)^{\boldsymbol{\lambda}^{III}}\mathbf{E}^{III} + r\mathbf{G}^{III}$	IV	$\mathbf{A}_z^{IV} = \left(\frac{r}{R_4}\right)^{|\mathbf{K}|}\mathbf{D}^{IV} + \left(\frac{R_3}{r}\right)^{|\mathbf{K}|}\mathbf{E}^{IV}$
V	$\mathbf{A}_z^V = \mathbf{P}^V\left(\frac{r}{R_5}\right)^{\boldsymbol{\lambda}^V}\mathbf{D}^V + \mathbf{P}^V\left(\frac{R_4}{r}\right)^{\boldsymbol{\lambda}^V}\mathbf{E}^V + r\mathbf{G}^V$	VI	$\mathbf{A}_z^{VI} = \left(\frac{r}{R_6}\right)^{|\mathbf{K}|}\mathbf{D}^{VI} + \left(\frac{R_5}{r}\right)^{|\mathbf{K}|}\mathbf{E}^{VI}$
VII	$\mathbf{A}_z^{VII} = \mathbf{P}^{VII}\left(\frac{r}{R_7}\right)^{\boldsymbol{\lambda}^{VII}}\mathbf{D}^{VII} + \mathbf{P}^{VII}\left(\frac{R_6}{r}\right)^{\boldsymbol{\lambda}^{VII}}\mathbf{E}^{VII}$	VIII	$\mathbf{A}_z^{VIII} = \mathbf{P}^{VIII}\left(\frac{r}{R_8}\right)^{\boldsymbol{\lambda}^{VIII}}\mathbf{D}^{VIII} + \mathbf{P}^{VIII}\left(\frac{R_7}{r}\right)^{\boldsymbol{\lambda}^{VIII}}\mathbf{E}^{VIII} + r^2\mathbf{F}$
IX	$\mathbf{A}_z^{IX} = \left(\frac{r}{R_9}\right)^{|\mathbf{K}|}\mathbf{D}^{IX} + \left(\frac{R_8}{r}\right)^{|\mathbf{K}|}\mathbf{E}^{IX}$	X	$\mathbf{A}_z^X = \left(\frac{r}{R_9}\right)^{|\mathbf{K}|}\mathbf{E}^X$

Where $\mathbf{D}^I, \mathbf{D}^{II}, \ldots, \mathbf{D}^{IX}$ and $\mathbf{E}^I, \mathbf{E}^{II}, \ldots, \mathbf{E}^X$ are (N*1) vector.

The expressions of **X**, **S**, **T** are:

$$\mathbf{X} = \begin{bmatrix} (\mathbf{D}^I)^T & (\mathbf{D}^{II})^T & (\mathbf{E}^{II})^T & (\mathbf{D}^{III})^T & (\mathbf{E}^{III})^T & \cdots & (\mathbf{D}^{IX})^T & (\mathbf{E}^{IX})^T & (\mathbf{E}^X)^T \end{bmatrix}^T \tag{A4}$$

$$\mathbf{S} = \begin{bmatrix}
\mathbf{K}_{1,1} & \mathbf{K}_{1,2} & \mathbf{K}_{1,3} & 0 & 0 & 0 & 0 & 0 & 0 & 0 & 0 & 0 & 0 & 0 & 0 & 0 & 0 & 0 \\
\mathbf{K}_{2,1} & \mathbf{K}_{2,2} & \mathbf{K}_{2,3} & 0 & 0 & 0 & 0 & 0 & 0 & 0 & 0 & 0 & 0 & 0 & 0 & 0 & 0 & 0 \\
0 & \mathbf{K}_{3,2} & \mathbf{K}_{3,3} & \mathbf{K}_{3,4} & \mathbf{K}_{3,5} & 0 & 0 & 0 & 0 & 0 & 0 & 0 & 0 & 0 & 0 & 0 & 0 & 0 \\
0 & \mathbf{K}_{4,2} & \mathbf{K}_{4,3} & \mathbf{K}_{4,4} & \mathbf{K}_{4,5} & 0 & 0 & 0 & 0 & 0 & 0 & 0 & 0 & 0 & 0 & 0 & 0 & 0 \\
0 & 0 & 0 & \mathbf{K}_{5,4} & \mathbf{K}_{5,5} & \mathbf{K}_{5,6} & \mathbf{K}_{5,7} & 0 & 0 & 0 & 0 & 0 & 0 & 0 & 0 & 0 & 0 & 0 \\
0 & 0 & 0 & \mathbf{K}_{6,4} & \mathbf{K}_{6,5} & \mathbf{K}_{6,6} & \mathbf{K}_{6,7} & 0 & 0 & 0 & 0 & 0 & 0 & 0 & 0 & 0 & 0 & 0 \\
0 & 0 & 0 & 0 & 0 & \mathbf{K}_{7,6} & \mathbf{K}_{7,7} & \mathbf{K}_{7,8} & \mathbf{K}_{7,9} & 0 & 0 & 0 & 0 & 0 & 0 & 0 & 0 & 0 \\
0 & 0 & 0 & 0 & 0 & \mathbf{K}_{8,6} & \mathbf{K}_{8,7} & \mathbf{K}_{8,8} & \mathbf{K}_{8,9} & 0 & 0 & 0 & 0 & 0 & 0 & 0 & 0 & 0 \\
0 & 0 & 0 & 0 & 0 & 0 & 0 & \mathbf{K}_{9,8} & \mathbf{K}_{9,9} & \mathbf{K}_{9,10} & \mathbf{K}_{9,11} & 0 & 0 & 0 & 0 & 0 & 0 & 0 \\
0 & 0 & 0 & 0 & 0 & 0 & 0 & \mathbf{K}_{10,8} & \mathbf{K}_{10,9} & \mathbf{K}_{10,10} & \mathbf{K}_{10,11} & 0 & 0 & 0 & 0 & 0 & 0 & 0 \\
0 & 0 & 0 & 0 & 0 & 0 & 0 & 0 & 0 & \mathbf{K}_{11,10} & \mathbf{K}_{11,11} & \mathbf{K}_{11,12} & \mathbf{K}_{11,13} & 0 & 0 & 0 & 0 & 0 \\
0 & 0 & 0 & 0 & 0 & 0 & 0 & 0 & 0 & \mathbf{K}_{12,10} & \mathbf{K}_{12,11} & \mathbf{K}_{12,12} & \mathbf{K}_{12,13} & 0 & 0 & 0 & 0 & 0 \\
0 & 0 & 0 & 0 & 0 & 0 & 0 & 0 & 0 & 0 & 0 & \mathbf{K}_{13,12} & \mathbf{K}_{13,13} & \mathbf{K}_{13,14} & \mathbf{K}_{13,15} & 0 & 0 & 0 \\
0 & 0 & 0 & 0 & 0 & 0 & 0 & 0 & 0 & 0 & 0 & \mathbf{K}_{14,12} & \mathbf{K}_{14,13} & \mathbf{K}_{14,14} & \mathbf{K}_{14,15} & 0 & 0 & 0 \\
0 & 0 & 0 & 0 & 0 & 0 & 0 & 0 & 0 & 0 & 0 & 0 & 0 & \mathbf{K}_{15,14} & \mathbf{K}_{15,15} & \mathbf{K}_{15,16} & \mathbf{K}_{15,17} & 0 \\
0 & 0 & 0 & 0 & 0 & 0 & 0 & 0 & 0 & 0 & 0 & 0 & 0 & \mathbf{K}_{16,14} & \mathbf{K}_{16,15} & \mathbf{K}_{16,16} & \mathbf{K}_{16,17} & 0 \\
0 & 0 & 0 & 0 & 0 & 0 & 0 & 0 & 0 & 0 & 0 & 0 & 0 & 0 & 0 & \mathbf{K}_{17,16} & \mathbf{K}_{17,17} & \mathbf{K}_{17,18} \\
0 & 0 & 0 & 0 & 0 & 0 & 0 & 0 & 0 & 0 & 0 & 0 & 0 & 0 & 0 & \mathbf{K}_{18,16} & \mathbf{K}_{18,17} & \mathbf{K}_{18,18}
\end{bmatrix} \tag{A5}$$

$$\mathbf{T} = \begin{bmatrix} 0 & 0 & (R_2\mathbf{G}_1)^T & (R_2\boldsymbol{\mu}_{II}\mathbf{G}_1)^T & (-R_3\mathbf{G}_1)^T & (-R_3\boldsymbol{\mu}_{IV}\mathbf{G}_1)^T & (R_4\mathbf{G}_2)^T & (R_4\boldsymbol{\mu}_{IV}\mathbf{G}_2)^T & (-R_5\mathbf{G}_2)^T \\ (-\boldsymbol{\mu}_{VI}R_5\mathbf{G}_2)^T & 0 & 0 & (R_7^2\mathbf{F})^T & (2R_7^2\mu_{open,\theta}\mathbf{F})^T & (-R_8^2\mathbf{F})^T & (-2R_8^2\boldsymbol{\mu}_{IX}\mathbf{F})^T & 0 & 0 \end{bmatrix}^T \tag{A6}$$

where:

$$\begin{bmatrix} \mathbf{K}_{1,1} & \mathbf{K}_{1,2} & \mathbf{K}_{1,3} \end{bmatrix} = \begin{bmatrix} \mathbf{I} & -\left(\frac{R_1}{R_2}\right)^{|\mathbf{K}|} & -\mathbf{I} \end{bmatrix} \tag{A7}$$

$$\begin{bmatrix} \mathbf{K}_{2,1} & \mathbf{K}_{2,2} & \mathbf{K}_{2,3} \end{bmatrix} = \begin{bmatrix} \boldsymbol{\mu}_{II} & -\boldsymbol{\mu}_I\left(\frac{R_1}{R_2}\right)^{|\mathbf{K}|} & \boldsymbol{\mu}_I \end{bmatrix} \tag{A8}$$

$$\begin{bmatrix} \mathbf{K}_{3,2} & \mathbf{K}_{3,3} & \mathbf{K}_{3,4} & \mathbf{K}_{3,5} \end{bmatrix} = \begin{bmatrix} \mathbf{I} & \left(\frac{R_1}{R_2}\right)^{|\mathbf{K}|} & -\mathbf{P}^{III}\left(\frac{R_1}{R_2}\right)^{|\mathbf{K}|} & -\mathbf{P}^{III} \end{bmatrix} \quad (A9)$$

$$\begin{bmatrix} \mathbf{K}_{4,2} & \mathbf{K}_{4,3} & \mathbf{K}_{4,4} & \mathbf{K}_{4,5} \end{bmatrix} = \begin{bmatrix} \mu_{PM,\theta} & -\mu_{PM,\theta}|\mathbf{K}|\left(\frac{R_1}{R_2}\right)^{|\mathbf{K}|} & -\mu_{II}\mathbf{P}^{III}\boldsymbol{\lambda}^{III}\left(\frac{R_2}{R_3}\right)^{\boldsymbol{\lambda}^{III}} & \mu_{II}\mathbf{P}^{III}\boldsymbol{\lambda}^{III} \end{bmatrix} \quad (A10)$$

$$\begin{bmatrix} \mathbf{K}_{5,4} & \mathbf{K}_{5,5} & \mathbf{K}_{5,6} & \mathbf{K}_{5,7} \end{bmatrix} = \begin{bmatrix} \mathbf{P}^{III} & \mathbf{P}^{III}\left(\frac{R_2}{R_3}\right)^{\boldsymbol{\lambda}^{III}} & -\left(\frac{R_3}{R_4}\right)^{|\mathbf{K}|} & -\mathbf{I} \end{bmatrix} \quad (A11)$$

$$\begin{bmatrix} \mathbf{K}_{6,4} & \mathbf{K}_{6,5} & \mathbf{K}_{6,6} & \mathbf{K}_{6,7} \end{bmatrix} = \begin{bmatrix} \mu_{IV}\mathbf{P}^{III}\boldsymbol{\lambda}^{III} & -\mu_{IV}\mathbf{P}^{III}\boldsymbol{\lambda}^{III}\left(\frac{R_2}{R_3}\right)^{\boldsymbol{\lambda}^{III}} & \mu_{PM,\theta}|\mathbf{K}|\left(\frac{R_3}{R_4}\right)^{|\mathbf{K}|} & \mu_{PM,\theta}|\mathbf{K}| \end{bmatrix} \quad (A12)$$

$$\begin{bmatrix} \mathbf{K}_{7,6} & \mathbf{K}_{7,7} & \mathbf{K}_{7,8} & \mathbf{K}_{7,9} \end{bmatrix} = \begin{bmatrix} \mathbf{I} & \left(\frac{R_3}{R_4}\right)^{|\mathbf{K}|} & -\mathbf{P}^{V}\left(\frac{R_4}{R_5}\right)^{\boldsymbol{\lambda}^{V}} & -\mathbf{P}^{V} \end{bmatrix} \quad (A13)$$

$$\begin{bmatrix} \mathbf{K}_{8,6} & \mathbf{K}_{8,7} & \mathbf{K}_{8,8} & \mathbf{K}_{8,9} \end{bmatrix} = \begin{bmatrix} \mu_{Mod,\theta}|\mathbf{K}| & -\mu_{Mod,\theta}|\mathbf{K}|\left(\frac{R_3}{R_4}\right)^{|\mathbf{K}|} & -\mu_{IV}\mathbf{P}^{V}\boldsymbol{\lambda}^{V}\left(\frac{R_4}{R_5}\right)^{\boldsymbol{\lambda}^{V}} & \mu_{IV}\mathbf{P}^{V}\boldsymbol{\lambda}^{V} \end{bmatrix} \quad (A14)$$

$$\begin{bmatrix} \mathbf{K}_{9,8} & \mathbf{K}_{9,9} & \mathbf{K}_{9,10} & \mathbf{K}_{9,11} \end{bmatrix} = \begin{bmatrix} \mathbf{P}^{V} & \mathbf{P}^{V}\left(\frac{R_4}{R_5}\right)^{\boldsymbol{\lambda}^{V}} & -\left(\frac{R_5}{R_6}\right)^{|\mathbf{K}|} & -\mathbf{I} \end{bmatrix} \quad (A15)$$

$$\begin{bmatrix} \mathbf{K}_{10,8} & \mathbf{K}_{10,9} & \mathbf{K}_{10,10} & \mathbf{K}_{10,11} \end{bmatrix} = \begin{bmatrix} \mu_{VI}\mathbf{P}^{V}\boldsymbol{\lambda}^{V} & -\mu_{VI}\mathbf{P}^{V}\boldsymbol{\lambda}^{V}\left(\frac{R_4}{R_5}\right)^{\boldsymbol{\lambda}^{V}} & -\mu_{Mod,\theta}|\mathbf{K}|\left(\frac{R_5}{R_6}\right)^{|\mathbf{K}|} & \mu_{Mod,\theta}|\mathbf{K}| \end{bmatrix} \quad (A16)$$

$$\begin{bmatrix} \mathbf{K}_{11,10} & \mathbf{K}_{11,11} & \mathbf{K}_{11,12} & \mathbf{K}_{11,13} \end{bmatrix} = \begin{bmatrix} \mathbf{I} & \left(\frac{R_5}{R_6}\right)^{|\mathbf{K}|} & -\mathbf{P}^{VII}\left(\frac{R_6}{R_7}\right)^{\boldsymbol{\lambda}^{VII}} & -\mathbf{P}^{VII} \end{bmatrix} \quad (A17)$$

$$\begin{bmatrix} \mathbf{K}_{12,10} & \mathbf{K}_{12,11} & \mathbf{K}_{12,12} & \mathbf{K}_{12,13} \end{bmatrix} = \begin{bmatrix} \mu_{Open,\theta}|\mathbf{K}| & \mu_{Open,\theta}|\mathbf{K}|\left(\frac{R_5}{R_6}\right)^{|\mathbf{K}|} & -\mu_{VI}\mathbf{P}^{VII}\boldsymbol{\lambda}^{VII}\left(\frac{R_6}{R_7}\right)^{\boldsymbol{\lambda}^{VII}} & \mu_{VI}\mathbf{P}^{VII}\boldsymbol{\lambda}^{VII} \end{bmatrix} \quad (A18)$$

$$\begin{bmatrix} \mathbf{K}_{13,12} & \mathbf{K}_{13,13} & \mathbf{K}_{13,14} & \mathbf{K}_{13,15} \end{bmatrix} = \begin{bmatrix} \mathbf{P}^{VII}\boldsymbol{\lambda}^{VII} & \mathbf{P}^{VII}\left(\frac{R_6}{R_7}\right)^{\boldsymbol{\lambda}^{VII}} & -\mathbf{P}^{VIII}\left(\frac{R_7}{R_8}\right)^{\boldsymbol{\lambda}^{VIII}} & -\mathbf{P}^{VIII} \end{bmatrix} \quad (A19)$$

$$\begin{bmatrix} \mathbf{K}_{14,12} & \mathbf{K}_{14,13} & \mathbf{K}_{14,14} & \mathbf{K}_{14,15} \end{bmatrix} = \begin{bmatrix} \mu_{Slot,\theta}\mathbf{P}^{VII}\boldsymbol{\lambda}^{VII} & -\mu_{Slot,\theta}\mathbf{P}^{VII}\boldsymbol{\lambda}^{VII}\left(\frac{R_6}{R_7}\right)^{\boldsymbol{\lambda}^{VII}} & -\mu_{Open,\theta}\mathbf{P}^{VIII}\boldsymbol{\lambda}^{VIII}\left(\frac{R_7}{R_8}\right)^{\boldsymbol{\lambda}^{VIII}} & \mu_{Open,\theta}\mathbf{P}^{VIII}\boldsymbol{\lambda}^{VIII} \end{bmatrix} \quad (A20)$$

$$\begin{bmatrix} \mathbf{K}_{15,14} & \mathbf{K}_{15,15} & \mathbf{K}_{15,16} & \mathbf{K}_{15,17} \end{bmatrix} = \begin{bmatrix} \mathbf{P}^{VIII} & \mathbf{P}^{VIII}\left(\frac{R_7}{R_8}\right)^{\boldsymbol{\lambda}^{VIII}} & -\left(\frac{R_8}{R_9}\right)^{|\mathbf{K}|} & -\mathbf{I} \end{bmatrix} \quad (A21)$$

$$\begin{bmatrix} \mathbf{K}_{16,14} & \mathbf{K}_{16,15} & \mathbf{K}_{16,16} & \mathbf{K}_{16,17} \end{bmatrix} = \begin{bmatrix} \mu_{IX}\mathbf{P}^{VIII}\boldsymbol{\lambda}^{VIII} & -\mu_{IX}\mathbf{P}^{VIII}\boldsymbol{\lambda}^{VIII}\left(\frac{R_7}{R_8}\right)^{\boldsymbol{\lambda}^{VIII}} & \mu_{Slot,\theta}|\mathbf{K}|\left(\frac{R_8}{R_9}\right)^{|\mathbf{K}|} & \mu_{Slot,\theta}|\mathbf{K}| \end{bmatrix} \quad (A22)$$

$$\begin{bmatrix} \mathbf{K}_{17,16} & \mathbf{K}_{17,17} & \mathbf{K}_{17,18} \end{bmatrix} = \begin{bmatrix} \mathbf{I} & -\left(\frac{R_8}{R_9}\right)^{|\mathbf{K}|} & -\mathbf{I} \end{bmatrix} \quad (A23)$$

$$\begin{bmatrix} \mathbf{K}_{18,16} & \mathbf{K}_{18,17} & \mathbf{K}_{18,18} \end{bmatrix} = \begin{bmatrix} \mu_{X} & -\mu_{X}\left(\frac{R_8}{R_9}\right)^{|\mathbf{K}|} & \mu_{IX} \end{bmatrix} \quad (A24)$$

References

1. Liu, C. Emerging Electric Machines and Drives—An Overview. *IEEE Trans. Energy Convers.* **2018**, *33*, 2270–2280. [CrossRef]
2. Zhu, Z.Q.; Liu, Y. Analysis of Air-Gap Field Modulation and Magnetic Gearing Effect in Fractional-Slot Concentrated-Winding Permanent-Magnet Synchronous Machines. *IEEE Trans. Ind. Electron.* **2018**, *65*, 3688–3698. [CrossRef]
3. Atallah, K.; Rens, J.; Mezani, S.; Howe, D. A Novel "Pseudo" Direct-Drive Brushless Permanent Magnet Machine. *IEEE Trans. Magn.* **2008**, *44*, 4349–4352. [CrossRef]
4. Atallah, K.; Howe, D. A novel high-performance magnetic gear. *IEEE Trans. Magn.* **2001**, *37*, 2844–2846. [CrossRef]
5. Acharya, V.M.; Bird, J.Z.; Calvin, M. A Flux Focusing Axial Magnetic Gear. *IEEE Trans. Magn.* **2013**, *49*, 4092–4095. [CrossRef]
6. Linni, J.; Chau, K.T.; Gong, Y.; Jiang, J.Z.; Chuang, Y.; Wenlong, L. Comparison of Coaxial Magnetic Gears With Different Topologies. *IEEE Trans. Magn.* **2009**, *45*, 4526–4529. [CrossRef]
7. Holehouse, R.C.; Atallah, K.; Wang, J.B. Design and Realization of a Linear Magnetic Gear. *IEEE Trans. Magn.* **2011**, *47*, 4171–4174. [CrossRef]
8. Liu, C.T.; Chung, H.Y.; Hwang, C.C. Design Assessments of a Magnetic-Geared Double-Rotor Permanent Magnet Generator. *IEEE Trans. Magn.* **2014**, *50*, 1–4. [CrossRef]

9. Wang, L.L.; Shen, J.X.; Luk, P.C.K.; Fei, W.Z.; Wang, C.F.; Hao, H. Development of a Magnetic-Geared Permanent-Magnet Brushless Motor. *IEEE Trans. Magn.* **2009**, *45*, 4578–4581. [CrossRef]

10. Zhu, X.; Chen, L.; Quan, L.; Sun, Y.; Hua, W.; Wang, Z. A New Magnetic-Planetary-Geared Permanent Magnet Brushless Machine for Hybrid Electric Vehicle. *IEEE Trans. Magn.* **2012**, *48*, 4642–4645. [CrossRef]

11. Liu, C.; Chau, K.T. Electromagnetic Design of a New Electrically Controlled Magnetic Variable-Speed Gearing Machine. *Energies* **2014**, *7*, 1539–1554. [CrossRef]

12. Ho, S.L.; Wang, Q.; Niu, S.; Fu, W.N. A Novel Magnetic-Geared Tubular Linear Machine With Halbach Permanent-Magnet Arrays for Tidal Energy Conversion. *IEEE Trans. Magn.* **2015**, *51*, 1–4.

13. Liu, C.; Yu, J.C.; Lee, C.H.T. A New Electric Magnetic-Geared Machine for Electric Unmanned Aerial Vehicles. *IEEE Trans. Magn.* **2017**, *53*, 1–6. [CrossRef]

14. Zhao, H.; Liu, C.; Song, Z.; Liu, S. A Consequent-Pole PM Magnetic-Geared Double-Rotor Machine With Flux-Weakening Ability for Hybrid Electric Vehicle Application. *IEEE Trans. Magn.* **2019**, 1–7. [CrossRef]

15. Sun, L.; Cheng, M.; Zhang, J.W.; Song, L.H. Analysis and Control of Complementary Magnetic-Geared Dual-Rotor Motor. *IEEE Trans. Ind. Electron.* **2016**, *63*, 6715–6725. [CrossRef]

16. Bai, J.; Liu, J.; Zheng, P.; Tong, C. Design and Analysis of a Magnetic-Field Modulated Brushless Double-Rotor Machine—Part I: Pole Pair Combination of Stator, PM Rotor and Magnetic Blocks. *IEEE Trans. Ind. Electron.* **2019**, *66*, 2540–2549. [CrossRef]

17. Chan, C.C. The state of the art of electric, hybrid, and fuel cell vehicles. *Proc. IEEE* **2007**, *95*, 704–718. [CrossRef]

18. Miller, J.M. Hybrid electric vehicle propulsion system architectures of the e-CVT type. *IEEE Trans. Power Electron.* **2006**, *21*, 756–767. [CrossRef]

19. Liu, C.; Chau, K.T.; Zhang, Z. Novel Design of Double-Stator Single-Rotor Magnetic-Geared Machines. *IEEE Trans. Magn.* **2012**, *48*, 4180–4183. [CrossRef]

20. Chen, C.L.; Tsai, M.C. Kinematic and Dynamic Analysis of Magnetic Gear With Dual-Mechanical Port Using Block Diagrams. *IEEE Trans. Magn.* **2018**, *54*, 1–5. [CrossRef]

21. Chen, L.; Hopkinson, D.; Wang, J.; Cockburn, A.; Sparkes, M.; O'Neill, W. Reduced Dysprosium Permanent Magnets and Their Applications in Electric Vehicle Traction Motors. *IEEE Trans. Magn.* **2015**, *51*, 1–4.

22. Baloch, N.; Kwon, B.I.; Gao, Y.T. Low-Cost High-Torque-Density Dual-Stator Consequent-Pole Permanent Magnet Vernier Machine. *IEEE Trans. Magn.* **2018**, *54*, 1–5. [CrossRef]

23. Gao, Y.; Qu, R.; Li, D.; Li, J.; Zhou, G. Consequent-Pole Flux-Reversal Permanent-Magnet Machine for Electric Vehicle Propulsion. *IEEE Trans. Appl. Supercond.* **2016**, *26*, 1–5. [CrossRef]

24. Wang, H.T.; Fang, S.H.; Yang, H.; Lin, H.Y.; Wang, D.; Li, Y.B.; Jiu, C.X. A Novel Consequent-Pole Hybrid Excited Vernier Machine. *IEEE Trans. Magn.* **2017**, *53*, 1–4. [CrossRef]

25. Wang, Q.S.; Niu, S.X.; Yang, S.Y. Design Optimization and Comparative Study of Novel Magnetic-Geared Permanent Magnet Machines. *IEEE Trans. Magn.* **2017**, *53*, 1–4. [CrossRef]

26. Zhang, X.X.; Liu, X.; Chen, Z. Investigation of Unbalanced Magnetic Force in Magnetic Geared Machine Using Analytical Methods. *IEEE Trans. Magn.* **2016**, *52*, 1–4. [CrossRef]

27. Shin, K.H.; Cho, H.W.; Kim, K.H.; Hong, K.; Choi, J.Y. Analytical Investigation of the On-Load Electromagnetic Performance of Magnetic-Geared Permanent-Magnet Machines. *IEEE Trans. Magn.* **2018**, *54*, 1–5. [CrossRef]

28. Djelloul-Khedda, Z.; Boughrara, K.; Dubas, F.; Kechroud, A.; Tikellaline, A. Analytical Prediction of Iron-Core Losses in Flux-Modulated Permanent-Magnet Synchronous Machines. *IEEE Trans. Magn.* **2019**, *55*, 1–12. [CrossRef]

29. Lubin, T.; Mezani, S.; Rezzoug, A. Two-Dimensional Analytical Calculation of Magnetic Field and Electromagnetic Torque for Surface-Inset Permanent-Magnet Motors. *IEEE Trans. Magn.* **2012**, *48*, 2080–2091. [CrossRef]

30. Lubin, T.; Mezani, S.; Rezzoug, A. Analytical Computation of the Magnetic Field Distribution in a Magnetic Gear. *IEEE Trans. Magn.* **2010**, *46*, 2611–2621. [CrossRef]

31. Sprangers, R.L.J.; Paulides, J.J.H.; Gysen, B.L.J.; Lomonova, E.A. Magnetic Saturation in Semi-Analytical Harmonic Modeling for Electric Machine Analysis. *IEEE Trans. Magn.* **2016**, *52*, 1–10. [CrossRef]

32. Bai, J.G.; Zheng, P.; Tong, C.D.; Song, Z.Y.; Zhao, Q.B. Characteristic Analysis and Verification of the Magnetic-Field-Modulated Brushless Double-Rotor Machine. *IEEE Trans. Ind. Electron.* **2015**, *62*, 4023–4033. [CrossRef]

33. Chung, C.-T.; Wu, C.-H.; Hung, Y.-H. Effects of Electric Circulation on the Energy Efficiency of the Power Split e-CVT Hybrid Systems. *Energies* **2018**, *11*, 2342. [CrossRef]
34. Li, L. Use of Fourier series in the analysis of discontinuous periodic structures. *JOSA A* **1996**, *13*, 1870–1876. [CrossRef]
35. Articolo, G.A. *Partial Differential Equations and Boundary Value Problems with Maple*, 2nd ed.; Elsevier: Burlington, MA, USA, 2009; pp. 29–30.
36. Djelloul-Khedda, Z.; Boughrara, K.; Dubas, F.; Ibtiouen, R. Nonlinear Analytical Prediction of Magnetic Field and Electromagnetic Performances in Switched Reluctance Machines. *IEEE Trans. Magn.* **2017**, *53*, 1–11. [CrossRef]

Permissions

All chapters in this book were first published in MDPI; hereby published with permission under the Creative Commons Attribution License or equivalent. Every chapter published in this book has been scrutinized by our experts. Their significance has been extensively debated. The topics covered herein carry significant findings which will fuel the growth of the discipline. They may even be implemented as practical applications or may be referred to as a beginning point for another development.

The contributors of this book come from diverse backgrounds, making this book a truly international effort. This book will bring forth new frontiers with its revolutionizing research information and detailed analysis of the nascent developments around the world.

We would like to thank all the contributing authors for lending their expertise to make the book truly unique. They have played a crucial role in the development of this book. Without their invaluable contributions this book wouldn't have been possible. They have made vital efforts to compile up to date information on the varied aspects of this subject to make this book a valuable addition to the collection of many professionals and students.

This book was conceptualized with the vision of imparting up-to-date information and advanced data in this field. To ensure the same, a matchless editorial board was set up. Every individual on the board went through rigorous rounds of assessment to prove their worth. After which they invested a large part of their time researching and compiling the most relevant data for our readers.

The editorial board has been involved in producing this book since its inception. They have spent rigorous hours researching and exploring the diverse topics which have resulted in the successful publishing of this book. They have passed on their knowledge of decades through this book. To expedite this challenging task, the publisher supported the team at every step. A small team of assistant editors was also appointed to further simplify the editing procedure and attain best results for the readers.

Apart from the editorial board, the designing team has also invested a significant amount of their time in understanding the subject and creating the most relevant covers. They scrutinized every image to scout for the most suitable representation of the subject and create an appropriate cover for the book.

The publishing team has been an ardent support to the editorial, designing and production team. Their endless efforts to recruit the best for this project, has resulted in the accomplishment of this book. They are a veteran in the field of academics and their pool of knowledge is as vast as their experience in printing. Their expertise and guidance has proved useful at every step. Their uncompromising quality standards have made this book an exceptional effort. Their encouragement from time to time has been an inspiration for everyone.

The publisher and the editorial board hope that this book will prove to be a valuable piece of knowledge for researchers, students, practitioners and scholars across the globe.

List of Contributors

Lucia Frosini
Department of Electrical, Computer and Biomedical Engineering, University of Pavia, Via Ferrata 5, 27100 Pavia, Italy

Rajeev Thottappillil and Nathaniel Taylor
KTH – Royal Institute of Technology, 100 44 Stockholm, Sweden

Mallikarjun Kande
KTH – Royal Institute of Technology, 100 44 Stockholm, Sweden
ABB Power Generation, Wickliffe, OH 44092, USA

Alf J. Isaksson
ABB Corporate Research, 721 78 Västerås, Sweden

Shuo Chen, Xiao Zhang, Xiang Wu and Guojun Tan
School of Electrical and Power Engineering, China University of Mining and Technology, Xuzhou 221116, China

Xianchao Chen
Xuzhou Yirui Construction Machinery Co. Ltd., Xuzhou 221000, China

Zih-Cing You, Cheng-Hong Huang and Sheng-Ming Yang
Electrical Engineering, National Taipei University of Technology, Taipei 10608, Taiwan

Ahmed A. Zaki Diab, Abou-Hashema M. El-Sayed and Hossam Hefnawy Abbas
Electrical Engineering Department, Faculty of Engineering, Minia University, Minia 61111, Egypt

Montaser Abd El Sattar
El-Minia High Institute of Engineering and Technology, Minia 61111, Egypt

Pedro Gonçalves, Sérgio Cruz and André Mendes
Department of Electrical and Computer Engineering, University of Coimbra, and Instituto de Telecomunicações, Pólo 2-Pinhal de Marrocos, P-3030-290 Coimbra, Portugal

Ya Li, Hui Yang, Heyun Lin, Shuhua Fang and Weijia Wang
School of Electrical Engineering, Southeast University, Nanjing 210096, China

Chaelim Jeong
Department of Industrial Engineering, University of Padova, 35131 Padova, Italy

Dongho Lee
Sungshin Precision Global (SPG) Co., Ltd., Incheon 21633, Korea

Jin Hur
Department of Electrical Engineering, Incheon National University, Incheon 22012, Korea

Krzysztof Kolano
Faculty of Electric Drives and Machines, Lublin University of Technology, 20-618 Lublin, Poland

Markel Fernandez, Endika Robles, Iñigo Kortabarria, Edorta Ibarra and Jose Luis Martin
Derpartment of Electronic Technology, University of the Basque Country (UPV/EHU), Plaza Ingeniero Torres Quevedo 1, 48013 Bilbao, Spain

Andres Sierra-Gonzalez
Tecnalia Research and Innovation, C. Mikeletegi 7, 20009 Donostia, Spain

Jongwon Choi and Kwanghee Nam
Department of Electrical Engineering, Pohang University of Science and Technology, 77 Cheongam Ro, Nam-Gu, Pohang 37673, Korea

Hang Zhao, Chunhua Liu, Zaixin Song and Jincheng Yu
School of Energy and Environment, City University of Hong Kong, 83 Tat Chee Avenue, Kowloon Tong, Hong Kong, China
Shenzhen Research Institute, City University of Hong Kong, Nanshan District, Shenzhen 518057, China

Index

Signal Injection, 42, 59, 199-200, 207, 210-211, 215-218

Signum Function, 43-44

Six-phase Machines, 101, 103, 105, 108, 114, 137

Sliding-mode Gains, 42-44, 53, 55

Speed Range, 42-43, 49, 55, 58, 143, 153-154, 217

State Variables, 44

Static Operation Points, 190-191

Stationary Reference, 44, 59, 86, 89, 111, 114, 116, 120, 135

Stator Currents, 7-8, 12-15, 24, 44, 103, 200

Stator Resistance, 44, 78, 99, 145, 192

Stator Voltages, 44, 86, 89, 135, 182, 191

Stator Windings, 5, 7, 14, 21-22, 33, 103-105, 116, 139-140, 180, 219, 221, 223, 233-237

Stress Tensor, 156-157, 169, 229-230

Switching Frequency, 52, 111-112, 114, 116, 120-122, 124-125, 127, 129-132, 135, 140-142, 190, 192, 212

Switching Losses, 184, 188-191, 193-194

System Efficiency, 193-194

T

Tangent Function, 43, 49, 51, 58-59, 200

Tangent-based, 49, 51-52, 56-57

Topology, 52, 103, 133, 143-144, 147, 198, 211, 222

Torque Ripple, 14, 124-125, 127, 129-130, 132-133, 141-142, 148, 156, 169, 180

Transfer Function, 47, 66, 83-86

V

Vector Control, 52, 59, 80-81, 94, 98, 100, 217

Virtual Vectors, 114, 116, 120-121, 124-126, 129-131, 133-135, 141

Voltage Model, 86, 201

Voltage Vector, 61, 111-112, 120, 122-123, 125, 130-131, 135, 141, 182, 184, 188

W

Waveforms, 54, 75, 147-148, 235-238

Printed in the USA
CPSIA information can be obtained
at www.ICGtesting.com
JSHW062236071123
51533JS00031B/82